# The Nature and Power of Mathematics

# The Nature and Power of **Mathematics**

*Donald M. Davis*

PRINCETON UNIVERSITY PRESS · PRINCETON, NEW JERSEY

Davis, Donald M., 1945–
The nature and power of mathematics / Donald M. Davis.
p. cm.
Includes bibliographical references and index.
ISBN 0-691-08783-0 (alk. paper) – ISBN 0-691-02562-2
(pbk. : alk. paper)
1. Geometry, Non-Euclidian. 2. Number theory. 3. Fractals.
4. Cryptography. I. Title.
QA685.D25 1993
516.9-dc20 92-26744

This book has been composed in Lucida Bright Regular and Demibold using $T_EX$

Princeton University Press books are printed on acid-free paper
and meet the guidelines for permanence and durability of the
Committee on Production Guidelines for Book Longevity of
the Council on Library Resources

Printed in the United States of America

10 9 8 7 6 5 4 3 2

# Contents

# Preface

*The Nature and Power of Mathematics* is written for the student who wants to learn some mathematics that is both surprising and deep. The book was written for the liberal arts student, but it could also be read with benefit by a good high school student, a college freshman who is taking calculus and considering a possible major in mathematics, an engineering student who would like to see another side of mathematics, or by an educated "general reader" who is willing to do a little work while reading. The book presumes only a background in high school algebra and geometry, and even this knowledge is not an absolute prerequisite.

The book has more depth and less breadth than most other books directed at a similar audience. It deals essentially with three principal topics: non-Euclidean geometry, number theory/cryptography, and fractals. The theme that binds them together is the unexpected applications of pure mathematics. The mid-twentieth-century physicist Eugene Wigner spoke of the "unreasonable effectiveness of mathematics." Here we will see three fine examples of this process, plus a fourth as we look carefully at the influence of Greek mathematics on the astronomer Johannes Kepler nearly two thousand years later, and a fifth as we glimpse how Alan Turing's work in pure mathematics led to the development of computers.

Most students have no idea that there is any geometry other than Euclidean. The notion that questions about Euclid's fifth postulate led to a new form of mathematics that eventually influenced Albert Einstein and subsequent physicists is a revelation. But unless this information is taught thoroughly and carefully, students will not be convinced.

Most students are also surprised to learn that the factoring of large numbers is the basis for a method of cryptography used by corporations and the military. And, although we pay lip service to the current applications of fractals to physics and applied science, the unexpected "application" on which we focus here is their beauty. The mathematicians Julia and Fatou, who developed much of the requisite mathematics before 1920, would have been surprised to learn that seventy years later the sets they were studying would become popular works of art.

Such applications are not useful in one's daily life. At Lehigh University, a course of these subjects is a sister course to a more practical Finite Mathematics course, in which topics such as matrices, probability and statistics, and mathematics of finance are taught. From *The Nature and Power of Mathematics*, the student will learn to appreciate why people devote their lives to mathematics (because of the beauty of its ideas), and why governments support people to do mathematics (because of its unexpected applications, often far in the future). The *Nature* part of the title suggests that the proofs which permeate the text are showing the reader the essence of mathematics, while the *Power* part refers to the fact that pure mathematics, freed from real-world constraints, can lead to insights with profound real-world implications.

We have found that the topics discussed in this book can motivate the student to want to learn deep mathematics. It is not necessary to work the exercises to obtain an overview, but someone who neglects to do so will be missing much that can be gained from this book. Some of the exercises ask the student to prove (usually with hints) theorems that were stated without proof in the text. Others provide an opportunity for students to test their mastery of an important concept. And some introduce supplementary topics not covered in the text. The order of the exercises follows roughly the order of the text material, and their difficulty is indicated by a * (for difficult) and ** (for very difficult).

Each section within each chapter closes with either a "Focus" or an Aside. A "Focus" is a biographical sketch, while an "Aside" is a related topic not necessary for further development. A student or instructor who is pressed for time may ignore these parts. Some of the Asides, such as "Computational Complexity" and "Fractal Ferns," are a bit more difficult than the basic text, but everyone should enjoy the biographies of Archimedes, Kepler, Gauss, Einstein, Turing, and Newton. There is also a good bit of history interwoven into the text.

The student should finish this book with a great respect for proofs. In both the geometry and number theory sections, the student will learn by doing: that you can't really understand a concept until you have struggled with some proofs that involve it. In the geometry section, we learn that what you can prove depends on what you assume as axioms, and in the number theory section, that a mathematician is not satisfied to know that a statement is true in the first billion cases. The logical thought processes the student develops in struggling with these proofs can be important in everyday life.

It is best if the student has access to a computer with BASIC, so that he or she can write and run the programs discussed in the chapter on fractals. Because running BASIC can vary so much from system to system, the student may need some help in getting started. But these programs are good ones to motivate the student to want to run them.

There is enough material here for a full year's course. Covering the first two chapters one semester and the last three chapters another semester is a reasonable schedule, although even that may require some omissions.

I acknowledge with great appreciation the many people who have assisted me in the preparation of this book. Ken Monks produced the color plates of the Mandelbrot set, including both computer work and photography. He also reviewed the manuscript and printed some of the black and white fractals. Judy Vervoort did the professional artwork. Jerry King and Albert Wilansky also made helpful criticisms of the text, as did a number of students in my Math 5 classes at Lehigh University. Matt Fante helped with production of some fractals. Peter Hilton, David Kahn, Henrik Lenstra, and John Milnor lent their expertise in certain sections of the book.

# The Nature and Power of Mathematics

# 1. Some Greek Mathematics

## 1.1 π and Irrational Numbers: An Introduction to Greek Mathematics

The modern approach to mathematics began with the Greeks. This is one of the major themes of this book—the extent to which the mathematics developed by the ancient Greek civilization still heavily influences mathematics today. The time frame for this flowering of mathematics is best given by tabulating the approximate time of the work of most of the Greek thinkers whose work we will discuss. For many of these, the exact dates of birth and death are not known.

| | |
|---|---|
| Thales | 600 B.C. |
| Pythagoras | 550 B.C. |
| Plato | 400 B.C. |
| Euclid | 300 B.C. |
| Archimedes | 250 B.C. |
| Eratosthenes | 230 B.C. |
| Apollonius | 220 B.C. |
| Ptolemy | 150 A.D. |
| Diophantus | 250 A.D. |

Not all of these people lived in the small country that is Greece today. The center of their civilization was indeed Athens, but Pythagoras and his band of followers lived in what is now southern Italy, while Euclid worked and taught in Alexandria, now part of Egypt.

Throughout this book, we will encounter many mathematical ideas, especially in geometry and number theory, that were introduced by the Greeks. More important than the specific facts, though, was the introduction of mathematics as a deductive science. This idea is sometimes attributed to Thales. The historian Plutarch wrote that "Thales was apparently the only one whose wisdom stepped, in speculation, beyond the limits of practical utility."

Historical accounts of Thales are quite limited, but he apparently noted various general truths of geometry and how they were related to one another. This method was developed more thoroughly by the Pythagoreans fifty years later, and it was utilized to a very high degree by Euclid three hundred years after Thales. The difference between the empirical approach and the deductive approach is the difference between saying that vertical angles are equal because it looks like they are, and proving it in the following way, as Thales may have done.

PROPOSITION 1.1.1. *Vertical angles are equal.*

This theorem says that in Figure 1.1, $\angle AOD = \angle BOC$.

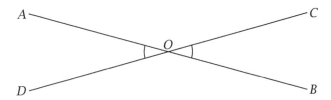

Figure 1.1: Vertical angles are equal.

*Proof.* If angle $AOC$ is added to either of the angles in question, a straight angle (180°) is obtained, in the first case along line $DC$, and in the second case along $AB$. Thus both angles are equal to 180° minus $\angle AOC$, and hence they are equal to each other. Q.E.D.[1]

In the next section we will discuss the assumptions underlying this proof. This is a simple example, but throughout this book we will see many examples of far-reaching proofs of nonobvious results. We will see other examples of statements that have been verified to be true in billions of cases, but which have defied all attempts to prove that they will necessarily be true in *every* case that anyone will ever be able to consider. This demand for a proof is what sets mathematicians apart from other scientists. In Section 2.6 we shall see how one of the most important laws of physics, Newton's Law of Gravity, had to be fine-tuned by Einstein 250 years after its discovery. On the other hand, a mathematical theorem such as Proposition 1.1.1 is a proven fact, and so will never need any modification.

In order to better appreciate Greek mathematics, we should compare it with the mathematics of the Egyptians, a great civilization that flourished from 2500 B.C. until it was conquered by the Macedonians[2] under Alexander the Great in 332 B.C. The Egyptians obviously had to have some knowledge of geometry and arithmetic to build their pyramids, but they were interested only in procedures that would

1. This abbreviation stands for *quod erat demonstrandum*, Latin for "that which was to be proved." It signifies the end of a proof.
2. The Macedonians lived in northern Greece until they began a series of conquests, which included Greece, Egypt, and regions to the east, in 352 B.C.

be accurate enough for their purposes. Most of what we know about Egyptian mathematics is obtained from the Rhind Papyrus, which dates back to 1650 B.C. This document was purchased from the Egyptians by the British in 1858 and is now located in the British Museum in London.

One of the Egyptian formulas[3] was for the area of a 4-sided figure with side lengths $a$, $b$, $c$, and $d$, as indicated in Figure 1.2.

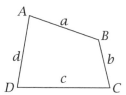

Figure 1.2. The quadrilateral for (1.1.2).

The formula is

(1.1.2)
$$\text{Area} = \frac{a+c}{2} \cdot \frac{b+d}{2}.$$

This is supposed to apply to any 4-sided figure, but it is actually correct only for rectangles. For one example in which it fails, consider the trapezoid in Figure 1.3, where $S$ stands for any length.

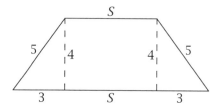

Figure 1.3. A quadrilateral for which the Egyptian formula is wrong by 25%.

The familiar Pythagorean 3–4–5 right triangle allows us to calculate the actual area of this as $12 + 4S$, while the Egyptian calculation for the area is $15 + 5S$, an error of 25%. There is no evidence in Egyptian documents that they were aware that their formulas were usually incorrect.

---

3. The Egyptians didn't use formulas in our sense of the word, but they used procedures that amounted to the same thing.

## THE NUMBER $\pi$

The Egyptian formula for the area of a circle, as given in the Rhind Papyrus, was

$$A = \left(\frac{8d}{9}\right)^2 = \frac{256}{81}r^2 \approx 3.1605r^2,$$

where $d$ is the diameter and $r$ the radius of the circle. We use the symbol "$\approx$" to mean "is approximately equal to." Since the actual formula is

$$A \approx 3.1416r^2,$$

this formula is better than their formula for the area of a quadrilateral, but it is still far from exact.

The number $\pi$ is usually *defined* to be the ratio of the circumference of a circle to its diameter, which will be the same for all circles. Even simply defining what is meant by the circumference of a circle is not a simple matter. Intuitively it is the length of string required to wrap around the circle, but that isn't a mathematical definition. A rigorous definition of length might involve approximating the circle by straight lines, and adding up the lengths of those lines, the length of a line being a fundamental notion. The actual length of the circle is the limit of the lengths of these polygonal approximations of the circle. Archimedes was the first to have a good understanding of this idea, one of the central ideas of calculus. Nearly two thousand years after Archimedes, calculus was developed into a subject of wide applicability by the British mathematician/physicist Isaac Newton[4] and the German mathematician/philosopher G. W. Leibniz in the seventeenth century, and it wasn't until the nineteenth century that the notion of a limit was placed on a firm foundation.

You will get some experience with this idea in Exercise 3, where you will study the diagram of a regular hexagon and a regular 12-sided polygon, each inscribed inside a circle of radius 1 (Fig. 1.4).

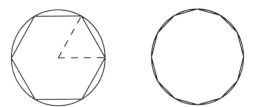

Figure 1.4. Regular hexagon and 12-gon.

4. See the "Focus" in Section 5.3.

You will show that the perimeter of the hexagon is 6, while the perimeter of the 12-gon is approximately 6.21. These numbers should give approximations to $2\pi$, whose value is approximately 6.28. By extending this method to polygons of 24, then 48, and finally 96 sides, Archimedes was able to prove that

$$(1.1.3) \qquad\qquad 3\frac{10}{71} < \pi < 3\frac{1}{7}.$$

This means that $\pi$ has a value between 3.1408 and $^{22}/_7 \approx 3.142857$. It is significant that Archimedes wasn't just guessing; he knew that he was just approximating, and how good his approximation was.

Some readers may be surprised to read that $\pi$ is less than $^{22}/_7$, for this approximation to $\pi$ is misrepresented as an equality in many high school courses. Since $^{22}/_7$ agrees with $\pi$ in the first three significant figures (3.14), this is a reasonably good approximation to $\pi$, but the numbers are not equal. In fact, it was proven by J. H. Lambert in 1767 that $\pi$ is an irrational number, which means that it cannot be written as a fraction. We will say more about irrational numbers later in this section.

One consequence of the irrationality of $\pi$ is that its digits never settle down to a repeating pattern. Many mathematicians over the centuries have tried to obtain better and better approximations to $\pi$ by computing more decimal places in its expansion. Archimedes' approximation (1.1.3) was good to three significant figures, since the 3.14 part was correct, but it did not determine whether the next decimal place is a 0, 1, or 2. Improvements were made over the ages, so that by the year 1699, $\pi$ was known to seventy-one decimal places, using the infinite series,

$$(1.1.4) \qquad \pi = 2\sqrt{3}\left(1 - \frac{1}{3\cdot 3} + \frac{1}{3^2\cdot 5} - \frac{1}{3^3\cdot 7} + \cdots\right).$$

The dots at the end of the line mean that the sum continues forever, with the terms following the pattern that has (hopefully) been established by the listed terms. This infinite series can be derived using standard methods of calculus.

The advent of computers greatly increased the number of decimal places to which $\pi$ is known. In June 1989, a couple of mathematician brothers at Columbia University, David and Gregory Chudnovsky, computed 480 million decimal places of $\pi$, and by early 1992 they had upped it to over 2 billion. They developed a new algorithm and ran it on two separate computers using different programs, to insure that there were no mistakes. This discovery was featured in many newspaper articles, not because it was more important than many other mathematical discoveries, but because it was an attention grabber and understandable to the general public. (See the Associated Press newspaper article.)

## New pi slice: 480 million decimal places

NEW YORK (AP) – Two researchers say they have cut themselves a record slice of pi: 480 million decimal places of one of the most famous numbers in mathematics.

That exceeds the mark of about 201 million decimal places produced last year in Japan, David Chudnovsky of Columbia University said yesterday.

Pi is the ratio of the circumference of a circle to its diameter. The infinite string of digits begins with 3.14159, and contains no obvious pattern.

The newly calculated number would stretch for 600 miles if printed, Columbia University said.

Chudnovsky and his brother Gregory, associate research scientists at Columbia, came up with the number after devising a new algorithm, which is a series of mathematical steps.

The algorithm lets researchers calculate more and more decimal places for pi by simply adding to an existing string, rather than having to start from scratch, Chudnovsky said.

It also has a self-correcting feature that can check whether the pi calculation is on the right track while the calculation is in progress, he said.

The researchers checked

their calculation by running the algorithm through two supercomputers with different hardware and software, Chudnovsky said. When the results were compared, "everything perfectly fitted together."

So what good is knowing this number?

"We can use it as a very stringent test for testing the performance of hardware and software," said Chudnovsky, who called pi calculations "the ultimate stress test."

And by calculating pi out to great lengths, researchers can investigate whether the string of digits is random or contains some hidden pattern, he said.

"We sort of see more of the tail of the dragon," he said.

Associated Press newspaper article: December 19, 1986

There are several reasons why highly regarded mathematicians would bother to do this calculation. One is that the development of new algorithms is important for many computer applications. A second is that such a mammoth calculation provides an excellent test for bugs in computer hardware. A third is competition—several times the Chudnovskys' record was bettered by a Japanese team. A fourth, and most important reason, is that it provides data for studying some unsolved questions about the digits of $\pi$. For example, it is believed, but not yet proved, that each digit will occur on average just as often as any other digit. Thus, out of the first billion digits, one would expect that each digit from 0 to 9 will occur approximately 100 million times.

In 1988 a paper giving a statistical analysis of the first 29,360,000 digits of $\pi$ was presented.[5] The most frequent digit was 4, which appeared 2,938,787 times, and the least frequent one was 7, which

5. D. H. Bailey, "The computation of $\pi$ to 29,360,000 decimal digits using Borweins' quartically convergent algorithm," *Mathematics of Computation* 50 (1988): 283–96.

occurred 2,934,083 times. This difference is not unreasonable for a truly random sequence. The longest string of consecutive digits was a string of nine consecutive 7's. One can compute that in a string of 29,360,000 random digits, there is a 29% chance that there will be a string of nine consecutive digits; therefore this string did not give reason to doubt the randomness of the digits.

For an in-depth look at the Chudnovskys and their calculation, see Preston (1992), which gives a fascinating account that discusses their difficulty emigrating from Russia, and their difficulty in finding a regular position in America. It explains, in their own words, why they do this work: "We are looking for the appearance of some rules that will distinguish the digits of $\pi$ from other numbers." So far, they haven't found any, but to them 2 billion digits is a mere warmup for the infinitely many that follow. They speculate that every possible sequence of digits occurs somewhere in $\pi$. There are ways of associating letters to combinations of digits. Their speculation would imply that everything that has ever been written or could be written occurs somewhere among the digits of $\pi$.

The ubiquity of the number $\pi$ makes the memorizing of the digits of $\pi$ a popular stunt for some people. To this end, mnemonics have been developed. These are strings of words whose number of letters in the successive words equals the successive digits of $\pi$. For example, a sentence such as "May I have a large container of coffee?" can help one remember that $\pi$ begins with 3.1415926. To remember fifteen decimal places, you might try the sentence, "How I want a drink, alcoholic of course, after the heavy lectures involving quantum mechanics." Cleverer yet are mnemonics whose literary content bears some relationship to the number $\pi$. The following poem about Archimedes by A. C. Orr, which yields the first thirty-one digits of $\pi$, rates very highly in this regard:

> Now I, even I, would celebrate
> In rhymes unapt, the great
> Immortal Syracusan, rivaled nevermore,
> Who in his wondrous lore,
> Passed on before,
> Left men his guidance
> How to circles mensurate.

Since the thirty-third digit of $\pi$ is 0, one would have to make a convention about handling 0's in order to go much farther. The use of punctuation marks other than periods to represent 0's was cleverly implemented by M. Keith in a 402-word mnemonic that told a story of memorizing digits.[6]

6. *Mathematical Intelligencer* 8 (1986): 56–57.

Speaking of ubiquity, an attempt was once made in Indiana in 1897 to legislate the value of $\pi$. A Dr. Goodwin claimed that he supernaturally learned the exact value of $\pi$, and offered his state the use of his discovery and its free publication in school textbooks; other interested persons, however, would have to pay a royalty. The bill, which passed the Indiana House but failed in the Senate, actually gave four values of $\pi$: 4, 3.3333, 3.2, and 3.55555.[7]

### IRRATIONAL NUMBERS

We saw earlier that $\pi$ is an irrational number and therefore its decimal expansion does not repeat. In the remainder of this chapter, I will explain why this is true.

A *rational number* is a fraction, a number that can be expressed as the ratio of two integers. Some examples of rational numbers are 3, $1/3$, $-22/7$, and 3.14 (= $314/100$). A principal tenet of the Pythagorean philosophy was that everything could be related to whole numbers or ratios of whole numbers. Pythagoreans believed that certain numbers were associated to various real-world qualities. For example, 2 was male, 3 was female, 5 was marriage, and 7 was health. They discovered that for musical strings under the same tension, a ratio of 2 to 1 in length will produce an octave, a ratio of 3 to 2 a fifth, and a ratio of 4 to 3 a fourth—some of the most pleasing harmonies. This discovery is considered to be the first result in mathematical physics.

The famous Pythagorean Theorem[8] implies that if in a right triangle the sides surrounding the right angle both have length 1, then the hypotenuse has length $c$ satisfying

$$c^2 = 1^2 + 1^2 = 2.$$

A well-known legend is that one of the Pythagoreans, Hippasus, discovered the following clever argument that $c$ cannot equal any fraction; for his effort he was rewarded by being thrown off a boat and drowned. This discovery caused some rethinking of the Pythagorean dictum that identified number with geometry.

THEOREM 1.1.5. *There is no rational number $c$ which satisfies $c^2 = 2$. In other words, $\sqrt{2}$ is irrational.*

---

7. See A. E. Hallerberg, "Indiana's squared circle," *Mathematics Magazine* 50 (1977): 136–40.
8. See Proposition 47 in Section 1.2.

*Proof.* Suppose there is a fraction $p/q$ reduced to lowest terms, such that

$$\left(\frac{p}{q}\right)^2 = 2.$$

Then $p^2 = 2q^2$, and so $p$ is an even number, since its square is even, and the square of an odd number is odd. Thus $p = 2a$ for some integer $a$. Hence,

$$2q^2 = (2a)^2 = 4a^2,$$

and so $q^2 = 2a^2$. Thus $q$ must also be an even number, by the same reasoning as above. The fact that $p$ and $q$ are both even numbers contradicts the assumption that the fraction $p/q$ was reduced to lowest terms, because both $p$ and $q$ can be divided by 2. Q.E.D.

The Greeks preferred to work purely in geometrical terms, rather than assigning numerical values to length and area. They might rephrase Theorem 1.1.5 to read "In an isosceles right triangle, the hypotenuse is incommensurable with either of the equal sides." This means that there is no segment into which both of these sides can be evenly divided. To understand that this says the same thing as 1.1.5, note that if the hypotenuse were equal to $p/q$ times another side, then if that side were divided into $q$ equal parts, exactly $p$ of them would fill the hypotenuse. That lines should be commensurable seems like a natural truth, and was believed by the Greeks until Hippasus's discovery.

Throughout this book, we will see many examples of the method of proof by contradiction, also known as *reductio ad absurdum*, which was used in proving Theorem 1.1.5. You assume that the desired conclusion is false, and show that a contradiction is obtained. Thus the desired conclusion must be true, since it can't be false.

A key ingredient in the above proof was the statement that the square of an odd number is odd. This is such a basic fact that you are probably quite familiar with it. In exercise 7, you are asked to prove $\sqrt{3}$ is irrational. Here you will have to use the fact that if $p^2$ is a multiple of 3, then so is $p$. This may not seem quite so obvious. It is a consequence of the uniqueness of the factorization of numbers into prime factors, a topic that will be discussed in more detail in Section 3.1. The factors of $p^2$ are just the factors of $p$ taken twice. If a 3 is included as one of the factors, it must have come from one of the $p$'s; that is, 3 must be a factor of $p$.

Given the unique factorization of integers into primes, a more direct argument for the nonexistence of integers $p$ and $q$ satisfying $(p/q)^2 = 2$ can be given. Cross-multiplying yields $p^2 = 2q^2$. No matter how many factors of 2 are in $p$, there will be twice that many, an even number, in $p^2$. Similarly, there will be an odd number of factors of 2 in $2q^2$, since $q^2$ contributes an even number of factors, and the 2 contributes one more. Unique factorization of integers, which we will discuss more carefully in Section 3.1, implies that an integer with an even number of factors of 2 cannot equal an integer with an odd number of factors of 2. Thus $\sqrt{2}$ cannot be written as $p/q$.

The reader should not fail to note what a change such ideas are from the practical thinking of the Egyptians. The discovery that $\sqrt{2}$ is irrational may not have had any immediate practical applications, but it led to a much better understanding of the number system. This factor was important in the development of calculus, which has had a myriad of applications.

The number $\sqrt{2}$ can be approximated as accurately as desired by rational numbers. The decimal expansion of $\sqrt{2}$ begins 1.4142136. This means that the rational numbers

$$\frac{14}{10}, \quad \frac{141}{100}, \quad \frac{1414}{1000}, \quad \frac{14142}{10000}, \dots$$

yield better and better approximations to $\sqrt{2}$. To illustrate, we tabulate these numbers and their squares:

| $x$ | $x^2$ |
|---|---|
| 1.4 | 1.96 |
| 1.41 | 1.9881 |
| 1.414 | 1.99940 |
| 1.4142 | 1.99996 |

*REPEATING DECIMALS*

The reader is certainly familiar with the repeating decimal,

$$\frac{1}{3} = 0.333333333\dots.$$

Somewhat less familiar is

$$\frac{1}{7} = 0.142857142857142857\dots.$$

We write this as $0.\overline{142857}$, the line indicating the repeating digits. One can discover the repeating decimal for any fraction by performing the long division required to write the decimal expansion of the fraction, and noting that when one remainder is repeated, then all successive remainders will be repeated.

We illustrate with the fraction $^{137}/_{1110}$. We perform the long division as below, and note that the successive remainders are 260, 380, 470, and 260.

$$
\begin{array}{r}
0.12342 \\
1110\overline{)137.00000} \\
\underline{111.0} \\
26.00 \\
\underline{22.20} \\
3.800 \\
\underline{3.330} \\
4700 \\
\underline{4440} \\
2600 \\
\underline{2220} \\
380
\end{array}
$$

As the remainder 260 has been repeated, the rest of the division will just repeat the three remainders 260, 380, and 470, ad infinitum. One such additional step has been included above. This proves that

$$\frac{137}{1110} = 0.1\overline{234}.$$

Note that the initial 1 is not part of the repeating group. The numerator 137 should really have been considered as the first remainder, since with it as well as with all the actual remainders, the next division is performed by placing a 0 after the number and dividing 1110 into this number.

The remainders will necessarily eventually repeat when any fraction is calculated. This is because the only possibilities for them are the numbers from 0 to $D - 1$, where $D$ is the divisor. After $D + 1$ steps, some remainder will have to have been repeated. Often these repeated remainders will come much more quickly than $D + 1$ steps. In the above example, we got a repeat in the fifth step, whereas it could have taken as many as 1111 steps.

Note that if a 0 is obtained as a remainder, all successive remainders will also be 0's, as will all successive quotients. Thus, for example, $^3/_{20} = 0.15$, which you may not think of as a repeating decimal, actually has an infinite string of repeating 0's: $^3/_{20} = 0.150000. \ldots$

The above argument demonstrates that any fraction can be expressed as a repeating decimal. How about going the other way?

**Example.** What fraction equals $0.2454545 \ldots = 0.2\overline{45}$?
Let $X = 0.2\overline{45}$. Because the repeating group has 2 digits, we note that $100X$ and $X$ will have their repeated blocks lined up. Indeed,

$$100X = 24.54545\ldots$$

$$X = \phantom{2}.24545\ldots$$

If we subtract these equations, the lined-up 45's cancel out, leaving

$$99X = 24.3.$$

After multiplying this by 10 to turn the 24.3 into an integer, we obtain

$$X = \frac{243}{990} = \frac{27}{110}.$$

In the last step, we reduced to lowest terms by dividing numerator and denominator by 9.

This method will always work. It leads to an alternative method in which you don't actually have to write out the equations as we did above. The above method will show, for example, that

$$0.\overline{1342} = \frac{1342}{9999},$$

and similarly, for any other group of digits repeating immediately after the decimal point, the repeating decimal equals the fraction that has the repeating group in the numerator and a number with the same number of digits, all of which are 9, in the denominator. If the repeating digits do not begin immediately to the right of the decimal point and/or if they are preceded by some nonrepeating digits, the nonrepeating digits should be subtracted off and taken into account individually, and the fraction with the 9's in the denominator should be multiplied by an appropriate power of 10 to account for where it begins.

**Example.** Write $2.1\overline{543}$ as a fraction.

Think of it as $2.1 + .0\overline{543}$. The fact that the 543 block has one 0 separating it from the decimal point means that the $543/999$ will have to be multiplied by $1/10$. We obtain

$$\frac{21}{10} + \frac{543}{10 \cdot 999}.$$

From here on, it is just a matter of arithmetic, putting the fractions over a common denominator, and reducing to lowest terms, for completeness, yielding

$$\frac{21}{10} \cdot \frac{999}{999} + \frac{543}{9990} = \frac{21522}{9990} = \frac{3587}{1665}.$$

Although this method of inserting 9's in the denominator seems to lead more directly to an answer, the equation method probably involves less work. It would yield $999X = 1000X - X =$

$$\frac{\begin{array}{r} 2154.3543\ldots \\ 2.1543\ldots \end{array}}{2152.2}$$

Hence, $9990X = 21522$.

The methods we have learned for going back and forth between fractions and repeated decimals give a constructive proof of the following result.

THEOREM 1.1.6. *A number is rational if and only if its decimal representation is eventually repeating.*

Although $\sqrt{2}$ was historically the first number proved to be irrational, Theorem 1.1.6 allows one easily to construct many irrational numbers. One example is

(1.1.7) .1010010001000010000010000001000000010000000001…,

which has progressively more 0's between the 1's. This number is clearly never repeating, and hence is irrational. The Greeks wouldn't have known about this method of constructing irrational numbers because they did not use decimal notation for fractions. This usage did not become common in Europe until a book by Simon Stevin advocated it in 1579. Lambert's proof that $\pi$ is irrational used a different method—continued fractions.

*FOCUS: ARCHIMEDES*

Archimedes is frequently ranked with Newton (1642–1727) and Gauss (1777–1855) as one of the three greatest mathematicians of all time. We have already mentioned Newton's law of gravity and development of calculus, and in the succeeding chapters you will see numerous references to Gauss, a German. However, of the Greek mathematicians, we will focus mostly on Euclid, and you may begin to wonder why Archimedes is considered to be greater than Euclid.

As you will see in the next section, many of the results in Euclid's famous book, the *Elements*, were proved by earlier Greek mathematicians. Euclid did a masterful job of putting the results together, and undoubtedly produced many of the proofs himself, but a lack of documents from that time makes it difficult to determine who was the originator of most of the theorems and proofs. Archimedes, who lived shortly after Euclid, was a mathematician who clearly developed a number of important new mathematical techniques. One of his great contributions was the method of calculating an area enclosed by a curve, such as a circle, by the method of exhaustion, that is, by approximating the area from a sum of areas of triangles, and allowing the triangles eventually to exhaust the region by finer and finer approximations. This is essentially the method that became the integral calculus nineteen hundred years later. Archimedes also calculated volumes by a similar method.

Archimedes, about to be killed

One of Archimedes' great discoveries in physics is the first law of hydrostatics—that, when immersed in a fluid, a body is buoyed up by a force equal to the weight of the displaced fluid. A famous tale surrounds his discovery of this law. King Hieron had ordered a crown of pure gold, but suspected that some of the gold might have been replaced by a cheaper and less dense material. The king asked Archimedes to make a test, and Archimedes discovered the solution one day while in the public baths. He was so excited that he ran out of the bath naked, shouting "Eureka!" which translates to "I have found it!" The test he concluded was to balance the crown with an equal weight of gold, and then put the balance, with crown and gold, under water. The crown indeed contained some less dense material, and therefore took up more space than the pure gold. As a result, a more buoyant force pushed up on the crown, and it rose.

Another famous story about Archimedes concerns his death at the age of seventy-five. His town of Syracuse, on the island of Sicily, had been invaded by the Romans, led by General Marcellus. The townspeople withstood the siege for three years, in large part due to various contraptions devised by Archimedes, such as catapaults and large burning glasses to set the Romans' ships afire. Finally, the Romans did invade the city. Marcellus had great respect for Archimedes and ordered his soldiers not to harm him. As he often did, Archimedes was reputedly drawing geometrical diagrams in the sand and ordered a Roman soldier to stand clear of his diagram. Such arrogance was too much for the soldier, who speared him to death. But Archimedes was highly honored by both Greeks and Romans. His tomb, which has a mathematical diagram on it, was rediscovered in 1965, during excavations for a hotel in Syracuse.

## Exercises

1. Show that the formulas in the text regarding the actual area and Egyptian formula for the area of the trapezoid in Figure 1.3 are as claimed.
2.* Prove that the Egyptian area formula (1.1.2) is correct only for rectangles and overestimates the area the rest of the time by following these hints. Draw a diagonal $AC$ and, by considering the two triangles formed, show that

$$\text{Area} \leq \frac{cd}{2} + \frac{ab}{2},$$

with equality if and only if the figure is a rectangle. To do this, you will want to note that the side $AD$ is greater than or equal to the altitude of triangle $ADC$ dropped from $A$. Redraw the figure, and perform a similar analysis of the triangles on the two sides of the diagonal $BD$. If you add the two inequalities, you obtain an inequality directly comparable to a multiplied-out version of (1.1.2).

3. A regular polygon is a polygon all of whose sides are equal and all of whose angles are equal.

   a. Explain why the perimeter of a regular hexagon inscribed in a circle of radius 1 is 6. (Hint: The central angles in Figure 1.4 divide the total 360° into six equal parts. Note also that a triangle with three 60° angles is equilateral.)

   b. Let $P$ be a regular polygon with $n$ sides inscribed in a circle of radius 1, and let $Q$ be a regular polygon with $2n$ sides inscribed in a circle of radius 1. It can be proved that if the length of each side of $P$ is $S$, then the length of each side of $Q$ is

   $$\sqrt{2 - \sqrt{4 - S^2}}.$$

   (You need not prove this.) Apply this formula to obtain the length of each side and then the perimeter of a regular 12-gon inscribed in a circle of radius 1. You will probably want to use a calculator. Apply the formula again to determine the perimeter of a regular 24-gon inscribed in a circle of radius 1.

4. Find an approximate value of $\pi$ by using (1.1.4), including one more term than is listed there. Use your calculator.

5. Write a new mnemonic for $\pi$.

6. Write out the details of the following sketched argument, similar to one made by Archimedes, that the area of a circle is $\pi r^2$. If a regular polygon is inscribed in a circle, define the apothem of the polygon to be the length of the line from the center of the circle to the center of one of its edges. Prove that the area of the polygon equals one-half times its perimeter times its apothem. Note that with regular polygons that have a very large number of sides, the area, perimeter, and apothem of the polygons approach, as a limit, the area, circumference, and radius of the circle, respectively, as the number of sides becomes infinite. Now recall that $\pi$ is defined to be the ratio of the circumference to the diameter, and deduce the formula for the area of the circle.

7. Mimic the proof of Theorem 1.1.5 to show that $\sqrt{3}$ is irrational. You will need to use the fact that if $p^2$ is a multiple of 3, then so is $p$.

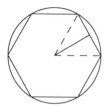

Figure 1.5. The apothem of a regular hexagon.

8. Prove that $\sqrt[3]{2}$ is irrational.
9. Show that the decimal expansion of $^2/_7$ is $0.\overline{285714}$. In fact, this group of six repeating digits may be started at any of its six positions and yield $^1/_7$, $^2/_7$, $^3/_7$, $^4/_7$, $^5/_7$, and $^6/_7$. Knowing this fact, tell by inspection what is the decimal expansion of $^3/_7$.
10. What is the decimal expansion of $^1/_{13}$? Work it out by long division until remainders start to repeat.
11. Use both the equation-method and the method of 9's to show that
$$0.7\overline{2014} = \frac{72007}{99990}.$$
12. Write $1.3\overline{721}$ as a reduced fraction.
13. Give an example of a nonrepeating decimal by completely describing all of its digits according to some pattern. Make it as different from (1.1.7) as you can.

## 1.2   Euclidean Geometry

The plane geometry that you studied in high school was probably a modernized version of the famous book, the *Elements*, written by the Greek mathematician Euclid around 300 B.C. This book is said to have been read by more people than any other book except the Bible. It has been considered by many to be the archetype of a rigorous deductive system. The *Elements* contains thirteen books (more like what we would call chapters) and deals with number theory as well as geometry. We will discuss some of Euclid's contributions to number theory in Chapter 3.

Euclid was not the first Greek to write a book entitled *Elements* which surveyed what was considered to be basic mathematics; however, he did such a good job of it that all earlier efforts have been lost. It is not known how many of the theorems are original to Euclid,

Euclid

although probably not very many of them are, since the deductive approach to geometry had already been going on for three hundred years. A serious historian of mathematics might worry about attribution of these theorems, but we will just call them theorems of Euclid, except for a few special cases such as the Pythagorean Theorem.

The words "theorem" and "proposition" have the same meaning, namely, a statement that has been rigorously deduced from a set of assumptions. In contemporary mathematics, a slight distinction is often made: a *theorem* is considered to be a more important result than a *proposition*. Euclid called all of his results propositions, and we will refer to Euclid's propositions using the numbers he gave them. Aside from those, we will use our

### *Chapter.Section.Number*

notation, and the words "theorem" or "proposition" to imply a measure of importance. Other words that are used for proved statements are *lemma*, for a result whose main purpose is to help prove another result, and *corollary*, for a result that is easily deduced from another.

Not much is known about Euclid's life except for an anecdote that suggests that he would probably not have much patience with many job-oriented students of today. A student studying geometry under Euclid asked what he would get from learning the subject, to which Euclid ordered a slave to give the student a penny, "since he must make gain from what he learns."

Euclid's approach, which has been the approach to deductive science ever since, is to begin with a few definitions and assumptions (called *postulates* or *axioms*) and then to deduce a string of propositions from them, in each step using only the assumptions, rules of logic, and previously proved propositions. Many nineteenth- and twentieth-century mathematicians have felt that Euclid's proofs are somewhat lacking in rigor. We will discuss many of these criticisms as we proceed, and the reader should not be dismayed by such appraisals. We can say four things in support of Euclid:

1. His presentation of his deductive system was deemed perfectly satisfactory for over two thousand years, and probably became the most influential scientific document ever written.
2. Although he may have allowed intuition to enter into some of his proofs, none of his theorems is wrong.
3. A completely rigorous treatment of Euclidean geometry can be written, in which all of Euclid's propositions are proved in a way that is completely divorced from the intuition.[9]
4. Not only in geometry, but also in other branches of mathematics, standards of rigor have evolved with the passage of time.

### *LOGIC*

Before we proceed with details of Euclid's treatment, we must learn a few terms from logic, most notably *converse* and *contrapositive*.

9. The famous German mathematician David Hilbert undertook this task. We will discuss his *Foundations of Geometry*, published in 1899, in Section 2.1.

First we note that the statements "If $P$, then $Q$" and "$P$ implies $Q$" have the same logical content. They mean that whenever $P$ is true, then $Q$ must necessarily be true. Here $P$ and $Q$ refer to statements that are either true or false. For example, if $P$ is the statement "$x$ is a boy" and $Q$ is the statement "$x$ is human," then the statement "If $P$, then $Q$" is true regardless of what $x$ is, as is the statement "$P$ implies $Q$." Indeed, in words these true statements become

$$\text{If } x \text{ is a boy, then } x \text{ is human,}$$

and

$$x \text{ is a boy implies that } x \text{ is human.}$$

The statements make no claim about what happens if $x$ is not a boy. Then $x$ might be human (e.g., a girl) or might not (e.g., a dog).

The *converse* of "If $P$, then $Q$" is "If $Q$, then $P$." There is no relationship between the truth of a statement and its converse. For example, if $P$ and $Q$ are as above, then the converse of the true statement "If $P$, then $Q$" is the statement,

$$\text{If } x \text{ is a human, then } x \text{ is a boy,}$$

which is of course not true, since not *all* humans are boys. It is not enough that some humans are boys, and it is not correct to say that the above statement is sometimes true. A conditional statement is either true or false, and it is true if and only if the "then" part is *always* a consequence of the "if" part. On the other hand, if $P$ is the statement "$x$ is a rational number" and $Q$ is the statement "$x$ is a number with an eventually repeating decimal expansion," then, by Theorem 1.1.6, both the statement "If $P$, then $Q$" and its converse are true.

The *contrapositive* of the statement "If $P$, then $Q$" is the statement "If not $Q$, then not $P$," or "If $Q$ is false, then $P$ is false." The contrapositive of a statement has the same logical content as the statement, that is, a statement is true if and only if its contrapositive is. In our above example, the contrapositive of the statement "If $P$, then $Q$" is the true statement "If $x$ is not human, then $x$ is not a boy." In order to prove a statement true, one can just as well prove the contrapositive.

This is closely related to proving something by contradiction, but it's not quite the same thing. For example, Theorem 1.1.5 could have been stated as

$$\text{If } c^2 = 2, \text{ then } c \text{ is irrational.}$$

The contrapositive of this is:

$$\text{If } c \text{ is rational, then } c^2 \neq 2.$$

Our proof showed that the statements "$c^2 = 2$" and "$c$ is a rational number in reduced form" lead to a contradiction.[10] This argument can be interpreted as proving either of the above statements, by saying that if the hypothesis is true, then the conclusion cannot be false, and hence the conclusion must be true.

### EUCLID'S DEFINITIONS AND POSTULATES

We begin our discussion of Euclid's *Elements* by listing some of his first definitions:

> *Definition* 1. A *point* is that which has no part.
> *Definition* 2. A *line* is breadthless length.
> *Definition* 3. The extremities of a line are points.
> *Definition* 4. A *straight line* is a line which lies evenly with the points on itself.

These definitions clearly leave something to be desired. They are not very useful in and of themselves. A point is meant to be some sort of infinitely small dot, but that really isn't what the definition says. If you were asked, before reading Euclid, for a single word for "breadthless length," it seems unlikely that "line" is the word that first comes to mind. Even if it did, what kind of line would you envision—straight or curved? As Euclid's Definition 4 makes clear, "line" could include either possibility. Finally, even if you were able to figure this out, you were using some intuition or prior knowledge about the word "length."

In a system of definitions, it is virtually impossible to avoid circularity. Checking a college edition of Webster's dictionary for the meaning of the word "straight" yields:

> **straight:** having the same direction throughout its length
> **direction:** the line leading to a place
> **line:** the path of a moving point, having length but not breadth, whether straight or curved

This seems to be going nowhere, and in fact has led us from "straight" to possibly curved. Let's try again, using the second definition of "straight":

---

10. Being in reduced form is no additional hypothesis, since every rational number can be written in this way.

**straight:** not bent
**bent:** not straight, curved
**curve:** a line having no straight part

Quite clearly, we are unable to avoid circularity here.

The solution is to begin with a few undefined terms, and then define everything in terms of them. In mathematics, you can attach meaning to undefined terms by specifying in the postulates every property that is supposed to be true of these terms. Modern rigorous formulations of geometry are conducted this way, with "point," "straight line," "lies on," and "between" as the major undefined terms. We will study this approach in more detail in Section 2.1.

Let's make two more comments on these definitions: (i) Definition 3 basically says that lines are composed of points; if "point" and "line" were undefined, this would be a postulate. (ii) Definition 4 may be trying to say that a straight line is the shortest distance between points on it. Euclid never said this explicitly; Archimedes did.

There are nineteen more definitions in Book 1 of the *Elements*. Here are a few of the most important ones. Note that words such as "right angle" and "circle" are defined in terms of earlier words such as "point" and "line" and do not need to be taken as undefined.

> *Definition* 10. When a straight line set up on a straight line makes the adjacent angles equal to one another, each of the equal angles is *right.*
>
> *Definition* 15. A *circle* is a plane figure contained by one line such that all the straight lines falling upon it from one point among those lying within the figure are equal to one another.
>
> *Definition* 23. *Parallel* straight lines are straight lines which, being in the same plane and being produced indefinitely in both directions, do not meet another in either direction.

These definitions make no claim about the existence of the defined notions. Existence will be asserted in postulates or proved in the propositions. Euclid makes no mention of degrees. One can get by perfectly well with the notion of "right angle" or "straight angle" and fractions thereof. That seems a bit overly formal for a book such as ours, and we will adopt the customary definition of a *degree* as the size of an angle $1/90$ of a right angle. We will not use the radian measure of angles used in calculus and other forms of higher mathematics.

Now for the postulates, which Euclid divided into two groups.

There are five postulates, which are assumptions quite specific to geometry, and will be the focus of most of our attention. Then there are five Common Notions, or Axioms, which underlie all of mathematics.

### *Postulates*

*Postulate* 1. (It is possible) to draw a straight line from any point to any point.

*Postulate* 2. (It is possible) to produce a finite straight line continuously in a straight line.

*Postulate* 3. (It is possible) to describe a circle with any center and distance.

*Postulate* 4. All right angles are equal.

*Postulate* 5. If a straight line falling on two straight lines makes the interior angles on the same side less than two right angles, the two straight lines, if produced indefinitely, meet on that side on which are the angles less than the two right angles.

Before we discuss these postulates, let's take a look at the Common Notions, which will not require further comment, except to say that they can apply to numbers, lengths, or angles.

### *Common Notions*

*CN1.* Things which are equal to the same thing are also equal to one another.

*CN2.* If equals be added to equals, the wholes are equal.

*CN3.* If equals be subtracted from equals, the remainders are equal.

*CN4.* Things which coincide with one another are equal to one another.

*CN5.* The whole is greater than the part.

In our statements of the postulates, rather than trying to modernize the phraseology we have followed, for the most part, the standard translations into English of Euclid's text.[11] However, we have added the phrase "It is possible" to several of the postulates, and will do likewise to some of the propositions.

The term "a straight line" in Postulate 1 is meant to say "exactly one straight line." This becomes clear in several of the early proofs. Postulate 2 means that a straight line can be extended beyond any

---

11. Heath 1956.

prescribed length. This harmless-sounding assumption will turn out to be a crucial difference between Euclidean geometry and one of the forms of non-Euclidean geometry that we will study in Chapter 2.

The word "distance" in Postulate 3 clearly means "radius." Postulate 4 is a statement about the uniformity of space. It says that it is impossible for space to bend in such a way that, whereas straight lines and right angles are, for the most part, the way you think they should be, in some far-away place you could have straight lines such as $PQR$ in Figure 1.6, so that the right angles $PQS$ and $RQS$ differ from the right angle $ABD$. A postulate such as this is necessary before it makes sense to talk about a degree as $1/90$ of a right angle.

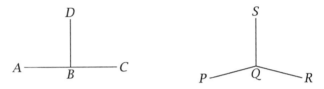

Figure 1.6. Hypothetical unequal right angles.

Postulate 5 is clearly much more complicated than the first four postulates. It says that if in Figure 1.7 the two indicated angles add up to less than 180°, then the two lines will eventually meet on the right side. It does *not* say what happens if the sum of the indicated angles equals 180°.

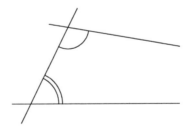

Figure 1.7. Euclid's fifth postulate.

Postulate 5 probably seems quite compatible with one's intuition. But, like many mathematicians during the twenty-one centuries following Euclid, you may feel that this is really too complicated a statement to be called a postulate, and should rather be proved as a proposition. We shall discuss attempts to do this in Section 2.2. It is quite likely that in your high school geometry course Postulate 5 was stated differently—something about one and only one parallel line. This, too, will be discussed in Section 2.2. Here we want to stay true to Euclid's presentation. When we discuss Euclid's ordering of his propositions later in the section, we will see that he showed a special feeling toward Postulate 5 by delaying its use in his proofs as long as possible.

### THE FIRST THIRTY-TWO PROPOSITIONS

Book 1 of the *Elements* contains forty-eight propositions, culminating in the Pythagorean Theorem, which we shall discuss in the next subsection. First we shall look at many of the first thirty-two propositions, including proofs when they serve a purpose. Before commencing, we point out that Book 1 is not the only book of the *Elements* that deals with plane geometry. For example, circles, area, and proportional figures are covered in detail in other books. Here we will concentrate on Book 1. We should study many of these propositions in detail to be able to contrast them with non-Euclidean geometry, the subject of Chapter 2.

PROPOSITION 1. *On a given finite straight line (it is possible) to construct an equilateral triangle.*

The proofs we write will generally be paraphrases of Euclid's proofs. Euclid wrote his proofs in prose style, unlike the column-style Statement and Reason method used in many high school geometry texts. We will follow Euclid's style, which is the style used by virtually all contemporary mathematicians in their work.

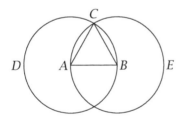

Figure 1.8. Euclid's Proposition 1.

*Proof.* Let *AB* be the given finite straight line. Draw circle *BCD* with center *A* and radius *AB*, and draw circle *ACE* with center *B* and radius *BA*. [Postulate 3] Let *C* be the point where the circles cut one another, and draw the straight lines *AC* and *BC*. [Postulate 1] Now, *AC* = *AB* and *BC* = *BA* by Definition 15. Thus the triangle is equilateral by CN1. Q.E.D.

Although this proof may, at first glance, appear very clear and obvious, it is one which is highly criticized for over-reliance on intuition or the diagram. The question is: How do we know that the circles intersect? Assumptions are being made here in several ways—for example, how do we know that the circles might not have tiny little holes in them so that the one could slip through a hole in the other? There is nothing in the postulates to prevent this occurrence. A new postulate is required here, such as:

> A circle separates the points of the plane not on the circle into two regions called its inside and outside. Any line drawn from a point outside to a point inside intersects the circle.

Note that the word "line" is used here in the way Euclid uses it, to allow the possibility of curved lines. Here it would be applied to the arc of the second circle that connects a point inside the first circle with one outside it. This may seem pedantic, but to become comfortable with non-Euclidean geometry, you must divorce yourself from your intuition and just think logically. In modern mathematics, there are forms of geometry in which circles do have tiny holes and might not intersect.[12] In Section 2.1, we will point out how Hilbert's modern form of "plane geometry made rigorous" handles questions such as this.

Even with the added postulate above, there is another worry: How do we know that the second circle contains points both inside and outside the first circle? Worrying about this may be stretching your credulity; however, in one of the forms of non-Euclidean geometry (spherical) we will study, sometimes the second circle stays totally inside the first, so that there is no point of intersection, and hence no equilateral triangle.

For most of the purposes of this book, Euclid's level of rigor is perfectly appropriate, and we need not dwell on fine points, as we have done above. As noted earlier, all of Euclid's results are correct in the sense that they can be rigorously deduced from a set of postulates which extends his slightly. Many geometry texts (such as Prenowitz and Jordan 1965) illustrate how overreliance on diagrams can actually lead to incorrect conclusions in geometry. The "Aside" at the end of this section tells a story of how such overreliance led to a highly publicized incorrect proof of a famous conjecture in 1986.

Postulate 3 is the principal ingredient in the proofs of the next two propositions, whose proofs we omit. Exercise 3 asks you to prove Proposition 3; think how you would do it if you had a compass, and try to formalize it.

---

12. For example, geometry of points in the plane whose coordinates are rational numbers.

PROPOSITION 2. *(It is possible) to place at a given point (as an extremity) a straight line equal to a given straight line.*

PROPOSITION 3. *Given two unequal straight lines, (it is possible) to cut off from the greater a straight line equal to the less.*

Proposition 4 is the famous Side-Angle-Side Proposition:

PROPOSITION 4. *If two triangles have the two sides equal to two sides, respectively, and have the angles contained by the equal straight lines equal, they will be congruent.*

Euclid doesn't use the word "congruent," but rather includes in the statement of the proposition each pair of corresponding parts that are equal. This is one of the propositions whose proof receives a lot of criticism. The proof amounts to "applying" one triangle to the other so that they agree at one point and at the angle formed there, and noting that they must agree everywhere else. This is an argument by superposition, which assumes a uniformity of space. You may believe quite strongly now that there is no question that space (i.e., Euclid's plane) must have this homogeneity, but by the time you finish Chapter 2, you will probably realize that this is something that really should have been stated. We have already mentioned that Postulate 4 is an assumption about uniformity of space, but it is not as strong as the uniformity implied by Proposition 4. Propositions 2 and 3 could easily have been proved by superposition, but instead Euclid preferred to prove them by using circles, essentially by moving a compass. Hilbert, in his rigorous 1899 approach to Euclidean geometry, used a statement similar to Proposition 4 as one of his postulates.

Recall that an isosceles triangle is one with two equal sides.

PROPOSITION 5. *In an isosceles triangle, the angles at the base are equal to one another.*

The proof of this proposition, which we defer to Exercise 14, is considered to be one of the cleverest of Euclid's proofs. The interested reader is challenged to prove this proposition without referring to the hints in Exercise 14, using only the propositions that precede it. Proposition 6 is the converse of Proposition 5. (What would it say?) Proposition 8 is the Side-Side-Side Congruence Theorem.

Propositions 9 and 10 say that you can bisect an angle and a line, respectively. Propositions 11 and 12 say that you can construct perpendiculars to a given line either at a point on it or from a given point not on it. By "constructing," Euclid meant that you are only allowed to use a straightedge and compass; no measuring device such

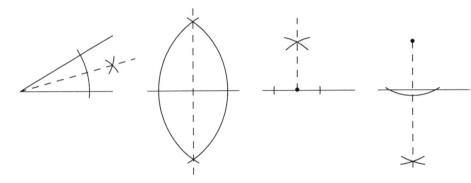

Figure 1.9. Euclid's Propositions 9 through 12.

as a ruler or a protractor is allowed. The reader will probably recall many of these constructions from a high school course. For those who have not seen them before, they are reproduced in Figure 1.9.

Proposition 15 says that vertical angles are equal. This proposition and its proof were discussed as our Proposition 1.1.1. The next result may seem technical, but it is an important tool in some of our later results about parallel lines.

PROPOSITION 16. *In any triangle, if one of the sides is produced, the exterior angle is greater than either of the opposite interior angles.*

This says that angle $ACD$ is greater than both angles $A$ and $B$ in Figure 1.10. We will prove $ACD$ is greater than $A$, and ask you, in Exercise 5, to prove it is greater than $B$.

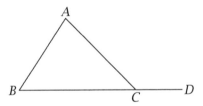

Figure 1.10. The setup for Proposition 16.

*Proof that* $\angle ACD > \angle A$. Let the point $E$ bisect $AC$. (See Figure 1.11.) Extend the segment $BE$ to an equal segment $EF$. Draw $CF$. Since $AE = EC$, $BE = EF$, and $\angle BEA = \angle CEF$, triangles $AEB$ and $CEF$ are congruent. Thus $\angle ECF = \angle A$. But $\angle ECD > \angle ECF$, and so the result follows, since $\angle ACD = \angle ECD > \angle ECF = \angle A$. Q.E.D.

We will need to refer to the following result in Chapter 2.

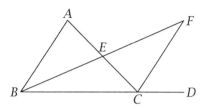

Figure 1.11. Proof of Proposition 16.

PROPOSITION 17. *Two angles of a triangle taken together are less than two right angles.*

In other words, the sum of two angles is less than 180°. You may be thinking, "Of course this is true because the sum of the angles of a triangle is 180°." However, there is a subtlety here. We haven't *proved* yet that the sum of the angles of a triangle is 180°; that doesn't come until Proposition 32. More to the point, the proof that the sum of the angles is 180° requires the use of Postulate 5, while the proof of Proposition 17 does not. The distinction between propositions that use Postulate 5 and those that do not will be of great importance to us in Chapter 2. Remember, *proofs are only allowed to use results that have preceded them.* You must exercise a certain amount of discipline to refrain from using results that you already learned in your high school course.

Proposition 20 says that any side of a triangle is less than the sum of the other two. This proposition was said to be so obvious that even an ass would know it: if fodder is placed at one vertex of a triangle and an ass at another, the ass does not, in order to get to its food, traverse the two sides of the triangle but only the side separating them. Obvious though it may be, it still requires proof, but we shall omit it here.

Proposition 23 allows construction of an angle equal to a given angle (see Exercise 7), and Proposition 26 is the Angle-Side-Angle Theorem. Now we are prepared to study the crucial sequence of results involving parallel lines.

PROPOSITION 27. *If a straight line falling on two straight lines makes the alternate interior angles equal to one another, then the straight lines are parallel to one another.*

This proposition says that if the indicated angles in Figure 1.12 are equal, then the lines are parallel. You are asked to prove this in Exercise 8.

Euclid's first twenty-eight propositions can be proved without us-

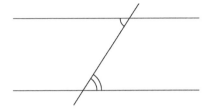

Figure 1.12. Proposition 27.

ing the fifth postulate. *Absolute geometry* is the collection of theorems that can be proved by using only Euclid's first four postulates (and his Common Notions). You will read a lot more about this in Chapter 2. Thus these first twenty-eight propositions are all theorems of absolute geometry. They are true in any system in which the first four postulates are true, regardless of whether or not the fifth postulate is true in that system. Euclid postponed using Postulate 5 just about as long as he could. Proposition 29, finally, does use the fifth postulate in its proof.

PROPOSITION 29. *A straight line falling on parallel lines makes the alternate interior angles equal.*

This is the converse of Proposition 27, but there is a big difference in that 27 is a result of absolute geometry, while Proposition 29 is not. This proposition says that if $\ell$ and $\ell'$ are parallel in Figure 1.13, then angles $A$ and $B$ are equal. In Exercise 9, you are asked to write out a proof of this proposition. It is an important proof because most of the applications of Postulate 5 are actually applications of Proposition 29.

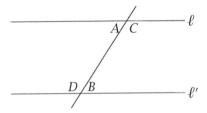

Figure 1.13. Proposition 29.

Note that saying $A = B$ is the same as saying $B + C = 180°$ or $A + D = 180°$. (Why?) The way to work Postulate 5 into the proof of (the contrapositive of) Proposition 29 is to say that if $B + C < 180°$, then lines $\ell$ and $\ell'$ intersect on the right side, whereas if $A + D < 180°$, then $\ell$ and $\ell'$ meet on the left side.

Proposition 29 (and hence Postulate 5) is used in the proof of the following result, whose proof is left to the reader as Exercise 10 (with hints there).

PROPOSITION 30. *Straight lines parallel to the same line are parallel to each other.*

Note that this proposition implies that there can be no more than one line through a given point parallel to a given line. Indeed, Proposition 30 says that it is impossible to have two distinct lines through a point parallel to a given line. Although Euclid does not make this explicit statement, we shall want to be able to refer to it, and so we record it as a corollary.

COROLLARY OF PROPOSITION 30. *There can be no more than one line through a given point parallel to a given line.*

The following proposition guarantees the existence of *at least* one line through a point parallel to a given line.

PROPOSITION 31. *Through a given point not on a given straight line, it is possible to draw a straight line parallel to the given straight line.*

Thus Propositions 30 and 31 together give the familiar statement about "one and only one parallel," which will play a major role throughout our remaining study of geometry. In Exercise 11, you are asked to write out the proof of Proposition 31, following the hints given there. You will see that this proof does not use Postulate 5 or anything that requires Postulate 5 in its proof. In other words, Proposition 31 is a proposition of absolute geometry. We can say that in absolute geometry parallel lines exist, but, as far as we know from what we have learned so far from this book, they may not be unique. Perhaps Euclid was inclined to list Proposition 31 before Proposition 29, in line with his apparent wish to postpone as long as possible those results that require Postulate 5. His apparent reason for postponing Proposition 31 until after Proposition 30 is so that he could talk about *a* straight line, and not have to worry about whether there might be more than one.

This subsection ends with the following result, which will also be a key point of comparison of Euclidean geometry with other forms of geometry. It is not obvious why having 180 degrees in triangles should have anything to do with uniqueness of parallel lines, but both are intimately related to Euclid's Postulate 5.

PROPOSITION 32. *The sum of the angles in a triangle equals two right angles* (180°).

The proof of this proposition is also relegated to Exercise 12. You should note carefully in your proof how Postulate 5 enters into the proof (indirectly).

*THE PYTHAGOREAN THEOREM*

The most famous theorem in plane geometry is certainly the Pythagorean Theorem, which appears in Book 1 of Euclid as Proposition 47. Some learned people claim it is the most famous and even the most important theorem in all of mathematics,[13] because it gives such a basic, beautiful, and nonobvious relationship between geometry and numbers. The Egyptians were aware of this relationship $c^2 = a^2 + b^2$ empirically, but they didn't bother to wonder *why* it was true.

PROPOSITION 47. *In right-angled triangles, the square on the side subtending the right angle is equal to the squares on the sides containing the right angle.*

Stated in this way, as Euclid did, it is a statement about areas rather than length squared. The proof in Euclid proceeds from this point of view, as well.

*Euclid's Proof of Proposition 47.* Refer to Figure 1.14. We will show that the area of the rectangle *BKLE* equals the area of the square *ABFG*. An analogous argument shows that the area of the rectangle

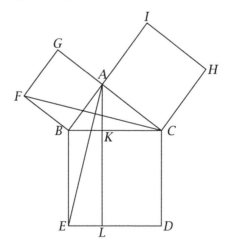

Figure 1.14. Euclid's proof of the Pythagorean Theorem.

13. Bronowski 1973, p. 160.

$KCDL$ equals that of the square $ACHI$. Since the area of the square $BCDE$ equals the sum of the rectangles $BKLE$ and $KCDL$, this will imply the theorem.

To prove the first claim about area, first note that triangles $ABE$ and $FBC$ are congruent by Side-Angle-Side. (Check this.) Thus they have the same area. But triangle $ABE$ and rectangle $BKLE$ have the same base $BE$ and the same height $BK$. Thus the area of $BKLE$ equals twice that of $ABE$. Similarly, the area of $ABFG$ equals twice that of $FBC$. (Check this.) Thus, using $|-|$ to denote area,

$$|BKLE| = 2|ABE| = 2|FBC| = |ABFG|,$$

as desired. Q.E.D.

You might wonder whether or not this is a result of absolute geometry. Did the fifth postulate, or anything that requires the fifth postulate in its proof, enter into this proof? The answer is yes. In fact, the existence of squares requires the fifth postulate. A proposition stating this fact should logically have preceded the Pythagorean Theorem, and did so in Euclid's treatment.

PROPOSITION 46. *On a given straight line, it is possible to describe a square.*

Here, as before, Euclid uses the word "describe" where we might use "construct," or "there exists." We sketch a proof which is similar to some that we will see in Section 2.2. Filling in of the details is left to Exercise 13. On the given line $AB$ in Figure 1.15, construct perpendiculars $AC$ and $BD$ with length equal to that of $AB$. Draw the line $CD$. Congruence theorems imply $\angle C = \angle D$. (To do this, draw in the diagonals, show that the bottom triangles are congruent, and then show that the top two triangles are congruent.) Now use a decomposition of the square into two triangles and Proposition 32 to deduce that angles $C$ and $D$ equal $90°$. Finally, use a congruence theorem to prove $CD = AB$.

Figure 1.15. Constructing a square.

A more visual proof of the Pythagorean Theorem uses Figure 1.16. Note that the two big squares have the same area, and both contain

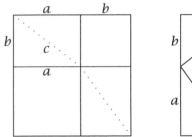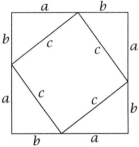

Figure 1.16. Another proof of the Pythagorean theorem.

four of the *a-b-c* right triangles. In addition to these, one has area $a^2 + b^2$, while the other has area $c^2$. Thus $c^2 = a^2 + b^2$.

Underlying this proof is Proposition 32, which tells us that the two non-right angles of the *a-b-c* triangles add up to 90°. This is necessary in order to form either of the two diagrams in Figure 1.16. In the left one it is needed to make the two triangles fit together to form a rectangle, while in the right one it is needed to make the line along the bottom straight.

There are *many* proofs of the Pythagorean Theorem. One is even attributed to President James A. Garfield.

You probably know that a triangle with sides of length 3, 4, and 5 is a right triangle, because

$$3^2 + 4^2 = 9 + 16 = 25 = 5^2.$$

The Egyptians knew this, too, and sometimes used it to construct perpendiculars. The general fact being used here is actually the converse of Proposition 47, and was Euclid's Proposition 48. Proposi-tio 47 says that if a triangle has a right angle, then $c^2 = a^2 + b^2$, while Proposition 48 says that if $c^2 = a^2 + b^2$, then the triangle has a right angle. Both are true, but logically they are different statements.

A *Pythagorean triple* is a triple of integers $(a, b, c)$ such that $a^2 + b^2 = c^2$. Simple algebra shows that for any positive integers $m$ and $n$,

(1.2.3)                     $(m^2 - n^2,\ 2mn,\ m^2 + n^2)$

is a Pythagorean triple. For example, letting $m = 2$ and $n = 1$ yields the triple (3,4,5).

Any multiple of a Pythagorean triple is also a Pythagorean triple. For example, multiplying (3,4,5) by 2 and 3 yields the Pythagorean triples $(6,8,10)$ and $(9,12,15)$. But this isn't very interesting. It is more interesting to find *primitive Pythagorean triples*, which are Pythagorean triples that are not a multiple of any other Pythagorean triple. Thus a Pythagorean triple $(a, b, c)$ is *primitive* if $a$, $b$, and $c$ have no common divisors (which are greater than 1). It is not difficult to see that the Pythagorean triple of the form given by (1.2.3) above is primitive if and only if $m$ and $n$ have no common divisors and have opposite parity[14]. It is not too much harder to prove that *all* primitive Pythagorean triples can be obtained from (1.2.3). We state this formally, but will not prove it.

THEOREM 1.2.4. *The set of all primitive Pythagorean triples equals the set of all triples of the form (1.2.3) such that m and n have no common factors and have opposite parity.*

You will apply this in the exercises.

### ASIDE: POINCARÉ CONJECTURE

In this brief aside, we discuss how overreliance on diagrams led to a highly publicized incorrect proof of a famous unsolved problem in pure mathematics in 1986. In 1904 the French mathematician Henri Poincaré, who was then one of the two most famous mathematicians in the world,[15] made a conjecture that arose in his work in topology. Since that time, many mathematicians have tried unsuccessfully to prove or disprove Poincaré's conjecture, and it has become one of the most famous unsolved problems in all of mathematics.

Topology is the study of properties of surfaces that are unaffected by stretching or bending. Thus length and curvature are not properties of interest to a topologist, but "number of holes" is. Most of the surfaces studied by contemporary topologists have more than two dimensions, and so they cannot be visualized directly. Surfaces of any number of dimensions can be defined by equations, and it is frequently through equations, or through analogy with visualizable surfaces, that they are studied. The Poincaré Conjecture deals with 3-dimensional manifolds. These are surfaces that locally look like our 3-dimensional space, but they usually require more than three dimensions for their global existence. See Section 2.6 for further discussion of the idea of more than three dimensions. For a simple analogue, a circle is a 1-dimensional manifold because it is locally 1-dimensional, even though 2 dimensions are required to

---

14. This means that one is even and one is odd.
15. Along with the German, David Hilbert, who has already been mentioned

see the whole thing. The Poincaré Conjecture states that any simply connected 3-dimensional manifold must be deformable to a 3-dimensional sphere in a manner that allows no cutting or gluing. "Simply connected" means that any loop in it can be contracted in it, and a 3-dimensional sphere is like a basketball, but it is one dimension higher.

A similar statement was proved for 2-dimensional manifolds in the early part of the twentieth century. Later workers realized that a modified version could be considered for higher dimensional spaces. This was proved to be true for spaces of dimensions 5 or more by Stephen Smale of the University of California at Berkeley in 1959, and for spaces of dimension 4 by Michael Freedman of the University of California at San Diego in 1982. So it is known to be true in every dimension except the dimension in which Poincaré conjectured it. For their efforts, both Smale and Freedman received the Fields Medal, the mathematical equivalent of the Nobel Prize. (A story that is widely circulated among mathematicians gives the purported reason why there is no Nobel Prize in mathematics.[16] It claims that Alfred Nobel had a great personal dislike for the great Swedish mathematician G. M. Mittag-Leffler, and he wanted to avoid the possibility of Mittag-Leffler winning a Nobel Prize.)

In 1986 the British mathematician Colin Rourke and his Portuguese colleague, Eduardo Rego, announced that they had proved the conjecture. Like most mathematicians who think they have proved a new result, they wrote out their proof and sent it to many experts in the field. But Rourke did something that is not usually done—he sent out a press release before his proof had been accepted as being correct by the mathematical community. Normally, officially designated referees and other interested experts study the proof to see if it is correct before it is published or announced to the public. Rourke's announcement received coverage in the *Manchester Guardian*, the *New York Times*, and *Science*. Several months later, Rourke held a series of seminars at Berkeley, in which a serious gap in his argument was uncovered, and most experts feel that there is no hope that his proof can be corrected. His delicate mistake had to do with an overreliance on intuition and diagrams.

(See the collage of a portion of the September 30, 1986, announcement in the *New York Times*, the headline from the December 19, 1986, announcement of the mistake in *Science*, and one of the diagrams from the paper that Rourke and Rego distributed to many mathematicians.)

16. Eves 1969, p. 130.

# One of Math's Major Problems Reported Solved

## Math Proof Refuted During Berkeley Scrutiny

*A highly publicized proof of a famous math problem—the Poincaré conjecture—has a gap, which might be unbridgeable*

O N Monday, 3 November, mathematician Colin Rourke of the University of Warwick got up in front of a cluster of mathematicians at the University of California at Berkeley to defend his claim that he and his colleague Eduardo Rego of the University of Oporto in Portugal had proved the Poincaré conjecture—a famous and difficult problem that has taunted mathematicians for 80 years. It was not an easy proof, and the mathematicians in attendance had already put in dozens of hours reading Rourke and Rego's work and trying to understand it. Now Rourke was about to start the first of several 3-hour seminars to explain the proof.

But the claim of a proof spurs a spirited debate.

**By JAMES GLEICK**

T WO mathematicians in England and Portugal have announced that they have proved one of the most important, longstanding problems in mathematics, the Poincaré conjecture — setting off a heated and sometimes emotional flurry of talks, seminar sessions and analysis that will occupy many of their colleagues for months to come.

If it survives the dissection, the proof will settle deep-seated questions about the possible curved or looped or twisted shapes that three-dimensional space can take, and the consequences will ripple through modern geometry and physics. But so far, American mathematicians who have studied preliminary versions of the proof remain skeptical.

"It's such an old and famous problem that there's a lot of speculation in the mathematics community, and there has already been creative work done in trying to analyze the proof," said Michael Freedman, a topologist at the University of California at San Diego. Nevertheless, he and other mathematicians said the first renderings had been too sketchy to confirm.

The proof's authors, Colin Rourke of the University of Warwick and Eduardo Rego of the University of Oporto, have circulated, withdrawn, revised and reissued several versions since January, and they mailed a newly clarified, 123-page version to mathematicians last week. Meanwhile, Dr. Rourke reported the proof in the British magazine New Scientist this month — to the annoyance of other mathematicians, who say he should satisfy his professional colleagues before publicizing the work.

Dr. Rourke considers such criticism unfair. "We're so excited about it, it seems silly not to tell people about it," he said. "As far as I'm concerned, it's not up in the air at all — it's cut and dried."

The effect of ribbon intersections on the standard picture is to give rise to situations such as those pictured in figure 27:

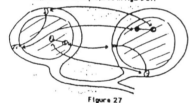

Figure 27

## Exercises

1. For each of the following five statements, write the converse and the contrapositive, and tell whether the statement, its converse, and its contrapositive are true.

   a. If the planet is closer to the sun than is the earth, then the planet is Mercury.

   b. If $x^2 = 9$, then $x = -3$.

   c. If you are in the most populous city in the United States, then you are in New York City.

   d. If a figure is a square, then it has four sides.

   e. If you attend Princeton University, then you are an engineer.

2. Try to write a better definition of *point* and *straight line* than Euclid did.

3. Prove Proposition 3.
4. Explain how to do at least two of the constructions in Propositions 9 to 12, and prove that they do what they claim to do. (You will probably want to use congruence theorems for triangles in your proofs.)
5. Write out the part of the proof of Proposition 16 that was omitted in the text, i.e., prove $\angle ACD > \angle B$. Your proof should mimic the part of the proof written out in the text. You should bisect $BC$ instead of $AC$. Indicate where each of the following is used in your proof: Postulate 2, Proposition 2, Proposition 4, Proposition 10, and Proposition 15. The proof here does require one step that was not present in the text—a use of vertical angles to make a direct comparison with the exterior angle.
6. Prove Proposition 17. You are being asked to prove that the sum of two angles of a triangle is less than 180°. Compare their sum with the sum of one of them plus an adjacent exterior angle.
7. Prove Proposition 23. Given angle $BAC$ and the line from $D$, strike an arc from $D$ equal to one from $A$. Then use a compass to make arc $EF$ equal to $C'B'$. What do you do from here, and why does it work?

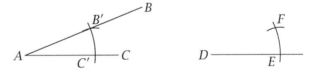

Figure 1.17. For Exercise 7.

8. Write the contrapositive of Proposition 27 and prove it, thus proving Proposition 27. The key ingredient in your proof should be Proposition 16. Note that if the lines in the diagram in Figure 1.17 are not parallel, then a triangle is formed on one side.
9. Prove Proposition 29, following the hints given after it in the text.
10. Prove Proposition 30, by applying Propositions 29, 15, and 27 to Figure 1.18, in which lines $\ell$ and $\ell'$ are both assumed to be parallel to $\ell''$.

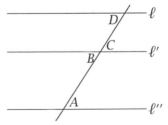

Figure 1.18. Proof of Proposition 30.

11. Prove Proposition 31 by using Proposition 23 to construct a line $\ell'$ making angles $A$ and $B$ equal in Figure 1.19. How do you deduce that the lines are parallel?

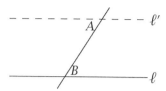

Figure 1.19. Proof of Proposition 31.

12. Prove Proposition 32, following the outline below. Indicate explicitly each time that you use one of the earlier propositions. In Figure 1.20, construct the line through $C$ parallel to $AB$. Explain why the sum of the angles of the triangle $ABC$ equals the straight angle $\angle 1 + \angle 2 + \angle 3$.

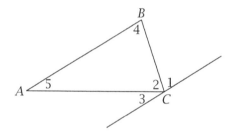

Figure 1.20. Proof of Proposition 32.

13. Write out the proof of Proposition 46, following the outline sketched in the text.

14. Prove Proposition 5, following the outline below. Suppose triangle $ABC$ has $AB = AC$. Extend $AB$ to $AD$, and $AC$ to $AE$, with $BD = CE$. Prove that triangles $ABE$ and $ADC$ are congruent, and then that triangles $BCE$ and $DBC$ are congruent. Now an analysis of the angles at $B$ and $C$ will yield the desired conclusion.

15. Write the primitive Pythagorean triples obtained from (1.2.3) by letting $(m, n)$ be $(3, 2)$, $(4, 1)$, $(4, 3)$, $(5, 2)$, and $(5, 4)$.

16. It is a fact that every integer greater than 2 appears in some Pythagorean triple (not necessarily primitive). Verify this for all integers less than 19. There will be one integer for which the list begun in Exercise 15 must be extended slightly.

17. Verify that $(m^2 - n^2, 2mn, m^2 + n^2)$ is a Pythagorean triple for every value of $m$ and $n$.

## 1.3   Greek Mathematics and Kepler

In this section we will discuss two major achievements of Greek geometry and their influence on the great German astronomer Johannes Kepler nearly two thousand years later. The first of these is the complete analysis of regular polyhedra, which was contained in Euclid's *Elements* but had been discovered by Plato's school one hundred years earlier, and the second is the book *Conic Sections* written by Apollonius around 200 B.C., one hundred years after Euclid.

### REGULAR POLYHEDRA

A *polyhedron* (plural, *polyhedra*) is a solid with plane faces and straight edges, arranged so that every edge joins two vertices and is the common edge of two faces. A polyhedron is called *convex* if any two points in it can be joined by a line segment lying completely inside it. A convex polyhedron is *regular* if all its faces are regular polygons congruent to one another, and the same number of faces meet at each vertex.[17] Another way of saying that all faces are regular polygons congruent to one another is to say that the edges of all the polygons are equal and the angles of all the polygons are equal.

Figure 1.21 shows four polyhedra, none of which is regular. Three of them are convex, but the fourth one isn't, since the line connecting the points $P$ and $Q$ does not lie inside the polyhedron. Note that the faces of the three convex polyhedra are all regular polygons (squares, equilateral triangles, or regular octagons). The first two also satisfy the property that the same number of faces (three) meet at each vertex but do not have all faces congruent to one another. All faces of the fourth are congruent equilateral triangles, but three faces meet at the top and bottom vertices, while four faces meet at the three middle vertices.

Figure 1.22 shows five regular polyhedra. It is an amazing fact that these five are the *only* possible regular polyhedra. They are sometimes called the Platonic solids, because they were discovered and revered by Plato's school. One member of that school, Theaetetus (c. 380 B.C.), is generally credited with proving that these are the only possible regular polyhedra. The last book of Euclid's *Elements* gives a detailed description of them, as well as a proof that there can be no others.

---

17. Recall that a polygon is regular if all its sides are equal and all its angles are equal.

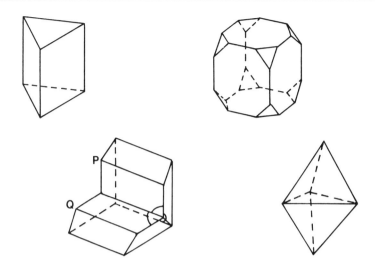

Figure 1.21. Several kinds of polyhedra.

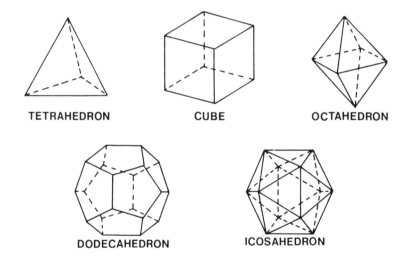

TETRAHEDRON    CUBE    OCTAHEDRON

DODECAHEDRON    ICOSAHEDRON

Figure 1.22. The five regular polyhedra.

Euclid's proof that there are only five regular polyhedra relied on two facts from earlier books of the *Elements*. The first was the following proposition, which he didn't state explicitly.

PROPOSITION 1.3.1. *In a polygon with n sides, the sum of the angles is $(n-2)180°$. The number of degrees in each angle of a regular polygon with n sides is $(n-2)180/n$.*

Note the difference between regular polygons, which are figures in the plane, and regular polyhedra, which are 3-dimensional. A regular polygon exists with any number of sides equal to or greater than 3. Proposition 1.3.1 is easily proved by dividing the polygon into $(n-2)$ triangles, as illustrated in Figure 1.23, and then using the proposition proved in Section 1.2 (Euclid's Proposition 32), that the sum of the angles of a triangle is 180°. Note that this is true for any polygon, not just a regular one, but the second part of Proposition 1.3.1 requires that all the angles be equal. The information we will use from this proposition is that the number of degrees in each angle of a regular polygon with three, four, five, or six edges is 60, 90, 108, and 120, respectively, while a regular polygon with more than six sides has angles of more than 120°.

Figure 1.23. 7-gon is composed of five triangles.

The second fact we need comes from Euclid's Book XI.

PROPOSITION 1.3.2. *In a convex polyhedron, the sum of the vertex angles at any vertex is less than 360°.*

This proposition need not be true in a nonconvex polyhedron. For example, in the nonconvex polyhedron in Figure 1.21c, the sum of the four angles at the indicated vertex is greater than 360°. One can make a convenient plausibility argument for Proposition 1.3.2 from one's intuition. If some polygons fit together to give a flat figure, the sum of the angles at each vertex will equal 360°. In order to form a convex polyhedron, the angles must be decreased somewhat, making the sum less than 360°.

This proof is somewhat unsatisfying, and it is not the way Euclid did it. His proof only addressed the situation where three faces meet at the vertex, where one can can cut off the figure and obtain a tetrahedron $ABCD$ (not necessarily regular, as illustrated in Figure 1.24). By considering the angles of the three triangles coming down from $A$, one sees that the sum of the three angles at $A$ equals $540 - S$, where $S$ is the sum of the six angles at $B$, $C$, and $D$ that are not in the triangle $BCD$. At each of $B$, $C$, and $D$, the sum of the two angles not in the base triangle is clearly greater than the angle in the base triangle. Thus $S$ is greater than the sum of the angles in the base tri-

Figure 1.24. A vertex of a polyhedron.

angle, which equals 180. Since $S > 180$, we deduce that our desired value, $540 - S$, is less than $540 - 180 = 360$. Note how the sense of the inequality $S > 180$ is changed when the terms are negated. This argument is easily generalized to a vertex at which any number of faces meet.

Now we are ready to prove the famous result, which appears without a number at the end of Euclid's Book XIII, but was probably proved by Theaetetus one hundred years earlier.

THEOREM 1.3.3. *There are only five regular polyhedra.*

*Proof.* There must be at least three faces that meet at each vertex. This is really just a property of what it means to be a vertex; if only two faces met there, only two edges would meet, the picture would be as in Figure 1.25, and the vertex could be removed.

Figure 1.25. A removable vertex.

If exactly three faces meet at each vertex, these faces can have three, four, or five sides, and they cannot have six or more sides. If there were six or more sides, the sum of the angles at the vertex would be at least $3 \cdot 120$ by Proposition 1.3.1, but Proposition 1.3.2 will not allow the sum of the angles to be this large. If four faces meet at each vertex, these faces cannot have four or more sides, since the sum of the angles would be at least $4 \cdot 90$. Similarly, if five faces meet at each vertex, it is possible that they might be triangles, since $5 \cdot 60 = 300$, but if they have four or more sides, then the sum of the angles will be at least $5 \cdot 90$, which is too large. Finally, six or more faces cannot meet at each vertex, for even if they are triangles, the sum of the angles will be $6 \cdot 60 = 360$, which is not allowed according to Proposition 1.3.2. Thus the only five possibilities are to have at each vertex (i) 3 triangles, (ii) 3 squares, (iii) 3 pentagons,

(iv) 4 triangles, or (v) 5 triangles. Look at Figure 1.22 to see how these possibilities occur. Q.E.D.

This really completes the proof. However, you might wonder why there couldn't be two distinct regular polyhedra with, say, three pentagons coming together at each vertex. This is really a rigidity result that you note at the same time that you show that they actually exist by constructing them. After all, it isn't obvious that there would have to be any such polyhedron. You can cut a bunch of congruent regular pentagons out of cardboard and tape three of them together at a vertex; there is only one way to do this (see Fig. 1.26). You now have three free vertices where two of the pentagons meet. There is only one way to tape a pentagon at each of these places. As you continue this process, the rigidity becomes clear; there is certainly only one possible polyhedron that can be formed in this way. What isn't so obvious is that there should be any polyhedra at all. Why will this procedure necessarily close on itself to give a regular polyhedron?

Figure 1.26. Starting to build a dodecahedron.

The answer is basically just luck! It works in each of the five cases. For example, to make a regular dodecahedron, cut out two copies of the pattern in Figure 1.27a, and fit them together as indicated in Figure 1.27b. It is clear that they will mesh. The constructibility of the other four regular polyhedra is even easier, and is left to the reader as Exercise 1.

For a regular polyhedron, it is standard to use the following letters:

$p$ = number of edges in each face;

$q$ = number of edges meeting at each vertex

    = number of faces meeting at each vertex;

$v$ = total number of vertices;

$e$ = total number of edges;

$f$ = total number of faces.

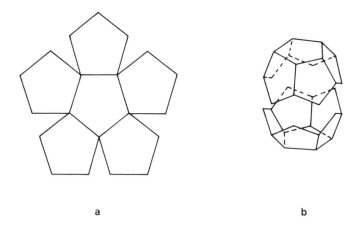

a                                                                    b

Figure 1.27. Building a dodecahedron.

Think for a moment about why the two quantities in the definition of $q$ are equal; as you move around the vertex, each edge separates two faces, and each face separates two edges, so the same number of faces and edges must meet at that vertex. Note that $v$, $e$, and $f$ can be defined for any polyhedron, but $p$ and $q$ make sense only for certain polyhedra that include the regular polyhedra. We can tabulate all the numbers for the regular polyhedra (see Table 1.1).

TABLE 1.1. The Geometric Numbers for the Regular Polyhedra.

|  | $p$ | $q$ | $v$ | $e$ | $f$ |
|---|---|---|---|---|---|
| Tetrahedron | 3 | 3 | 4 | 6 | 4 |
| Cube | 4 | 3 | 8 | 12 | 6 |
| Dodecahedron | 5 | 3 | 20 | 30 | 12 |
| Octahedron | 3 | 4 | 6 | 12 | 8 |
| Icosahedron | 3 | 5 | 12 | 30 | 20 |

*EULER'S FORMULA*

The reader can easily check that for all of the regular polyhedra in the table, $v - e + f = 2$. This is true not only for regular polyhedra but for *all* convex polyhedra, and in fact for a wider class of polyhedra called *simply connected*.[18]

This very famous formula was discovered by Leonhard Euler (1707–1783). Euler (pronounced "oiler") was the most prolific math-

18. This concept was mentioned in the "Aside" at the end of Section 1.2.

ematician of all time, having published 886 books and papers. He is generally ranked just below the triumvirate of Archimedes, Newton, and Gauss as one of the greatest mathematicians of all time. He was nearly blind for the last seventeen years of his life, but this disability did not slow his productivity: "Euler calculated without any apparent effort, just as men breathe and as eagles sustain themselves in air."[19] He worked within a wide range of mathematics, as we shall see throughout this book. A native of Switzerland, he is pictured on a 10-Swiss-franc note, but he lived most of his life in Russia and Germany.

Euler's formula is considered to be one of the first theorems of topology even though this field was not seriously developed until the time of Poincaré, around 1900. We mentioned topology in the "Aside" at the end of Section 1.2; this theorem fits into topology because $v - e + f$ measures a property of polyhedra that is not affected by the kinds of changes allowed by topologists, such as subdividing the faces into more faces. Direct generalizations, called the "Euler characteristic," are used frequently by contemporary topologists. Now we will formally state Euler's formula and sketch a proof.

THEOREM 1.3.4. *For any convex polyhedron, $v - e + f = 2$.*

*Proof.* Think of the edges of the polyhedron as rods, and shine a light from just outside one of the faces to a screen behind the polyhedron, away from the light. The convexity is required to avoid unwanted overlap of the projections of the edges.

This projection yields a connected network in the plane that has the same number of vertices and edges as the polyhedron, and if we include the exterior region as one of the regions, the number of regions, $r$, of the network equals the number of faces of the polyhedron. In Figure 1.28, the six faces of the cube project to the square region $E'F'H'G'$ in the middle, the four trapezoidal regions, such as $A'B'F'E'$, surrounding it, and the infinite region exterior to $A'B'D'C'$. So it will suffice to show that if the plane is divided into regions (including the infinite one) by a connected network, then $v - e + r = 2$.

Working from the outside, remove one edge at a time. If the edge bounds a region, as in Figure 1.29a, then both $e$ and $r$ are decreased by 1, so that $v - e + r$ is unchanged. If the edge sticks out, as in Figure 1.29b, then both $e$ and $v$ are decreased by 1, so that $v - e + r$ is left unchanged. This simplification procedure can be repeated, without changing $v - e + r$, until only a single edge, with two vertices and one infinite region, remains. Here we have $v - e + r = 2 - 1 + 1 = 2$,

19. Attributed to François Arago in Eves 1990, p. 435.

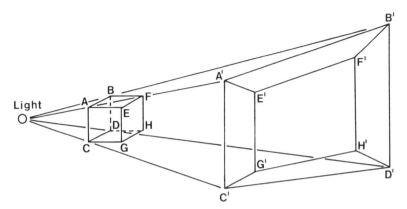

Figure 1.28. Projecting a polyhedron onto a screen.

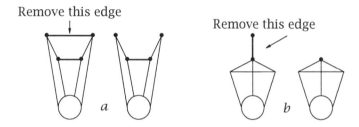

Figure 1.29. The reduction procedure in Euler's formula.

and since this quantity hasn't changed throughout our reduction, it must have equaled 2 for the original network. Q.E.D.

We can give an alternate proof that there are no more than five regular polyhedra by using Euler's formula. It was not the one used by the Greeks, since they apparently didn't know Euler's formula. Its character is totally different than that of the Greek one given earlier: that one was geometry, while Euler's is topology.

Recall that $p$ and $q$ denote the number of edges in each face and the number of edges that meet each vertex, respectively. An *edge-face intersection* is a place where an edge meets a face. Since every edge meets exactly two faces, the number of edge-face intersections equals $2e$. On the other hand, since each of the $f$ faces has $p$ edges, the number of edge-face intersections equals $fp$. Thus,

(1.3.5)                              $$2e = fp.$$

We can define and count edge-vertex intersections similarly, and deduce that

(1.3.6)                                    $2e = vq.$

We rewrite these equations as $2/p = f/e$ and $2/q = v/e$. Now take Euler's formula and divide all terms by $e$. This yields

$$\frac{v}{e} - 1 + \frac{f}{e} = \frac{2}{e},$$

and substituting the previous equations, we get

$$\frac{2}{q} - 1 + \frac{2}{p} = \frac{2}{e}.$$

Finally, we manipulate this result into the form

(1.3.7)                          $$\frac{1}{q} + \frac{1}{p} = \frac{1}{e} + \frac{1}{2},$$

which shows that $p$ and $q$ must be chosen so that $\frac{1}{p} + \frac{1}{q}$ is greater than $1/2$.

Remember that $p$ and $q$ must both be at least 3, because you can't have a 2-sided face or a vertex with only two edges meeting it. There aren't very many ways to choose integers $p$ and $q$ satisfying $p \geq 3$, $q \geq 3$, and $1/p + 1/q > 1/2$. In fact, at least one of them must equal 3, and the other must be 3, 4, or 5. To see this, note that both $1/4 + 1/4$ and $1/3 + 1/6$ equal $1/2$, and this isn't quite large enough. Thus we get the same values of $p$ and $q$ that we obtained from the geometrical analysis and tabulated earlier. Q.E.D.

One of the advantages of this method is that it enables us to determine $v$, $e$, and $f$ from the equations, whereas in the earlier approach we had to draw the figure and count the vertices, edges, and faces. For example, in the regular polyhedron where five triangles meet at each vertex, $p = 3$ and $q = 5$. By (1.3.7),

$$\frac{1}{e} = \frac{1}{5} + \frac{1}{3} - \frac{1}{2} = \frac{6 + 10 - 15}{30} = \frac{1}{30},$$

and so $e = 30$. Then the values of $f$ and $v$ can be determined from (1.3.5) and (1.3.6); indeed, they say $60 = f \cdot 3$ and $60 = v \cdot 5$.

### CONIC SECTIONS

Another major contribution of Greek mathematics was the eight-book treatise, *Conic Sections*, written by Apollonius around 220 B.C. We will learn later in this chapter how this work influenced Johannes Kepler eighteen hundred years later.

*Conic sections* are the curves obtained when a right circular cone is cut by a plane. These curves can be ellipses, parabolas, and hyperbolas, depending on the angle the plane makes with the cone (see Fig. 1.30). The names are attributed to Apollonius. Circles are not part of the list because they are considered to be special cases of ellipses.

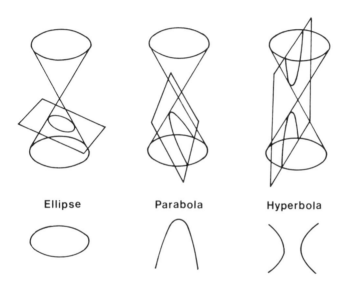

Figure 1.30. The conic sections.

These curves are usually studied today in analytic geometry courses, using equations. But analytic geometry, which establishes the link between geometry and algebra, was not invented until the 1600s by Descartes. Apollonius effectively gave all of the equations involving conic sections and their applications in purely geometrical terms. Because it was Apollonius's results for ellipses which were important to Kepler, we will restrict our attention to them.

Two of the most basic geometric properties of ellipses are summarized in the following result.

PROPOSITION 1.3.8. *Inside an ellipse are two points called foci such that*
  *i. The sum of the distances from the foci is the same for all points on the ellipse.*
  *ii. The lines from the foci to any point on an ellipse make equal angles with the tangent line to the ellipse at the point.*

The first part says that in Figure 1.31, where $A$ and $B$ are the foci, the sum $AP + PB$ is the same at all points on the ellipse. It can be used to draw an ellipse by holding a loose string at two points (the foci) and moving a pencil so that the string is always stretched, as suggested in Figure 1.31b.

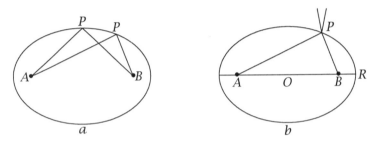

Figure 1.31. The sum of the distances from the foci is constant on an ellipse.

The *eccentricity* of an ellipse is a measure of how skinny it is. It is the ratio $OB/OR$ in Figure 1.31b, in which $B$ is a focus.[20] If the eccentricity is very close to zero, the foci will be very close to the center, and the ellipse is almost circular. If the eccentricity equals zero, then the two foci become one, and the ellipse is a circle.

The second part of Proposition 1.3.8 says that the indicated angles in Figure 1.32 are equal. On a billiard table with walls in the shape of an ellipse, every ball shot from one focus will, after hitting one wall, go to the other focus. This is because balls bounce so that the indicated angles are equal. Another application of (the 3-dimensional analogue of) this property is in whispering galleries, such as in the Statuary Hall of the Capitol Building in Washington, D.C. The ceiling is an ellipsoid, so if one whispers from one focus all the sound waves will be reflected to the other focus, and someone standing at the other focus will hear the sound distinctly, even when the room is quite crowded.

A more obscure result about ellipses, which is quoted from Apollonius in Kepler's *Astronomia Nova*, is the following, which we include

---

20. Focus is the singular of foci.

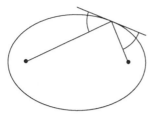

Figure 1.32. The reflecting property of an ellipse.

in order to give a sample of the material from Apollonius that Kepler found useful:

PROPOSITION 1.3.9. *If an ellipse be inscribed within a circle so that its longer axis touches the circle at opposite points, and a diameter be drawn through the center and the points of contact, and perpendiculars be dropped onto the said diameter from other points on the circumference, then all these perpendiculars will be divided by the circumference and the ellipse in the same ratio. Moreover, the area of the ellipse thus inscribed in the circle bears the same ratio to that of the circle as the ratio in which the abovesaid lines are divided.*

This says that in Figure 1.33 the ratio $RQ/PQ$ will be the same for perpendiculars dropped from all points $P$ on the circle, and this ratio will also equal the ratio of the area of the ellipse to the area of the circle.

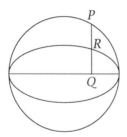

Figure 1.33. Proposition 1.3.9.

This is an easy exercise for a modern student of analytic geometry. The equations of the circle and ellipse are

$$x^2 + y^2 = a^2 \quad \text{and} \quad \frac{x^2}{a^2} + \frac{y^2}{b^2} = 1,$$

respectively, where $a$ is the radius of the circle. Solving for $y$, we see that $y = \sqrt{a^2 - x^2}$ in the first and $y = \frac{b}{a}\sqrt{a^2 - x^2}$ in the second, so that the $y$ value on the ellipse is always $b/a$ times the $y$ value on the circle. The easiest way to understand the statement about area is that the area under the curve $y = \frac{b}{a}f(x)$ equals $b/a$ times the area under the curve $y = f(x)$, for any function $f$. This statement should be clear to anyone who has had some calculus, but calculus isn't necessary: think of the area under the curve as being approximated by the sum of the areas of skinny vertical rectangles under the curve, as in Figure 1.34. If you multiply their heights by a number, then you multiply the sum of their areas by that number. The Greeks had a good grasp on this idea, which is fundamental to calculus.

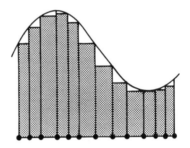

Figure 1.34. Approximating the area under a curve.

The final result about ellipses that we will note played an important role in Kepler's quest to understand planetary orbits, as we shall see later in this chapter. It is similar to some of Apollonius's results.

PROPOSITION 1.3.10. *If an ellipse is inscribed in a circle of radius $r$ so that the principal axis of the ellipse and the diameter of the circle coincide, then*

$$d = r - c\cos(\beta),$$

*where*
    *$d$ = distance from focus to a point on the ellipse,*
    *$c$ = distance from center of circle to focus of ellipse, and*
    *$\beta$ = angle at the center of the circle which the diameter makes with the line out to the point on the circle directly above the point on the ellipse.*
*Moreover, this equation characterizes the ellipse.*

In Figure 1.35, where $F$ is the focus and $R$ the point on the ellipse, $d = FR$, $c = OF$, and $\beta = \angle POQ$. The proof of this result is relegated to Exercise 8.

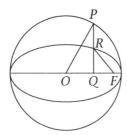

Figure 1.35. Proposition 1.3.10.

### PRE-KEPLERIAN THEORIES OF PLANETARY MOTION

Johannes Kepler (1571–1630) discovered the laws of planetary motion. The theme of Section 1.3 is the way in which he was influenced in his quest by the Greek results about regular polyhedra and conic sections. Before we discuss these events in any detail, we will review the major theories of planetary motion that were popular before Kepler.

The prevailing opinion from 150 A.D. until 1543 A.D. was the one postulated by Ptolemy, a Greek. According to this theory, the sun, moon, and stars all follow circular paths around the earth, while the planets follow complicated paths around the earth called *epicycles*. An epicycle is the path followed by a point that moves around a small circle, which in turn moves around a larger circle. Thus in Figure 1.36 the center $C$ of the smaller circle moves along the larger one, and at the same time the point $M$ moves around the smaller circle. The resulting motion of the point $M$ is as indicated by the looping curve.

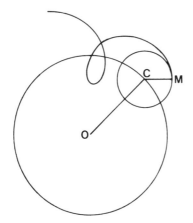

Figure 1.36. M traces an epicycle.

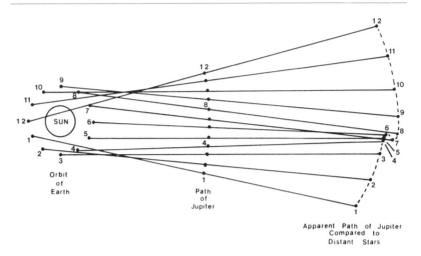

Apparent Path of Jupiter
Compared to
Distant Stars

Figure 1.37. Why Jupiter's path appears to back up.

Such a description was required in order to explain the oscillating paths of the planets in the sky with respect to the fixed stars. The apparent path of Jupiter over the course of a year is the dashed curve in Figure 1.37. In this figure we can see the actual cause of the backtracking in the path: while the earth travels completely around the sun, Jupiter covers only $1/12$ of its orbit. The lines going out from the earth past Jupiter at any of the months of the year show how Jupiter would be viewed from the earth in relation to the distant stars. Thus Jupiter appears to us hardly to move during months 4, 5, and 6. Then its orbit seems to back up for a month before reversing again at month 7, after it continues in its original direction.

Ptolemy's epicycles explained this reasonably well, although for many of the planets more than two circles were required in order to get reasonably good agreement with observation. Ptolemy required forty circles in all. Another problem had to do with the speed of the planets. As we will see when we study Kepler's laws, the planets do not travel at constant speed. In order to explain this, Ptolemy used *equants*, which are points displaced from the center of the circle. Ptolemy's theory was that the planets swept out angle from the equant at constant rate, rather than the more natural notion of sweeping out angle from the center of the circle at constant rate. Thus the farther a planet is from the equant, the faster it must move. For example, in Figure 1.38 the two angles from the equant $E$ are equal, and so the planet will have to pass $CD$ in the same time it takes it to pass $AB$.

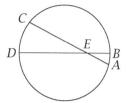

Figure 1.38. If $E$ is an equant, then $CD$ must be passed in the same time as $AB$.

The flexibility of choosing equants gave Ptolemy's theory even a better chance of agreeing with observations. There was a great philosophical prejudice in favor of circular motion and constant speed, and so Ptolemy's awkward system prevailed for fourteen hundred years. The idea that the earth is at the center of the universe was very appealing to the religious people of the Middle Ages.

At least one Greek was a proponent of the heliocentric (sun-centered) solar system—Aristarchus, who lived around 280 B.C., long before Ptolemy. He put forward the theory that the earth and other planets all follow circular orbits, with the sun at the center. His theory correctly explained why the planets do some backtracking in their orbits, as shown in Figure 1.37. However, as we shall see from Kepler's laws, the orbits are actually ellipses with the sun at one focus. For many of the planets, the ellipses are quite closely approximated by circles, except for Mercury, where the eccentricity is 0.2. This means that the sun is displaced along the major axis 20% from the center of the orbit of Mercury. This fact and other matters, such as the nonconstant speed of the planets, caused Aristarchus's theory to be given only slight attention, since it deviated too much from observed data. Even he was not convinced enough to write a book about it, and we know about his theory only from passing remarks made by other Greek writers such as Archimedes and Plutarch.

The Polish astronomer Nicholas Copernicus (1473–1543) is generally credited with being the first successful proponent of the heliocentric theory. He distributed a brief manuscript to a few scholars in 1514 stating his belief that the planets revolved around the sun, but he did not present much evidence for it in that manuscript. His theory developed an underground reputation and finally enticed Joachim Rheticus, a German, to study under him. It was only after much prodding from Rheticus that Copernicus allowed a much more detailed manuscript, *De Revolutionibus Orbium Cœlestium*, to be published, just before his death.

Although he was correct about the center of the solar system, Copernicus was incorrect about the nature of the orbits. He was still too fixed with the idea of circular orbits, and in order to get agreement with observations, he had to resort to using epicycles—almost as many as Ptolemy.

Copernicus's book did not have much immediate impact, either on the scientific community or on the church: it took the church seventy-three years to get around to banning it. One reason why Copernicus's book was given little attention was due to its preface, written (but not signed) by a theologian named Osiander, in a way that made it look as if it had been written by Copernicus himself. Osiander was in charge of the printing of the book, and it is unknown whether Copernicus ever even saw this preface, which diminished the revolutionary philosophical impact of the text. For example, it stated: "These hypotheses are not necessarily true or even probable, but if they provide us with a method of calculation which is consistent with observation, this alone is enough."

### KEPLER AND THE REGULAR POLYHEDRA

As we will see in the next subsection, Johannes Kepler eventually gave the correct explanation of planetary motion; but prior to his correct discovery, he had proposed several incorrect ones. The most spectacular of these was based on the five regular polyhedra, which he discussed in detail in his first book, *Mysterium Cosmographicum* (Cosmic Mystery), published in 1596.

Kepler was an avid spokesman for Copernicus's heliocentric theory, in large part because it gave a better explanation of certain qualitative aspects of planetary motion than Ptolemy's theory did, such as *why* the apparent motion of the inferior planets (Mercury and Venus) was so different than that of the superior planets (Mars, Jupiter, and Saturn), and both were so different from that of the sun and moon. But he was dissatisfied with Copernicus's use of epicycles and sought a simpler explanation that was consistent with observations. Like virtually everyone else at the time, Kepler was deeply religious and felt that God must have had good reasons for making the universe the way he did.

Along these lines, Kepler was concerned with questions such as *why* are there exactly six planets (Uranus, Neptune, and Pluto had not yet been discovered). He seriously considered, but then dismissed, the theory that there were six because 6 was the first perfect number.[21] When teaching one day, he drew on the blackboard

---

21. See Section 3.1 for a discussion of perfect numbers.

Figure 1.39. The figure Kepler drew on the board.

an equilateral triangle with inscribed and circumscribed circles, as shown in Figure 1.39.

Kepler immediately noticed that the ratio of the radii of the circles appeared to be the same as the ratio of the radii of the orbits of Jupiter and Saturn.[22] He immediately concluded that this could not be a coincidence, and extended the drawing as in Figure 1.40, with a square inscribed in the circle, then another circle and inscribed pentagon, and so on, convinced that the circles would have to agree with the orbits of the planets.

Figure 1.40. Kepler's expansion of Figure 1.39 to include more planets.

Kepler was disappointed when he still didn't get good agreement, but he didn't let this shortcoming stop him. He next concluded that since the universe is 3-dimensional, he should be using the regular polyhedra instead of flat polygons, and then he was struck that his theory must be correct, because the five regular polyhedra would determine six spheres from the outer one to the inner one, and this *must* be the reason that there are six planets. It couldn't be just a coincidence that there are five regular polyhedra and six planets! There are 120 ways to arrange the five regular polyhedra in order, and one of these should yield spheres that agree quite well with the orbits of the planets. The ordering that worked best is the one pic-

22. In Exercise 9, you will compare these ratios.

tured in Figure 1.41, which shows the plans for a physical model he never constructed. The order was Sphere of Saturn, Cube, Sphere of Jupiter, Tetrahedron, Sphere of Mars, Dodecahedron, Sphere of Earth, Icosahedron, Sphere of Venus, Octahedron, Sphere of Mercury. In his "Preface to the Reader" in *Mysterium Cosmographicum*, he wrote:

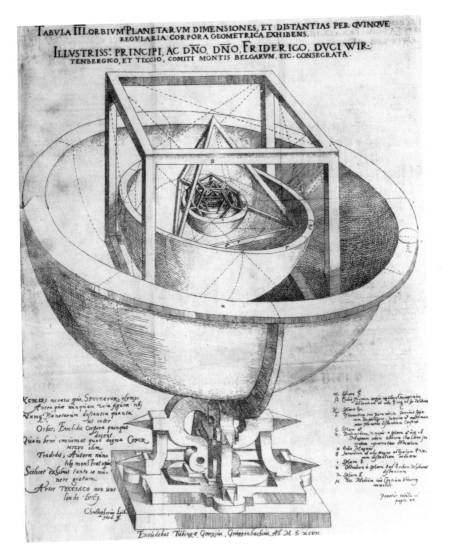

Figure 1.41. A model, never constructed, of Kepler's theory.

It is amazing! Although I had as yet no clear idea of the order in which the perfect solids had to be arranged, I nevertheless succeeded ... in arranging them so happily that later on, when I checked the matter over, I had nothing to alter. Now I no longer regretted the lost time; I no longer tired of my work; I shied from no computation, however difficult. Day and night I spent with calculations to see whether the proposition that I had formulated tallied with the Copernican orbits or whether my joy would be carried away by the winds .... Within a few days everything fell into place. I saw one symmetrical solid after the other fit in so precisely between the appropriate orbits, that if a peasant were to ask you on what kind of hook the heavens are fastened so they don't fall down, it will be easy for thee to answer him.[23]

The agreement this model gives for orbital radii is not great. Some of the planets were off by about 20%, which Kepler attributed in part to inaccuracies in the tabulated data and in part to the need for more refinement. Although he maintained an affinity for this idea throughout his life, luckily he moved on to others.

*KEPLER'S LAWS*

Kepler's fame rests not with his far-fetched theory relating regular polyhedra to planetary orbits but rather to the three laws of planetary motion he described in his books *Astronomia Nova* (New Astronomy, 1609) and *Harmonice Mundi* (Harmony of the World, 1618). The laws state the following:

1. The path of each planet is an ellipse with the sun at one focus.
2. The speed of a planet varies so that it sweeps out area at a constant rate.
3. The square of the period of revolution for any planet is proportional to the cube of its average distance from the sun.

The second law, while maintaining a form of constant speed, implies that a planet will move faster when it is closer to the sun so that the area swept out per unit of time is as large as it is when the planet is farther from the sun. In Figure 1.42, the areas of the two wedges are equal, so the time required to travel from $P_1$ to $P_2$ will equal the time required to travel from $P_1'$ to $P_2'$.

---

23. All of our Kepler quotations are taken from Koestler, *The Sleepwalkers*, which has, in turn, taken them from Kepler's *Collected Works*. This one is from Koestler 1959, p. 253.

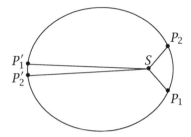

Figure 1.42. Kepler's first and second laws.

Kepler did not really *prove* that these laws had to be true. That proof did not come until sixty years later, when Isaac Newton, using his newly developed calculus, proved that Kepler's laws were a consequence of Newton's law of gravity. Newton was happy to have Kepler's laws as a verification of his law of gravity, and honored Kepler with the famous words: "If I have seen farther than others, it is because I stand on the shoulders of giants."

Kepler merely demonstrated that his laws gave better agreement with observed data than previous theories, and were much simpler than the previous theories, for they eliminated the need for the epicycles used by Ptolemy and Copernicus. He also had better data to work with, for he spent several key years as assistant to Tycho Brahe, a wealthy Danish astronomer who had privately compiled the best tables of astronomical data using the finest pretelescope observational devices.

Kepler did make some attempts at what we would call proof today, and this is one of his great advances—looking for *reasons* rather than just agreement with observations. He had a primitive idea of force, which was not really treated satisfactorily until Newton. His idea was akin to the sun sweeping the planets around in their orbits, as shown in Figure 1.43, taken from his *Astronomia Nova*.

Using this notion and facts about ellipses taken from Apollonius, such as 1.3.9, he did present some arguments for his theory, but the aesthetic appeal was much more convincing than his mathematical arguments. En route from circular to elliptical orbits, Kepler spent some time convinced that the orbits must be oval-shaped, a shape somewhat like an ellipse but more egg-shaped. In a letter to a friend in 1603, he wrote: "If only the shape were a perfect ellipse, all the answers could be found in Archimedes' and Apollonius' work."[24] But when he realized that the orbit of Mars appeared to satisfy the prop-

24. Koestler 1959, p. 335.

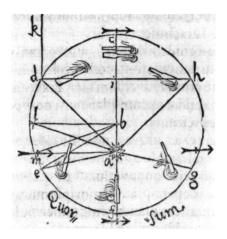

Figure 1.43. Kepler's idea of force.

erty described in our Proposition 1.3.10, he at first failed to realize that this property characterized an ellipse. He took many mis-
steps along the way, but his prodigious capacity for work carried
him through. Typically, he wrote in *Astronomia Nova*: "If thou, dear
reader, art bored with this wearisome method of calculation, take
pity on me who had to go through at least seventy repetitions of it,
at a very great loss of time; nor wilst thou be surprised that by now
the fifth year is nearly past since I took on Mars."[25]

It is safe to say that without the extensive work on ellipses performed by Apollonius eighteen centuries earlier, Kepler would have
been much less likely to achieve his discovery. (For a fascinating
nontechnical discussion of Kepler's circuitous route to his discovery, as well as an expansion on the aspects of his life, beliefs, and
personality that we will discuss in the "Focus" that follows, see *The
Sleepwalkers*.)

*FOCUS: KEPLER*

Kepler formed the boundary between the era of mysticism and religion and the era of scientific thought. He clung to many of the ideas
of the earlier period while initiating the ideas of the later one. One of
his mystical fascinations was with the regular polyhedra. Plato had
associated the four easily constructed polyhedra—the tetrahedron,
octahedron, icosahedron, and cube—with the four "elements" of fire,
air, water, and earth, respectively, while the dodecahedron was asso-

25. Koestler 1959, p. 325.

Johannes Kepler

ciated with the universe. Kepler explained this as follows. The cube is associated with the earth because it is most stable. The octahedron is easily spun, and so is air. The volume/surface area ratio is smallest for the tetrahedron and (he thought, incorrectly) largest for the icosahedron. This ratio is related to wetness, hence the associations of these with fire and water. Finally, the dodecahedron is associated with the universe because its twelve faces correspond to the twelve signs of the Zodiac. He pictured the polyhedra as shown in Figure 1.44.

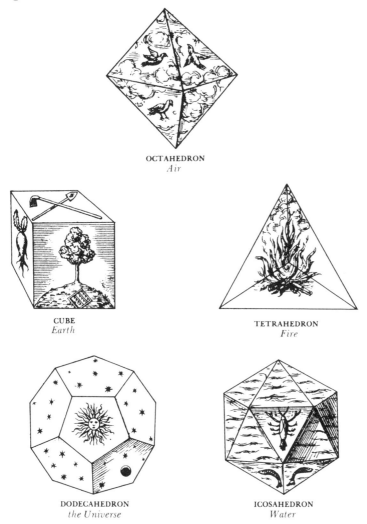

Figure 1.44. Kepler's drawings of the regular polyhedra.

Another strange idea was his "music of the spheres." He suggested that each planet corresponds to a sequence of musical notes (Fig. 1.45). The pitches of the notes are directly related to the speed of the planets (so that the closer, faster, planets have higher notes), and the range of the notes is related to the eccentricity of the orbits (so that Venus, with the roundest orbit emits only one note, while Mercury, with the flattest, emits the most).

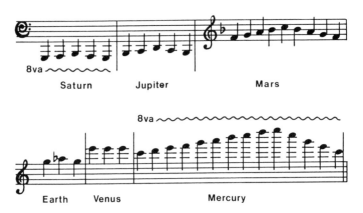

Figure 1.45. Kepler's tunes for the planets.

Early in his career, Kepler was a mathematics teacher. In this capacity he was also expected to serve as astrologer, which he did not enjoy. He wrote: "A mind accompanied to mathematical deduction, when confronted with the faulty foundations of astrology, resists a long, long time, like an obstinate mule, until compelled by beating and curses to put its foot into that dirty puddle."[26] He made some lucky predictions early on; for example, he once predicted cold weather, and indeed it got so cold that "it is reliably reported that when [people] arrive home and blow their noses, the noses fall off."[27] These predictions qualified him for the position of court astrologer to the Duke of Wallenstein.

Kepler's great accomplishments seem even greater when one reads the miserable details of his life. His health was terrible: at age 4 he almost died of smallpox, which caused permanent damage to his hands and eyes; at age 14-15 he suffered continually from skin ailments; at age 19-20 he suffered from the mange. A description of his relatives reads like a page out of the worst of Dickens: his parents

26. Koestler 1959, p. 245.
27. Koestler 1959, p. 244.

quarreled incessantly; his father was a drunkard who turned from being a soldier of fortune to keeping a tavern; at one point his father made a serious threat to sell one of his brothers. In addition, his mother was accused of witchcraft in 1615, after Kepler had already become quite famous, and he spent portions of four years defending her against the charges. The case was initiated because of a quarrel his mother had with a woman, who then accused her of trying to poison her with a potion, and because of a girl who felt a pain in her arm whenever the mother passed by.

His first marriage, which lasted fourteen years, was without joy. His wife did not appreciate his work, and he in turn seemed to have no respect for her. (He wrote, "She gave birth with difficulty. Everything else is in the same vein."[28]) Naturally, they quarreled. Their first child was born with genitals so deformed that "their composition looked like a boiled turtle in its shell,"[29] which Kepler attributed to the fact that his wife liked to eat turtles. This baby soon died, of cerebral meningitis, as did two of their other five children. His wife died at the age of 37. His second marriage, two years later, turned out much better, it seems; perhaps his good fortune was due to the fact that he had performed an almost scientific analysis of the attributes of each of the eleven candidates his friends had selected as a potential wife for him.

While a schoolboy studying to be a priest, Kepler was already intensely disliked, being considered an intolerable egghead. When living with Tycho Brahe, he was in constant battle with him to gain access to data. Finally, he was persecuted for being Protestant in a predominantly Catholic region. However, he was never persecuted for his scientific discoveries, as was Galileo, who lived at approximately the same time.

## Exercises

1. Make physical models of the five regular polyhedra. Use the model for the dodecahedron given in Figure 1.27, the model in Figure 1.46 for the icosahedron, and construct your own models for the rest.
2. Tabulate $v$, $e$, $f$, and $v - e + f$ for each of the four nonregular polyhedra sketched in Figure 1.21.
3. Draw the projections onto a plane of two of the other regular polyhedra, similarly to what was done for the cube in Figure 1.28.

28. Koestler 1959, p. 275.
29. Koestler 1959, p. 276.

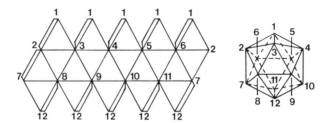

Figure 1.46. A model for the icosahedron.

4. Apply the reduction procedure used in the proof of Euler's formula (1.3.4) to the network in Figure 1.28, showing the values of $v$, $e$, $r$, and $v - e + r$ at each step.
5. Use equations (1.3.5), (1.3.6), and (1.3.7) to find $v$, $e$, and $f$ for a regular polyhedron with $p = 4$ and $q = 3$.
6.* Show that any convex polyhedron whose faces consist of some squares and some regular hexagons must have exactly six squares. Use the following hints. Using angle size show that there must be exactly three faces, and hence exactly three edges, meeting at each vertex. Thus you can use (1.3.6) with $q = 3$. Let $f_4$ denote the number of squares and $f_6$ the number of hexagons. Then you can use Euler's formula with $f$ replaced by $f_4 + f_6$, and you can use (1.3.5) with $fp$ replaced by $4f_4 + 6f_6$. These three equations should allow you to show $f_4 = 6$. Your procedure should be to eliminate $v$'s, then eliminate $e$'s, and the $f_6$'s will luckily cancel out.
7.* Problem 6 did not give enough information to deduce the number of hexagons. In fact, there might have been none. (What figure would that have been?) But suppose in addition that each square is surrounded by hexagons, and each hexagon is surrounded by three squares and three hexagons. How many hexagons, edges, and vertices are there then? (Hint: Count the number of hexagon-square intersections, and deduce $4f_4 = 3f_6$.)
8. Prove Proposition 1.3.10. You will need to know that $\cos(\beta) = x/r$ by the definition of the cosine, and that the equation of the ellipse can be written as

$$\frac{x^2}{r^2} + \frac{y^2}{r^2 - c^2} = 1.$$

9. The ratio of the average radius of Saturn to that of Jupiter is 1.827. What is the ratio of the radii of the circles in Figure 1.39, which Kepler thought equal to this?

# 2. Non-Euclidean Geometry

## 2.1 Formal Axiom Systems

In non-Euclidean geometry, words are sometimes used in a way that seems nonintuitive at first. To prepare the reader to prove theorems about unfamiliar entities, we now introduce Formal Axiom Systems.

A *Formal Axiom System* consists of some Undefined Terms, possibly some Defined Terms (defined in terms of the undefined terms), some statements called Axioms, which are assumed to be true without justification, and other statements called Theorems, which can be logically deduced from the axioms. The axioms are, in effect, telling you everything you need to know about the undefined terms.

We begin with an example that shows how theorems about nonintuitive concepts can be proved.

AXIOM SYSTEM 2.1.1. *The undefined terms are* pron, lem, *and* flunt. *The axioms are:*
>  Ax P1.  *There is at least one pron.*
>  Ax P2.  *For any two distinct lems, there is one and only one pron that flunts both of them.*
>  Ax P3.  *Each pron flunts at least two lems.*
>  Ax P4.  *For any pron, there is at least one lem which it does not flunt.*

Clearly, prons and lems are things, and flunting is something that prons do to lems. We have labeled the axioms with the letter P (for pron), and will label the theorems of this axiom system similarly. We will use different letters for different axiom systems. In Ax P2, "any" means "every," "distinct" means "not the same," and the pron that flunts the two lems in question might also flunt some other lems.

Now we can prove our first theorem.

THEOREM P1. *There are at least three lems.*

*Proof.* By Axiom P1, there is at least one pron. By Axiom P3, there are at least two lems that it flunts, and by Axiom P4, there is another lem that it doesn't flunt. Thus there are at least three lems. Q.E.D.

How did we come up with that proof? Mostly it was by studying the axioms and seeing what might yield the desired result. Both P3 and P4 were good in getting existence of lems, which is what we wanted, but in order to use them, we first had to have a way of asserting existence of prons, which is what P1 gave us. Note that if Axiom P1 had not been there, it would have been possible to have no prons and no lems. Axiom P3 by itself does not guarantee existence of a pron—it just says that if you have one, then it will flunt two lems.

To prove more complicated theorems, and to formulate theorems, it is helpful to have a way of picturing prons, lems, and flunting. The simplest way is probably to represent prons by the letter P, lems by the letter L, and flunting by a line from the pron to the lem. These pictures are not part of the proof, just as Euclid's proofs should really not have used pictures as anything more than a guide.

THEOREM P2. *If there are exactly three lems, then there are exactly three prons.*

*Proof.* Let L1, L2, and L3 be the three lems. Let P12 be the pron that flunts L1 and L2; P12 exists and is unique by Axiom P2. We can call this pron anything we want, and P12 seems like a good name, since it flunts L1 and L2. Axiom P2 does not prohibit P12 from also flunting L3, but Axiom P4 does. Since P12 does not flunt L3, there must be a different pron that flunts L2 and L3; call it P23. By Axiom P4 again, P23 does not flunt L1, and so a third pron is required to flunt L1 and L3; call it P13. There can be no prons other than the three just named, because if there were, it would have to flunt at least 2 lems by Axiom P3; but that would give us more than one pron flunting those two lems, in contradiction to Axiom P2. This completes the proof, which may be pictured as in Figure 2.1a. Q.E.D.

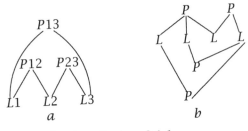

Figure 2.1. Models of Axiom System 2.1.1.

The example constructed in Figure 2.1a is what is called a model for Axiom System 2.1.1. A *model* for an axiom system is a way of interpreting the undefined terms in such a way that all the axioms are satisfied. Our interpretation is:

pron = letter P possibly followed by other symbols
lem = letter L possibly followed by other symbols
to flunt = to connect the letters representing a pron and a lem by a line

To establish that Figure 2.1a is a model of Axiom System 2.1.1, you must verify that each of the axioms is satisfied under the interpretation. For example, to verify Ax P3, you must check that each letter *P* in Figure 2.1a is connected by lines to at least two letter *L*'s.

Every theorem of an axiom system will necessarily be true in every model of the axiom system. This is the case because the axioms are true in the model, and the theorems are logical consequences of the axioms, and so they must be true, too. Another way of saying the same thing is that a statement which is false in some model of an axiom system cannot possibly be a theorem of that axiom system. For example, there is no hope of proving that "there are at least four lems" is a theorem of Axiom System 2.1.1 because it is not true in the above model.

On the other hand, a statement that is true of a model of an axiom system is not necessarily a theorem of that axiom system. In other words, it might not be provable from the axioms. For example, the statement, "There are exactly three lems," is true of the model of Axiom System 2.1.1 given in Figure 2.1a, but this statement is not a theorem of this axiom system, because Figure 2.1b gives another model of the axiom system in which this statement is not true.

Here is another axiom system.

AXIOM SYSTEM 2.1.2. *Undefined terms are* **corn** *and* **tweak**. *Axioms are:*
   Ax C1. *If A and B are distinct corns, then A tweaks B or B tweaks A. (The possibility of both happening is not excluded by this axiom.)*
   Ax C2. *No corn tweaks itself.*
   Ax C3. *If A, B, and C are corns such that A tweaks B and B tweaks C, then A tweaks C. (Some of A, B, and C here might be the same.) (This axiom is usually called a transitive law.)*
   Ax C4. *There are exactly four corns.*

We will prove two theorems about this axiom system.

THEOREM C1. *If A tweaks B, then B does not tweak A.*

*Proof.* Let $C = A$ in Axiom C3. (This is allowed.) It says "If *A* tweaks *B* and *B* tweaks *A*, then *A* tweaks *A*." Axiom C2 says that *A* cannot tweak itself. Thus, by the contrapositive of the version of Axiom C3 just cited, we deduce that we cannot have both *A* tweaking *B* and *B* tweaking *A*. Q.E.D.

The reader may be thinking, "Why didn't we just use the information of Theorem C1 as part of AxC1?" Certainly we could have done so; it would have eliminated our need to make this argument, which involves applying AxC3 with $C = A$, a step some readers may find awkward. The answer is that one generally tries to include the minimal amount of information in the axioms, leaving all redundant facts to be proved as theorems. The reader should not be overly concerned about this fine point, as it will not be used in our subsequent arguments.

THEOREM C2. *There is exactly one corn that tweaks all the others.*

*Proof.* We first note that there couldn't be more than one corn that tweaks all the others, for then they would tweak each other, in contradiction to Theorem C1. Thus it suffices to show that there is a corn that tweaks all the others.

Choose any two corns. By Axiom C1, one of them tweaks the other. We name it $C_1$, and name the other corn of that pair $C_2$. Now pick a third corn. By Axiom C1 again, either it tweaks $C_1$ or $C_1$ tweaks it. In the first case, it also tweaks $C_2$, by Axiom C3. So in either case, out of the three corns considered so far, there is one that tweaks both of the others. Let's name this champion of the three $C_0$. (In the second of the two cases being considered, $C_1$ must be renamed.)

Now consider the fourth corn. Either it tweaks $C_0$ or else $C_0$ tweaks it. In the second case, $C_0$ tweaks all the corns except itself. In the first case, the transitivity axiom implies that the fourth corn also tweaks the two corns tweaked by $C_0$, and so this fourth corn tweaks all the corns except itself. Thus, in either case, we have a corn that tweaks all the other corns. Q.E.D.

*CONSISTENCY*

An axiom system is said to be *inconsistent* if it implies a contradiction, that is, if it is possible to prove in it that some statement is both true and false. The statement being proved both true and false can frequently be chosen to be one of the axioms, which is trivially "proved" to be true. For example, consider the following axiom system.

AXIOM SYSTEM 2.1.3. *Undefined terms are X, Y, and* like. *Axioms are the following:*

    *Ax X1. There are exactly 2 X's.*
    *Ax X2. There are exactly 3 Y's.*
    *Ax X3. Each X likes exactly 2 Y's.*
    *Ax X4. No two X's like the same Y.*

Any one of these axioms can be proved to be false using the other axioms. Thus the axiom system is inconsistent. For example, axioms X1, X3, and X4 easily imply that there are at least 4 Y's, since one of the two X's will like two Y's, and the other X will like two different Y's. Thus the statement, "There are exactly 3 Y's." can be proved both true and false—true because it is assumed to be true, and false by the argument just presented.

Thus it is sometimes easy to prove an axiom system is inconsistent—we just need to find a contradiction. But how can we prove that an axiom system is consistent, that is, that it implies no contradictions? We cannot possibly list all the theorems it implies and then check that no two contradict each other. Luckily, there is an easier way to do this.

THEOREM 2.1.4. *If an axiom system has a model, then it is consistent.*

To see that this statement is true, we need merely to observe that all theorems provable in the axiom system must be satisfied by the model (we noted this earlier—the axioms are satisfied by the model and hence so are the logical consequences of them), and since a model is something that actually exists, there cannot be contradictory statements that are true of the model.

**Example.** Prove that the following axiom system is consistent.

AXIOM SYSTEM 2.1.5. *Undefined terms are* **post, bird,** *and* **mark.** *Axioms are:*

Ax M1. *There are at least two posts.*

Ax M2. *For any two distinct posts, there is exactly one bird that marks them both.*

Ax M3. *No bird marks every post.*

Ax M4. *Given any bird B and any post P that it does not mark, there exists exactly one bird that marks P but does not mark any of the same posts as B.*

We try to make a model. We start with two posts (Ax M1), a bird marking both (Ax M2), and a post that it doesn't mark (Ax M3), as illustrated in Figure 2.2.

Figure 2.2. A bird marking two posts.

Now, Ax M2 requires the addition of two more birds to mark the pairs P1-P3 and P2-P3 (Fig. 2.3).

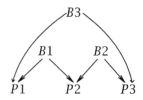

Figure 2.3. Two more birds.

This figure would satisfy the first three axioms, but it does not satisfy Ax M4. For example, we need a bird that marks P3 but does mark either of the posts (P1 and P2) marked by B1. We add a bird B4 which marks P3, and similarly B5 marking P2 (to satisfy Ax M4 for B2 and P2), and B6 marking P1 (see Fig. 2.4).

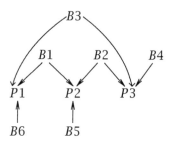

Figure 2.4. Six birds and three posts.

Cursory inspection of this figure might allow you to think it is a model. However, it fails Ax M4 for B4 and P2, since there are two birds (B1 and B5) that mark P2 but do not mark any of the posts marked by B4. The only way to remedy this situation would seem to be to add another post P4, and have both B4 and B5 mark it. Now we are left with a grouping such as in Figure 2.5 without the dashed

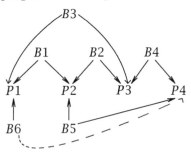

Figure 2.5. Six birds and four posts.

line, which still causes a problem for Ax M4 with B4 and P1, since P1 is marked by B1 and B6, neither of which mark any of the same posts as B4. However, if we make B6 also mark P4, then we obtain the situation shown in the completed Figure 2.5, which does satisfy all the axioms.

INDEPENDENCE

An axiom in a consistent axiom system is said to be *dependent* if it can be proved using just the other axioms. Thus a dependent axiom is essentially a superfluous axiom, since the same theorems can be proved regardless of whether or not the axiom is included in the list of axioms. An axiom is *independent* if it cannot be proved using just the other axioms.

In order to show that an axiom is dependent, you simply try to prove it, using only the other axioms. But how would you show that an axiom is independent? It is not easy to see how you would argue that this axiom cannot be deduced from the others. Again, models are useful. If you can construct a model in which the other axioms are all true, but the axiom in question is false, then the axiom will be independent. To see why this is true, we note that because all the other axioms are true in this model, every statement that can be deduced from the other axioms must also be true in this model. Since we are assuming that the "axiom in question" is not true in the model, it cannot be deduced from the other axioms and hence is independent.

For example, consider the following axiom system.

AXIOM SYSTEM 2.1.6. *Undefined terms are* **X, Y,** *and* **like.** *Axioms are:*
　*Ax L1. There are exactly three X's.*
　*Ax L2. There are exactly three Y's.*
　*Ax L3. Each X likes exactly two Y's.*
　*Ax L4. No two X's like exactly the same Y's.*
　*Ax L5. Each Y is liked by at least one X.*

We shall show that Axiom L5 is dependent, while all the other axioms are independent. First we note that the axiom system is consistent because it has the model pictured in Figure 2.6. Our remarks about independent and dependent axioms apply only to consistent axiom systems.

Now we prove that Ax L5 is dependent. Use Axioms L1, L2, and L3 to name the $X$'s and $Y$'s in such a way that the $X$'s and $Y$'s are named $X1$, $X2$, $X3$, $Y1$, $Y2$, and $Y3$, and $X1$ likes $Y1$ and $Y2$. Now, Ax L4 tells us that $X2$ cannot like both $Y1$ and $Y2$, and since it likes exactly

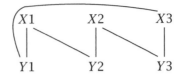

Figure 2.6. A model for Axiom System 2.1.6.

two $Y$'s, it must like one of $Y1$ or $Y2$, and also like $Y3$. Thus all three $Y$'s are liked, and so Ax L5 has been deduced simply by using the other four.

We show each of the other axioms is independent by constructing a model in which the other four axioms are true, but the axiom in question is false. We list our models in Figure 2.7, labeled by the axiom which is false in them.

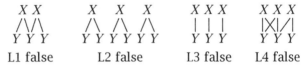

Figure 2.7. Models which show that axioms are independent.

EQUIVALENT AXIOMS

Our final concept involving axiom systems is that of equivalent axioms. Two axiom systems involving the same undefined terms are said to be *equivalent* if the same theorems can be deduced in each of them. This will be true if and only if each of the axioms of either one of the systems can be proved as theorems of the other, for if you can deduce all of the axioms of the other system, then you can deduce all consequences of them. Some axioms may be common to the two systems, but you don't have to worry about deducing them since they are assumed to be true.

It sometimes happens that the two axiom systems agree in all but one of the axioms. The two differing axioms are then said to be *equivalent* if the axiom systems are equivalent, which is to say that the differing axiom of each system can be deduced as a theorem of the other system. Equivalent axioms can be thought of as substitute axioms, for one can be substituted for the other without changing the content of the axiom system. The equivalence of two axioms depends

not only on the two axioms in question, but also on the common axioms of the two axiom systems.

**Example.** We will show that in Axiom System 2.1.1, Ax P1 is equivalent to the axiom,

Ax P1′. There are at least two prons.
If Ax P1′ is substituted for Ax P1 in 2.1.1, then it is a simple matter
to prove Ax P1. Indeed, since we are assuming that there are at least
two prons, it is certainly true that there is at least one pron. On the
other hand, if we assume Ax P1 along with the rest of the axioms
of 2.1.1, then we can deduce that there are at least two prons, as
follows.

Let P be a pron guaranteed by Ax P1. Then Ax P3 guarantees that
there is a lem L1 that is flunted by P, and Ax P4 guarantees that there
is another lem L2 that is not flunted by P. By Ax P2, there is a pron
that flunts both L1 and L2, and it cannot be P since P does not flunt
L2. Thus there are at least two prons.

Thus Ax P1′ can be substituted for Ax P1 in 2.1.1 without changing
the content of the axiom system. Many widely differing statements
can be used as axioms equivalent to Euclid's fifth postulate; this will
play an important role in the next section.

### HILBERT'S FOUNDATIOMS OF GEOMETRY

The main reason we consider formal axiom systems is to prepare
ourselves for the break from intuition required in order to study non-
Euclidean geometry. The notions of consistency, independence, and
equivalence of axioms will play a role in this study. However, as we
noted in Section 1.2, Euclidean geometry, as set out by Euclid, isn't
really a formal axiom system. He did not utilize undefined terms, and
he used intuition heavily in his proofs, for example when he asserted
that circles intersect to prove Proposition 1 and when he superposed
triangles to prove the Side-Angle-Side Theorem. Such criticisms were
already being made in the 1500s. It was not until the discovery of
non-Euclidean geometry in the 1800s that the distinction between
geometry as a formal axiom system, and geometry as the study of
the physical universe, was made.

The great German mathematician Gauss, who, as we shall see in
the next section, was one of the prediscoverers of non-Euclidean geometry, complained in a letter in 1832 that such intuitive notions as
the meaning of "the inside of a triangle" and "between" must be formalized. Morris Kline summarized the problem with Euclid's treatment nicely with the statement, "Euclidean geometry was supposed

to have offered accurate proofs of theorems suggested intuitively by figures, but actually it offered intuitive proofs of accurately drawn figures."[1]

David Hilbert (1862–1943) succeeded in making Euclidean geometry rigorous by twentieth century standards in his book *Foundations of Geometry*, first published in 1899. Hilbert was, along with Poincaré, one of the two most famous mathematicians in the world at the time. At the International Congress of Mathematicians in Paris in 1900, he gave a famous lecture in which he stated what he considered to be the twenty-three most important problems in mathematics. Some of those have been solved, and others are still being studied. In 1974 a major conference was held at Northern Illinois University to discuss the status of Hilbert's problems.

We present a somewhat abbreviated list of Hilbert's axioms for plane geometry, mainly to give the reader a flavor of what is required to do Euclidean geometry really rigorously. We will not make a detailed study of Hilbert's presentation, but will want to be able to say things such as "Lobachevsky's non-Euclidean geometry can be presented formally in a way similar to Hilbert's treatment of Euclidean geometry."

The undefined terms are *point, line, lies on (or is incident with or is on), congruence* (sometimes written ≡), and *between.* The idea of congruence of lines and congruence of angles as being undefined terms will be very important when we look at some models of hyperbolic geometry in Section 2.5. When working with these words, you must try to divorce yourself from your intuitive ideas about them. Hilbert wrote, "One must be able to say at all times—instead of points, straight lines, and planes—tables, chairs, and beer mugs."

The axioms are divided into five groups.

### Group 1: Axioms of Incidence
Ax1. Two distinct points $A$ and $B$ are incident with one and only one line, which we sometimes write as $AB$ or $BA$.

Ax2. Each line is incident with at least two points. There are at least three points, not all incident with the same line.

### Group 2: Axioms of Order
Ax3. If point $B$ is between points $A$ and $C$, then $A$, $B$, and $C$ are three distinct points on a line, and $B$ is also between $C$ and $A$.

Ax4. If $A$ and $C$ are two points on a line, then there is at least one point $B$ between $A$ and $C$.

---

1. Kline 1972, p. 1007

Ax5. Of any three points on a line, there is always one and only one which is between the other two.

Now that the properties of "between" have been spelled out, we can define a few new words. The *segment AB* consists of the points which lie on the line *AB* and are between *A* and *B*. The *ray* from *A* through *B* consists of the points *C* on the line *AB* such that *A* is not between *B* and *C*. Any point *A* on line $\ell$ determines two rays, which have only *A* in common. Two points different from *A* in the same ray from *A* are said to be *on the same side of A on line* $\ell$. Two rays *h* and *k* having a common end point *A* but not lying in the same line are said to form an *angle*, written $\angle(h, k)$. The terms *same side of a line in the plane, interior point of an angle*, and *triangle* are also defined.

Ax6. Let *A*, *B*, and *C* be three points not on the same line, and let $\ell$ be a line not passing through any of the points *A*, *B*, and *C*. Then, if line $\ell$ passes through a point of the segment *AB*, it will also pass through a point of the segment *BC* or a point of the segment *AC*.

### Group 3: Axioms of Congruence

Ax7. If *A* and *B* are two points on a line $\ell$, and if *A′* is a point on the same or another line $\ell'$, then on a given side of *A′* on the line $\ell'$, we can find one and only one point *B′* so that the segment *AB* is congruent to ($\equiv$) the segment *A′B′*.

Ax8. If $AB \equiv A'B'$ and $A'B' \equiv A''B''$, then $AB \equiv A''B''$.

Ax9. Let *AB* and *BC* be two segments of a straight line $\ell$ which have no points in common aside from the point *B*, and let *A′B′* and *B′C′* be two segments of the same or another straight line $\ell'$ having no other point than *B′* in common. Then, if $AB \equiv A'B'$ and $BC \equiv B'C'$, we have $AC \equiv A'C'$.

Ax10. Let an angle $(h, k)$, a ray *h′* along a line $\ell'$, and a prescribed side of the line $\ell'$ be given. Then there is one and only one ray *k′* such that the angle $(h, k)$ is congruent to the angle $(h', k')$ and at the same time all interior points of the angle $(h', k')$ lie upon the given side of $\ell'$.

Ax11. If, in the two triangles *ABC* and *A′B′C′*, the congruences $AB \equiv A'B'$, $AC \equiv A'C'$, and $\angle BAC \equiv \angle B'A'C'$ hold, then the congruence $\angle ABC \equiv \angle A'B'C'$ also holds.

### Group 4: Euclid's Parallel Postulate

Ax12. Through any point *A*, not incident with a line $\ell$, one and only one line can be drawn which does not intersect line $\ell$.

*Group 5: Archimedean Axiom*

Ax13. Let $A_1$ be any point upon a line between points $A$ and $B$. Take the points $A_2, A_3, \ldots$ so that $A_1$ lies between $A$ and $A_2$, $A_2$ between $A_1$ and $A_3$, $A_3$ between $A_2$ and $A_4$, etc. Moreover, let the segments $AA_1, A_1A_2, A_2A_3, \ldots$ be congruent to one another. Then, in this sequence of points, there always exists a certain point $A_n$ such that $B$ lies between $A$ and $A_n$.

Note that Axiom 11 contains the essence of the Side-Angle-Side Theorem. It tells that the other angles are equal.[2] That the other sides are equal requires proof, which we shall omit.

### ASIDE: RUSSELL'S PARADOX AND SET THEORY

Along with the efforts of mathematicians to formalize Euclid's geometry were efforts to make other areas of mathematics equally rigorous. One of these areas was set theory, which had been developed into a major area of mathematics by Georg Cantor in the 1870s, with his discovery that some infinite sets are bigger than others. The word to use when comparing sizes of sets is *cardinality.*

Sets $S$ and $T$ are said to *have the same cardinality* if there is a one-to-one correspondence between the elements of each. A one-to-one correspondence is a way of associating elements of $T$ to elements of $S$ so that each element of $T$ corresponds to exactly one element of $S$, and vice versa. The set $T$ has greater cardinality than $S$ if $S$ can be placed in one-to-one correspondence with some subset of $T$ but not with all of $T$. Cantor showed that the set of all real numbers, represented by their decimal expansions, has greater cardinality than the set of all positive integers. His argument is sketched in Exercise 12, and you are asked to fill in the details.

At the time of Cantor's work, a set was thought of as any collection of objects that could be described. In 1902 the British philosopher Bertrand Russell (1872–1970) discovered that this interpretation can lead to a paradox. First note that some sets are members of themselves, while most are not. For example, the set of all abstract concepts is an abstract concept, and the set of all sets is a set, while the set of all apples is not an apple. Let $H$ denote the set of all sets that are not members of themselves. Thus the set of all apples is a member of $H$, while the set of all ideas is not in $H$. Now you might ask, is $H$ a member of $H$? If so, then, by the definition of $H$, $H$ is

---

2. We use "equal" as a synonym for "congruent."

not a member of itself, that is, it is not a member of $H$. On the other hand, if $H$ is not a member of $H$, then the definition of $H$ implies that $H$ is a member of itself, and hence it is in $H$. Either way, we are led to a contradiction.

In 1918 Russell posed a more down-to-earth paradox of a similar type. In a certain town there is only one barber, and he claims that he shaves every man who does not shave himself, but does not shave any man who shaves himself. The question is, does this barber shave himself? You are quickly led to a paradox that he shaves himself if and only if he doesn't shave himself. The answer to this paradox is quite easy—no barber can make this claim. A similar answer to the paradox in set theory would say that we are not allowed to talk about a set such as $H$, and that is pretty much the way it was resolved. Russell and A. N. Whitehead gave an elaborate formulation of the concept of a set in a way that prohibited anything involving all members of a set from being a member of the set. Their book which did this, *Principia Mathematica*, required its first 362 pages to prove that $1 + 1 = 2$.

A famous question arising from Cantor's work on cardinality of infinite sets was known as the Continuum Hypothesis, which can be defined as the conjecture that every infinite set of real numbers has the same cardinality as either the set of positive integers or the set of all real numbers. In other words, there can be no set of intermediate size. This problem was famous enough for Hilbert to list it as Problem #1 in his list of twenty-three problems. In 1940 the Austrian logician Kurt Gödel, by then firmly ensconced at the Institute for Advanced Study in Princeton, New Jersey, proved that the Continuum Hypothesis was consistent with the accepted axioms for set theory. This means that he gave a model for set theory in which the Continuum Hypothesis was true, but it does not mean that he *proved* the continuum hypothesis using the axioms of set theory. There was now no further hope of proving that the Continuum Hypothesis is false. A simple analogy would be that the model we gave early in this section showed that the statement, "There are exactly 3 lems," is consistent with Axiom System 2.1.1. But you can't prove that there have to be exactly 3 lems because there are also models of 2.1.1 in which there are more than 3 lems.

Gödel's result made it all the more tantalizing to try to prove the Continuum Hypothesis. Therefore it came as quite a shock in 1963 when Paul Cohen, a former assistant of Gödel's who was by then at Stanford University, proved that the Continuum Hypothesis is independent of the usual axioms of set theory by constructing a model of set theory in which the Continuum Hypothesis is false. So the situation is similar to the number of lems in Axiom System 2.1.1. There, the axioms don't give enough information to decide how many lems there are, while here the usual axioms of set theory do not give enough information to tell whether there are any sets whose cardinality is greater than that of the set of positive integers but less than that of the set of all real numbers. Some set theorists now believe that the axioms of set theory should be changed so that a basic question such as this one can be answered.

### Exercises

1. a. Prove that the following theorem is true in Axiom System 2.1.1. If there are exactly 4 lems, then the number of prons is either 4 or 6. (This means that you must prove that each of these possibilities can occur, and that they are the only possibilities.) (Hint: Divide into cases as to whether or not there is a pron which flunts 3 lems.)
   b.* How many prons can there be if there are exactly 5 lems?
   c.** Try to formulate and prove some general relationships between the number of prons and the number of lems. For example, must the number of prons be at least as large as the number of lems?
2. In Axiom System 2.1.2, replace the 4 in Ax C4 by a 5; that is, assume there are 5 corns instead of 4. State and prove a theorem identical to Theorem C2. (Hint: You may find it convenient to isolate 4 of the 5 corns, and apply to them Theorem C2 proved in the text.)
3. In Axiom System 2.1.2, prove that the 4 corns have a "pecking order." In other words, show that they can be ordered so that each corn tweaks those that come after it in the ordering.
4. Consider the axiom system in which the undefined terms are *cat*, *twit*, and *bug*, and the axioms are:
   Ax B1. There is at least one cat.
   Ax B2. Every cat twits exactly one bug.
   Ax B3. Every cat twits the same bug.
   Ax B4. There is at least one bug that no cat twits.
   State and prove a theorem about how many bugs there are.
5.* Prove the following results about Axiom System 2.1.5. It is not enough to note that they are true in the model given in the text. You must write out an argument that shows that they must always be true (in every model). We will use the following definition: Two birds are *distant* if there is no post that they both mark.

Th 1. Two distinct birds each distant from a third bird are distant from each other.

Th 2. Every bird marks at least one post.

Th 3.** Every bird marks at least two posts.

Th 1 should be useful in proving the others.

6. Tell whether the following axiom system is consistent. Explain your reasoning. Undefined terms are $X$, $Y$, and *related*. Axioms are:

Ax XY1. There are exactly five $X$'s.

Ax XY2. Each $X$ is related to at least two $Y$'s.

Ax XY3. No $X$ is related to every $Y$.

7. Tell whether the following axiom system is consistent. Explain your reasoning. Undefined terms are *dot*, *lion*, and *meet*.

Ax D1. There are exactly four dots.

Ax D2. There are exactly four lions.

Ax D3. For every pair of dots, there is exactly one lion which both meet.

Ax D4. Every lion is met by exactly two dots.

8. In the following axiom system, tell which of the axioms are independent and which are dependent. Explain your reasoning. Undefined terms are *boy*, *girl*, and *is related to*. Axioms are:

Ax BG1. There are at least three boys.

Ax BG2. There are at least four girls.

Ax BG3. Every boy is related to at least one girl.

Ax BG4. No two boys are related to the same girl.

Ax BG5. There is at least one girl who is related to no boys.

9. Given the axiom system whose undefined terms are *bug, cow*, and *annoy*, with axioms:

Ax1. There is at least one bug.

Ax2. For every pair of cows, there is at least one bug that annoys both.

Ax3. Every bug annoys at least two cows.

Ax4. No bug annoys all the cows.

Prove that Ax1 is equivalent to

Ax1′. There are at least two bugs.

10. Explain, using your own words and a sketch, what is being said in Hilbert's Axiom 6, 9, and 13.

11. Show that the set of positive integers has the same cardinality as the set of even positive integers. Show that it has the same cardinality as the set of all integers.

12.* Write out a detailed discussion, perhaps including an illustrative example, of the following sketch of Cantor's argument that the set $R$ of all real numbers, represented by their decimal expansions, has greater cardinality than the set $N$ of positive integers. First you should show that $N$ can be put in one-to-one correspondence with a subset of $R$. That is easy. The tricky part is to show that $N$ cannot be put into one-to-one correspondence with all of $R$. If it could, let $d_1$ be the decimal number in $R$ corresponding to the integer 1, $d_2$ the decimal number corresponding to 2, etc. This listing of decimal numbers $d_1, d_2, \ldots$ cannot include all decimal numbers because one can construct a decimal number that differs from $d_1$ in the first digit, $d_2$ in the second digit, etc.

13. Discuss the following variant of Russell's paradox. Numbers can be described in words using letters and spaces (to separate the words). We will count the spaces along with the letters. For example, 729 can be described in twenty-five letters as "seven hundred twenty nine" and in twenty-one letters as "nine times eighty one." Now you might ask, what is the smallest number not expressible in less than seventy letters? There certainly must be such a number, because the number of such descriptions is less than $27^{70}$. (Explain.) But "the smallest number not describable in less than seventy letters" describes this number in less than seventy letters.

## 2.2   Precursors to Non-Euclidean Geometry

Many mathematicians over the centuries felt dissatisfied with Euclid's fifth postulate. They felt that it was so much more complicated and less intuitive than his other postulates that it shouldn't be included among them. Many attempts were made to prove that it was dependent, that is, that it could be deduced from the other four postulates. Many thought they had succeeded, but they were always making some hidden assumption. Other mathematicians stated that other, more palatable, postulates could be substituted for Euclid's fifth postulate. That is, they found axioms equivalent to Euclid's fifth postulate, and felt that their new axiom was a simpler statement that was more obviously true of space. In this section, we will discuss some of these attacks on Euclid's postulate. Our presentation will not be chronological.

### PLAYFAIR'S POSTULATE

The best-known axiom equivalent to Euclid's fifth postulate is the following.

STATEMENT 2.2.1. *Through a given point not on a given line can be drawn at most one line parallel to the given line.*

This statement is taken as the fifth postulate in many current high school geometry texts. It is called Playfair's Postulate, after the eighteenth-century Scottish mathematician John Playfair, who popularized it, although many mathematicians long before Playfair knew that it was equivalent to Euclid's postulate.

Statement 2.2.1 is a statement about uniqueness of parallels. Existence does not need to be postulated because it can be deduced from Euclid's first four postulates. We showed in Section 1.2 that Euclid's Proposition 31 asserts the existence of parallel lines, and does not require the use of the fifth postulate in its proof. Recall that absolute geometry is the set of statements that can be deduced from Euclid's first four postulates. Then existence of parallel lines is certainly a theorem of absolute geometry, while the question being addressed by most of the mathematicians discussed in this section is whether uniqueness of parallels is also a theorem of absolute geometry. The advantages of Playfair's statement over Euclid's are its somewhat less technical content and the ease of formulating alternatives. Statement 2.2.1 is often stated as "one and only one parallel." The only reason for stating it as we have is that one should try to postulate the minimal amount of information, and since we can prove existence of parallels from the other four postulates, we should not bother to postulate it.

THEOREM 2.2.2. *Statement 2.2.1 is equivalent to Euclid's fifth postulate.*

*Proof.* We must show that assuming either one of these along with Euclid's first four postulates enables us to prove the other. We saw in the corollary to Proposition 30 in Section 1.2 that Statement 2.2.1 could be deduced as a consequence of Euclid's five postulates. Now we want to work the other way. Assume Euclid's first four postulates and Statement 2.2.1, and try to deduce Euclid's fifth postulate as a consequence. We are, of course, allowed to use all the results that Euclid proved without using the fifth postulate. This includes the first twenty-eight propositions and Proposition 31.

To prove Euclid's Postulate 5, we assume in Figure 2.8a that $\angle PQR + \angle QPS < 180°$, and want to deduce that the lines will eventually meet on the right side. By Propositions 23 and 27, we can draw the line $UPT$ so that the indicated angles are equal and $UPT$ is parallel to $QR$. The lines $PT$ and $PS$ are not the same line. Of course, that seems clear from the drawing, but to prove it we note that

$$\angle PQR + \angle QPT = \angle QPU + \angle QPT = 180°,$$

and so, because of the assumption made at the beginning of this paragraph, angle $QPS < \angle QPT$, and so the line $PS$ must lie beneath the line $PT$. Now we can invoke Statement 2.2.1 to say that since $PT$ is parallel to $QR$, $PS$ cannot also be parallel to $QR$. Thus it must intersect $QR$. If it intersected $QR$ on the left side, that would mean that the line $PS$ has to drop below the line $PU$ at some point $V$ on the left side, as shown in Figure 2.8c, which would give two distinct lines passing throught the points $P$ and $V$, in violation of Euclid's Postulate 1. We conclude that the lines must meet on the right side. Q.E.D.

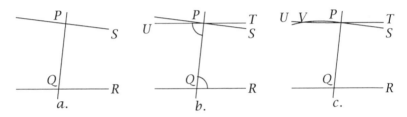

Figure 2.8. Proof of Theorem 2.2.2.

In Figure 2.8c, we have made our first non-Euclidean drawing. You may have objected that the line $SPV$ is not straight. And indeed it is not because it would contradict Euclid's Postulate 1. However, as we move on in our discussion, we will draw other so-called straight lines that look curved to you. Two points about this should be made here. First, the pictures may be greatly exaggerated so that the line $SPV$ doesn't meet $TU$ for a billion miles or more, in which case you would not notice that it doesn't seem straight. Second, no one really knows what space is like in the far reaches of the universe, and so the notion of "straight line" becomes rather nebulous. That is why it should be an undefined concept, satisfying only the properties given in the postulates. Several mathematicians in the late 1700s wrote about being bothered by the meaning of the word "straight." To provide a context for dates mentioned in this section, we note now that the official discovery of non-Euclidean geometry, on which we will focus in the next section, was in 1829.

### *ANGLE SUMS OF TRIANGLES*

Another statement that was long known to be equivalent to Euclid's fifth postulate is the statement that the sum of the angles of every triangle is 180°. This was known to the thirteenth-century Arabian, Nassir-Edin, who wrote a book entitled *Treatise That Heals the Doubt Raised by Parallel Lines.*

THEOREM 2.2.3. *The statement "All triangles have angle sum 180°" is equivalent to Euclid's fifth postulate.*

*Proof.* Euclid's Proposition 32 shows that the former statement can be deduced from the latter, and so we now assume that all triangles have angle sum 180°, and will deduce Playfair's Postulate as a consequence. This will imply Euclid's fifth postulate, by our Theorem 2.2.2.

This proof is a bit harder than the last one. Refer to Figure 2.9 for the first steps of the proof and 2.10 for the last steps. From the given point $P$, drop a perpendicular $PQ$ to the given line $\ell$, and construct the line $PA$ perpendicular to $PQ$. By Euclid's Proposition 27 (which, we emphasize, is a theorem of absolute geometry), $PA$ is parallel to $\ell$. Let $PR$ be any line that lies below $PA$ to the right of point $P$. We will show that $PR$ must meet $\ell$, and hence $PA$ is the only parallel, since a similar argument could be made about lines that lie below the extension of $PA$ to the left.

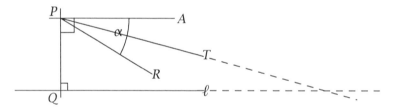

Figure 2.9. Beginning of proof of Theorem 2.2.3.

Let $\alpha$ be the angle that $PR$ makes with $PA$. The number $\alpha$ must be positive. We will prove that there is a line $PT$ that intersects $\ell$ and makes angle $TPA$ less than $\alpha$. The line $PR$ must stay below $PT$ (by Euclid's Postulate 1, as in the previous proof), and so it, too, must intersect $\ell$, as desired. Euclid would allow this final conclusion, that $PR$ intersects $\ell$ in Figure 2.9, on intuitive grounds, similarly to his proof of Proposition 1; to deduce it rigorously, one would need something like Hilbert's Axiom 6.

Thus it remains to construct $PT$. To do this, we construct a sequence of isosceles triangles, and perform an analysis of their angles, using our assumption about 180° and Euclid's Proposition 5, which states that the base angles of isosceles triangles are equal. Refer now to Figure 2.10. We begin by marking off $QQ_1$ equal to $PQ$.

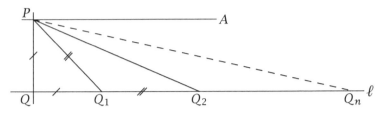

Figure 2.10. Construction of lines in proof of Theorem 2.2.3.

The angles in the isosceles triangle $PQQ_1$ are 45°. (Why?) The angle $Q_1PA$ is also 45°. (Why?) Next we draw $Q_1Q_2$ equal to $PQ_1$. Show that angles $Q_1PQ_2$, $Q_1Q_2P$, and $Q_2PA$ are all $22^1/_2$°. To do this you may first want to evaluate $\angle PQ_1Q_2$. This procedure of marking off segments along $\ell$, each equal to the previous diagonal line, can be continued indefinitely. Each successive step halves the angle at the base and the top. Eventually the angle $Q_nPA$ at the top will be less than $\alpha$. (Why?) The line $PT$ in Figure 2.9 is a portion of this line $PQ_n$. Q.E.D.

The statement that angle sums of triangles are 180° seems to have nothing to do with parallel lines, and yet, by the theorem we just proved, it amounts to the same thing. This at least provides a dramatic reformulation of the parallel postulate. But it doesn't help with the ultimate goal of deciding whether Euclid's fifth postulate is necessarily true of space, or whether it yields the only logically consistent geometry. In fact, the angle-sum formulation is probably less intuitive than the parallel-line formulation, but it certainly lends itself readily to the consideration of alternatives.

The history of attempts to prove the fifth postulate did not proceed as smoothly as the above theorems might suggest. Many mathematicians wrote books claiming to have proved the fifth postulate, but underlying their argument was usually some complicated assumption about behavior of lines or area as infinity is approached. For example, the famous French mathematician Adrien-Marie Legendre (1752–1833) wrote a very influential text in plane geometry that went through twelve editions. In most of these volumes, he presented an argument that seemed to prove Euclid's fifth postulate, but by the following edition, he would point out the flaw in the argument in the former one and present a new argument. Legendre's work was influential on subsequent workers, but he is most famous for work in number theory and analysis. Nassir-Edin, whom we mentioned above to have correctly understood the relationship of angle sums of triangles to uniqueness of parallels, came up with some subtle bogus proofs of Euclid's fifth postulate, assuming only the first four.

### SACCHERI'S APPROACH

An attempt by Gerolamo Saccheri (1667–1733) to prove that Euclid's fifth postulate was dependent almost resulted in the discovery of non-Euclidean geometry one hundred years before its actual discovery. Saccheri, who was a Jesuit priest and professor of mathematics at the University of Pavia, Italy, tried to prove the fifth postulate by the method of *reductio ad absurdum*. His book, *Euclid Freed of Every Flaw*, was published in the year of his death. As we shall discuss in detail, he derived many of the theorems of non-Euclidean geometry, but then rejected them on intuitive grounds.

In order to simplify the exposition, we will deviate a little bit from Saccheri's presentation; however, our approach is mathematically equivalent to his. See Exercises 5 to 7 for a presentation closer to Saccheri's method.

Saccheri's goal was to show, working only in absolute geometry, that the angle sum of every triangle is 180°. We saw in Theorem 2.2.3 that this would imply Euclid's fifth postulate. His method of doing this was going to be by eliminating the other possibilities. He began by showing that the sum could not be greater than 180°.

THEOREM 2.2.4. *Assume only absolute geometry. The sum of the angles of a triangle cannot be greater than* 180°.

You might be thinking, "We already proved in Section 1.2 that the sum of the angles must equal 180°, so why are we bothering to prove that they cannot be greater than 180°?" The answer is that in the proof of Euclid in Section 1.2 we were allowed to use the fifth postulate, but here we are not. The point here is that if you don't allow yourself to use Euclid's fifth, then you can still eliminate the possibility of angle sums greater than 180°. We note also that the proof we give here, as with some other proofs in this section, is not the one Saccheri gave, but rather a simpler one given by a later mathematician.

*Proof.* If a triangle has angle sum exceeding 180°, then it must exceed 180° by a certain amount, say $e$. We will show that there is a triangle with the same angle sum as the given triangle, but one of its angles is less than $e$. Thus the other two angles of this new triangle add up to more than 180°, contradicting Euclid's Proposition 17, which is, of course, a theorem of absolute geometry. This contradiction means that our assumption of a triangle with angle sum exceeding 180° must have been false.

Let $ABC$ be the given triangle, and suppose angle $A$ is its smallest angle, as in Figure 2.11. Let $D$ be the midpoint of $BC$, and extend $AD$ to $B_1$ so that $AD = DB_1$. The triangles $ABD$ and $B_1CD$ are congruent, and so in Figure 2.11 angles with the same label are equal. Thus triangles $ABC$ and $AB_1C$ have the same angle sum ($\alpha_1 + \alpha_2 + \beta + \gamma$). Moreover, since angle $A$ equals $\alpha_1 + \alpha_2$, at least one of the two angles $\alpha_1$ and $\alpha_2$ must be less than or equal to $^1/_2 \angle A$. Thus the minimal angle of triangle $AB_1C$ is less than or equal to half of that of triangle $ABC$. Now we repeat this process to triangle $AB_1C$, obtaining a new triangle with the same angle sum as $ABC$, but minimal angle less than one-fourth times that of triangle $ABC$. If we repeat this process often enough, we eventually get a triangle with the same angle sum as that of $ABC$, but with minimal angle less than $e$, as desired. Q.E.D.

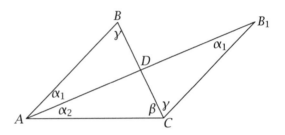

Figure 2.11. Proof of Theorem 2.2.4.

Next, Saccheri had to rule out the chaotic possibility that some triangles have 180°, while others have less than 180°. This is sometimes called the Three Musketeers Theorem ("United we stand . . . "). The proof we sketch for you in Exercises 9, 10, 11, and 12 is a later one due to Legendre; Saccheri's proof was more complicated.

THEOREM 2.2.5. *Assume only absolute geometry. Either all triangles have* 180° *or else all triangles have less than* 180°.

Finally, Saccheri hoped to show that the assumption that all triangles have less than 180° leads to a contradiction. We state and prove one of his most important consequences of this assumption.

THEOREM 2.2.6. *Assume absolute geometry and the statement, "All triangles have angle sum less than 180°." Then given any point P not on a line ℓ, there is more than one line through P which does not meet ℓ no matter how far extended.*

*Proof.* Refer to Figure 2.12. Draw $PQ$ perpendicular to $ℓ$, and $AP$ perpendicular to $PQ$. Then $AP$ is parallel to $ℓ$ by Proposition 27. Let

*PR* be any line meeting $\ell$, and, using Proposition 23, draw the line *PS* so that $\angle SPR = \angle QRP$. By Proposition 27, *PS* is parallel to $\ell$. We note that in Euclidean geometry $\angle QPS$ would equal 90°, and *PS* would coincide with *PA*. However, under our assumptions, we have

$$\angle QPS = \angle QPR + \angle RPS = \angle QPR + \angle QRP,$$

and this plus the right angle *PQR* equals the angle sum of triangle *PQR*, which by assumption is less than 180°. Thus angle *QPS* is less than 90°, and line *PS* does not coincide with line *PA*, so that both of them (and everything in between) are parallel to $\ell$. Q.E.D.

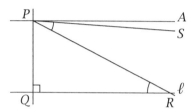

Figure 2.12. Proof of Theorem 2.2.6.

Figure 2.12 and Theorem 2.2.6 are distinctly non-Euclidean. You must resist your inclination to say that *PS* is going to meet $\ell$, for you just proved that, under certain hypotheses, it does not meet $\ell$. Keep in mind our comments earlier in this section about not knowing about the far reaches of the universe, and consequently not really having as good a feeling as we might assume for the word "straight." Also, keep in mind that this picture could be greatly exaggerated so that *PA* and *PS* might only diverge for less than one millimeter in a distance of more than a billion miles, in which case it doesn't seem quite so "obvious" that *PS* and $\ell$ are going to meet. As we prove theorems from now on, you must always keep in mind what assumptions underlie the theorem, and only allow yourself to use those facts about straight lines or angles which are implied by the assumptions.

Saccheri was bothered by Theorem 2.2.6 and other properties of parallel lines that he was able to deduce from the assumption that triangles had less than 180°. Many of these properties will be stated in Section 2.3 as theorems of hyperbolic geometry. Saccheri finally concluded that these properties were "repugnant" and "counter to the nature of a straight line." He concluded that this gave a contradiction to the assumption that triangles have less than 180°, and thus triangles must have exactly 180°. He then invoked Theorem 2.2.3 to conclude that he had proved Euclid's fifth postulate, assuming just the first four. Accordingly, he entitled his book *Euclid Freed of Ev-*

*ery Flaw*, because he felt that he had given the long-sought proof of Postulate 5.

What Saccheri should have said is that all the theorems he had derived under the hypothesis that triangles have less than 180° form a new kind of geometry, which makes as much logical sense as Euclidean geometry. Heath writes that "Saccheri . . . was the victim of the preconceived notion of his time that the sole possible geometry was the Euclidean, and he presents the curious spectacle of a man laboriously erecting a structure upon new foundations for the very purpose of demolishing it afterwards."[3]

## GAUSS'S WORK

Carl Friedrich Gauss (1777–1855) is ranked with Archimedes and Newton as one of the three greatest mathematicians of all time. We will learn a little bit more about him in the "Focus" at the end of this section, and will see consequences of his work pervading the rest of this book.

Gauss was probably the first person to realize that non-Euclidean geometry is every bit as valid as Euclidean geometry, both as a logical system and as a description of the universe. The next section will feature the acknowledged discoverers, Lobachevsky and Bolyai, who published their findings independently in 1829 and 1832, but from his diaries and letters it is clear that Gauss had already understood this fact by 1813. He is not given credit for being the discoverer because he did not publish his findings.

As we shall note in the "Focus" that follows, Gauss kept many findings to himself, possibly sharing them with a few correspondents. One reason for his reluctance to publish was his insistence on a perfectly rigorous and complete presentation. Another was his wish to avoid controversy. On the basis of what he had published, he was already the most famous mathematician in the world, so he did not bother to go to the trouble of carefully writing out all his additional ideas, and then having to persuade people that they were correct. This explains particularly why he did not publish his findings on non-Euclidean geometry. He wrote that he wanted to avoid the "clamor of the Boetians," a group of people whom the ancient Greeks had considered to be dull.

Gauss wrote an unpublished document speculating about non-Euclidean geometry at the age of 15. In a 1799 letter he wrote to Wolfgang Bolyai that his work had led him

3. Heath 1956, p. 211.

to doubt the validity of geometry. I have certainly achieved results which most people would look upon as a proof (of the parallel postulate), but which in my eyes prove almost Nothing; if, for example, one can prove that there exists a right triangle whose area is greater than any given number, then I am able to establish the entire system of geometry with complete rigor. Most people would certainly set forth this theorem as an axiom; I do not do so; though certainly it may be possible that, no matter how far apart one chooses the vertices of a triangle, the triangle's area still stays within a finite bound. I am in possession of several theorems of this sort, but none of them satisfy me.[4]

He is saying, in other words, that he has established the existence of triangles of arbitrarily large area as being equivalent to Euclid's fifth postulate, but that he sees no reason why there couldn't be a system of geometry in which there is a finite upper bound on the areas of all triangles. This is one of the stranger aspects of non-Euclidean geometry, but by the end of this chapter, it should seem quite natural to you.

In an 1817 letter, he wrote that "I keep coming closer to the conviction that the necessary truth of our geometry cannot be proved, at least by the human intellect for the human intellect. Perhaps in another life we shall arrive at other insights into the nature of space which at present we cannot reach. Until then we must place geometry on an equal basis, not with arithmetic, which has a purely *a priori* foundation, but with mechanics."[5] Here he is saying that the only way to tell whether Euclidean geometry is true of our universe is to go out and measure, and this is what he did. He was involved in a surveying project for the government in the early 1820s, and, using light signals and mirrors, he measured the sum of the angles in a triangle formed by three mountain peaks. The sides of the triangle were about 69, 85, and 107 kilometers. The sum of the angles exceeded $180°$ by about 15 seconds, that is, about $1/240$ of a degree. This was certainly within experimental error of possibly being exactly $180°$. He realized that his triangle was too small and his measurements too crude to detect a difference, if there was one. As we will see in the next section, if non-Euclidean geometry really describes space, then the bigger the triangle, the bigger is the deviation of its angle sum from $180°$.

We will also discuss in the next section Gauss's response to the

4. See Greenberg 1980, p. 144.
5. Faber 1983a, p. 156.

1829 publication of findings in non-Euclidean geometry. We close this subsection with more quotes from a letter he wrote in 1824:

> I have pondered it for over thirty years, and I do not believe that anyone can have given more thought to this than I, though I have never published anything on it. The assumption that the sum of the angles is less than 180° leads to a curious geometry, quite different from ours (the Euclidean), but thoroughly consistent, which I have developed to my entire satisfaction. . . . The theorems of this geometry appear to be paradoxical and, to the uninitiated, absurd; but calm, steady reflection reveals that they contain nothing at all impossible."[6]

A comparison of Gauss's and Saccheri's thinking here is illuminating. Both dealt with what happens if you assume that triangles have less than 180°. Saccheri felt that it contradicted his intuitive ideas about straight lines, while Gauss said it leads to a new form of geometry.

*FOCUS: GAUSS*

Gauss was called *Mathematicorum princeps*, the Prince of Mathematicians, and would certainly have been held in even greater esteem if he had made known the contents of his diaries, which were not disclosed until 1898, forty-three years after his death. There were results in the diary that anticipated much of the mathematics that was developed during the second half of the nineteenth century. We have already discussed his reasons—perfectionism and privacy—for not publishing all of his results.

We cannot begin to touch on all the areas of mathematics to which Gauss made fundamental contributions[7], but will mention one, a novel application of algebra to geometry, of which Gauss was particularly proud. The Greeks had been interested in what kinds of geometrical figures can be constructed using just a straightedge and compass. One of the most difficult of these problems was the regular polygons. The question is: For which values of $n$ can a regular polygon with $n$ sides be constructed (with straightedge and compass)? It is elementary to do this when $n = 3$ and 4, the equilateral triangle and square, and the Greeks knew how to do it when $n = 5$ and 6.

We point out the difference between saying that the regular $n$-gon exists and saying that it is constructible. They exist for all values

6. Greenberg 1980, p. 144.
7. But see Sections 3.3 and 5.3 for two major contributions, congruence arithmetic and the geometry of complex numbers.

Carl Friedrich Gauss

of $n$—in a circle, just draw angles of $360/n°$ out from the center. But that requires more than a straightedge and compass; it requires an infinitely accurate protractor. When in Section 1.3 we said there are only five regular polyhedra, that was a question of existence, not constructibility. There certainly are regular 7-gons (see Fig. 2.13), but they cannot be made with only a straightedge and compass.

At the age of nineteen, Gauss showed how to construct a regular 17-gon. This discovery was what convinced him to devote his life to mathematics rather than to philology. He wanted one of these on his tombstone, but the closest he came to his wish is a 17-pointed star on a statue erected in his honor; it was felt that the 17-gon would appear indistinguishable from a circle.

Figure 2.13. A regular 7-gon.

More important than the explicit construction of the regular 17-gon was Gauss's rule for exactly which regular $n$-gons can be constructed.

THEOREM 2.2.7. *The values of $n$ for which the regular n-gon can be constructed with straightedge and compass are those $n \geq 3$ which are a power of 2 or can be written as a product of one or more distinct primes of the form $2^{2^m} + 1$, multiplied perhaps by a power of 2.*

Thus 17 works because it is $2^{2^2} + 1$ and is prime. The primes of the form $2^{2^m} + 1$ are known as *Fermat primes*. There are only five such known numbers, corresponding to $m = 0, 1, 2, 3,$ and 4. It is known that for all values of $m$ from 5 through 21, the number $2^{2^m} + 1$ is not prime. The first Fermat number whose primeness is unknown is prime $2^{2^{22}} + 1$, a number with more than a million digits.

Gauss's proof of Theorem 2.2.7 involved reinterpreting the problem as which complex numbers can be obtained from integers using only simple arithmetic and square roots, and then solving that problem by studying roots of equations. This proof is one of the high points of the curriculum of an undergraduate mathematics major.

Gauss was a real prodigy. When he was two years old, he was playing on the floor while his father was paying wages to some workers, and he said, "Papa, you have made a mistake," and then named another figure. He was correct, of course.

When Gauss was about ten, his schoolteacher, who was a strict disciplinarian, asked the students to add the numbers from 1 to 100, expecting that this project would keep his pupils busy and give him a lot of time to relax. In just a few seconds Gauss wrote the answer, 5050, on his slate and laid it on the teacher's desk, saying, "*Ligget se*" (There it lies). While the other students toiled away, the teacher just glared at Gauss. What Gauss had noticed was the symmetry in the sum. $1 + 100 = 101$, $2 + 99 = 101$, and so on until $50 + 51 = 101$. Thus there are 50 sums of 101, and so the answer is $50 \cdot 101 = 5050$.

## *Exercises*

1. Answer the questions in the proof of 2.2.3. If the angle $\alpha$ had been $1°$, how many triangles $PQQ_n$ would you have had to construct?

2. Show that the axiom system that has Euclid's first postulates and Ax5', "All triangles have exactly $180°$," is equivalent to the axiom system that has Euclid's first four postulates and Ax5, "All quadrilaterals have exactly $360°$." (Hint: If you assume Ax5', then you can prove that a quadrilateral has $360°$ by dividing it into two triangles. On the other hand, if you assume Ax5'', then you can prove that a triangle has $180°$ by attaching a congruent triangle to it, forming a quadrilateral.)

3. Assume only absolute geometry. Prove that a quadrilateral cannot have more than $360°$. (Hint: Use 2.2.4.)

4. Recall that in Euclidean geometry, similar triangles are triangles whose corresponding angles are equal. Prove the following result, which implies that the existence of similar triangles is equivalent to Euclid's fifth postulate: "Assume only absolute geometry. If there is a pair of noncongruent similar triangles, then all triangles must have angle sum $180°$." (Hint: Follow these steps: Your assumption is that there are two noncongruent triangles $ABC$ and $A'B'C'$ with equal angles. Construct triangle $ADE$ inside the larger one and congruent to $A'B'C'$, as in Figure 2.14.

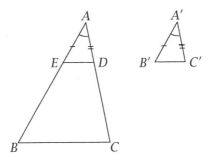

Figure 2.14. For Exercise 4.

Use the straight angles at $D$ and $E$ to determine the angle sum of the quadrilateral $BCDE$. Complete the proof by dividing the quadrilateral into two triangles and using 2.2.5.)

5. A *Saccheri quadrilateral* is a 4-sided plane figure that has two sides of equal length both perpendicular to a third side. (See Fig. 2.15.)

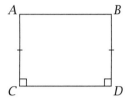

Figure 2.15. A Saccheri quadrilateral.

The angles at $A$ and $B$ in Figure 2.15 are called the *vertex angles* of the Saccheri quadrilateral. Prove that, assuming only absolute geometry, the vertex angles are equal to each other. (Hint: Draw in the diagonals, and apply the Side-Angle-Side proposition twice. To help you see which triangles are congruent, it might be helpful to draw the quadrilateral in an exaggerated non-Euclidean way, as in Figure 2.16.)

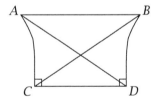

Figure 2.16. For Exercise 5.

6. Assume only absolute geometry. Prove that the vertex angles of a Saccheri quadrilateral cannot be greater than 90°. (Hint: Draw a diagonal in your Saccheri quadrilateral and apply 2.2.4 to the two triangles formed.)

7. Assume only absolute geometry. Prove that the statement, "All triangles have angle sum 180°," is equivalent to the statement, "All vertex angles of Saccheri quadrilaterals are right angles." (Hint: The implication in the forward direction is similar to Exercise 6. The backwards implication should be done in two steps. First prove that all right triangles have 180° by expanding a given right triangle into a Saccheri quadrilateral as in Figure 2.17a. See Exercise 8 for a discussion of the required congruence result. Then use Exercise 9.)

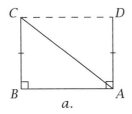 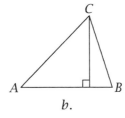

Figure 2.17. For Exercises 7 and 9.

8. In Exercise 7, you may want to make use of the fact that two right triangles with equal hypotenuses and a pair of equal legs must be congruent. It is not true that two triangles with a pair of equal angles and two pairs of equal sides are necessarily congruent. For example, in Figure 2.18, triangles $ABC$ and $ABD$ have the same angle at $A$, equal sides $AB$, and $BC = BD$, and yet are not congruent. Explain with a drawing why this cannot happen if the angle in which they agree is a right angle .

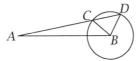

Figure 2.18. For Exercise 8.

9. Assume only absolute geometry. Prove that "all triangles have angle sum $180°$" if and only if "all right triangles have angle sum $180°$." (Hint: An arbitrary triangle can be divided into two right triangles. See Fig. 2.17b.)

10. Assume only absolute geometry. Prove that if in Figure 2.19 triangle $ADE$ has angle sum $180°$, then so does triangle $ABC$. (Hint: If $ABC$ has less than $180°$, what can you say about the angle sum of the quadrilateral $DBCE$?)

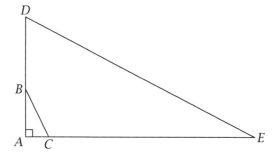

Figure 2.19. For Exercise 10.

11. Assume only absolute geometry. Prove that if there exists one right triangle with angle sum 180°, then there exists an arbitrarily large right triangle with angle sum 180°. (Hint: If, in Figure 2.20, *ABC* has angle sum 180°, then it can be completed to the rectangle *ABCD*, which can be used to form as large a rectangle as desired, from which one can form a large right triangle with angle sum 180°.)

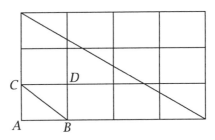

Figure 2.20. For Exercise 11.

12.* Prove Theorem 2.2.5. Use Exercises 9, 10, and 11 to show it is impossible to have one triangle with 180° and another with less than 180°.

13. Construct a regular hexagon using only straightedge and compass. Figure 2.21 should help you figure out how to do it.

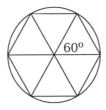

Figure 2.21. For Exercise 13.

14. Write the numerical values of the five known Fermat primes. Which regular $n$-gons are constructible for $n < 40$?

15. What is the sum of the integers from 1 to 1000? What is the sum of the integers from 100 to 400?

## 2.3   Hyperbolic Geometry

The previous section should have prepared you to give serious consideration to hyperbolic geometry, the first and most important form of non-Euclidean geometry. This is a form of geometry in which Euclid's first four postulates are true, but the fifth postulate is replaced by the statement that "there is more than one line through a point parallel to a line." The acknowledged discoverers of hyperbolic geometry are János Bolyai (1802–1860), a Hungarian, and Nikolai Lobachevsky (1793–1856), a Russian, who independently published in books in 1832 and 1829, respectively, demonstrations that this form of geometry has a collection of theorems that appear to be just as consistent as those of Euclidean geometry. Both were of the opinion that hyperbolic geometry could provide just as adequate a description of the space in which we live as does Euclidean geometry.

The significance of this discovery is immense. For over two thousand years people had felt that Euclid's geometry was necessarily the geometry of space. Mathematics and physics were wedded through this belief. Hyperbolic geometry showed that there are other conceivable descriptions of space. Physics became the study of physical space, while mathematics became a more abstract science. From this time on, it was clear that in mathematics one could begin with any set of postulates and study the abstract consequences of those postulates. The significance of this discovery is frequently compared with Copernicus's theory, which said that the earth is not the center of the universe (Section 1.3), or Einstein's theory that time is not a concept that is the same for all observers (see Section 2.6). The analogy of course is that each theory freed people from long-accepted modes of thought.

Another reason that the discovery of hyperbolic geometry is considered to be so important is that it led to yet more forms of geometry; we will discuss several of them later in this chapter. The subject of differential geometry was an outgrowth of this discovery, and it has been an essential tool for physicists, most notably Einstein, in their study of the physical universe. This is a prime example of the central theme of this book—topics in abstract mathematics that have turned out to have unexpected applications in the real world.

### BOLYAI AND LOBACHEVSKY

Neither Bolyai nor Lobachevsky received much credit for his discovery during his lifetime. János Bolyai's father, Wolfgang Bolyai, had been a classmate of Gauss, and throughout his life communicated with Gauss about mathematics, most notably about the parallel postulate. An excerpt of one such note from Gauss to Wolfgang Bolyai was included in the preceding section.

The older Bolyai warned his son János against working on the parallel postulate because of the frustration he had experienced in his own work. He wrote:

> You must not attempt this approach to parallels. I know this way to its very end. I have traversed this bottomless night, which extinguished all light and joy of my life. I entreat you, leave this science of parallels alone. You should detest it just as much as lewd intercourse; it can deprive you of all your leisure, your health, your rest, and the whole happiness of your life. This abysmal darkness might perhaps devour a thousand towering Newtons.[8]

However, János Bolyai succeeded in developing most of the results of hyperbolic geometry. Moreover, unlike Saccheri, he realized it was a form of geometry that should be given equal status with Euclid's, and, unlike Gauss, he was not afraid to publish his discovery. His enthusiasm is evident in the following portion of a letter he wrote to his father in 1823:

> I have now resolved to publish a work on the theory of parallels, as soon as I shall have put the material in order, and my circumstances allow it. . . . The goal is not yet reached, but I have made such wonderful discoveries that I have been almost overwhelmed by them, and it would be the cause of continual regret if they were lost. . . . *I have created a new universe from nothing.* All that I have sent you till now is but a house of cards compared to the tower.[9]

This enthusiasm was contagious and caused Wolfgang to change his tune completely. He urged his son to write up his findings as quickly as possible. The following excerpt of a letter from Wolfgang to János shows that Wolfgang correctly surmised that these ideas were prevalent enough for others to be publishing them soon:

> If you have really succeeded in the question, it is right that no time be lost in making it public, for two reasons: first, because ideas pass easily from one to another, who can anticipate its publication; and secondly, there is some truth in this, that many things have an epoch, at which they are found at the same time in several places, just as the violets appear on every side in spring. Also every scientific struggle is just a serious war, in which I cannot say when peace will arrive. Thus we ought to conquer when we are able, since the advantage is always to the first comer."[10]

8. First four sentences are from Gray 1979, p. 97; last two are from Coughlin and Zitarelli 1984, p. 240.
9. Faber 1983a, p. 161.
10. Gray 1979, p. 97.

Because János was also an army officer, it took him a few years to work out and write out all the details of his geometry. It was finally published in 1832 as a twenty-six-page appendix to his father's book, *Essay on the Elements of Mathematics for Studious Youths*. The title of the appendix was "Supplement Containing the Absolutely True Science of Space, Independent of the Truth or Falsity of Euclid's Axiom XI (that can never be decided a priori)."[11]

They sent a copy of the book to Gauss, who responded unenthusiastically, saying that he had been familiar with these ideas for a long time:

> If I commenced by saying that I am unable to praise this work, you would certainly be surprised for a moment. But I cannot say otherwise. To praise it would be to praise myself. Indeed the whole contents of the work, the path taken by your son, the results to which he is led, coincide almost entirely with my meditations, which have occupied my mind partly for the last thirty or thirty-five years. So I remained quite stupefied. So far as my own work is concerned, of which up to now I have put little on paper, my intention was not to let it be published during my lifetime. Indeed the majority of people have not clear ideas upon the question of which we are speaking, and I have found very few people who could regard with any special interest what I communicated to them on this subject. To be able to take such an interest it is first of all necessary to have devoted careful thought to the real nature of what is wanted and on this matter almost all are most uncertain. On the other hand it was my idea to write down all this later so that at least it should not perish with me. It is therefore a pleasant surprise for me that I am spared this trouble, and I am very glad that it is just the son of my old friend, who takes the precedence of me in such a remarkable manner."[12]

Wolfgang was satisfied with this response, but János was not. He never really believed that Gauss had done all this work earlier, and he never published any more mathematical writing. It seems unconscionable that Gauss would not have used his reputation to publicize the work of the son of his friend, even if he had previously done the work privately by himself. As it was, virtually no one paid attention to János's work until several years after his death.

Nikolai Ivanovich Lobachevsky was professor of mathematics and rector at the University of Kazan. He first presented his new geometry in 1826 in a talk at his university. In that talk, he discussed the

11. Common Notions were included along with the Postulates.
12. Gray 1979, p. 97.

Nikolai Ivanovich Lobachevsky

relationship between parallel lines and angle sums of triangles. He noted that huge triangles would be required before one might hope to notice any deviation of the angle sum of a triangle from 180°. Using the data supplied by the latest astronomical calendar, Lobachevsky showed that the angle sum of the triangle formed by the star Sirius and two diametrically opposed positions of the earth in its orbit differed from 180° by less than .0004 second. This was well within the limits of experimental accuracy, and so was inconclusive as to whether or not the angle sum of that triangle might be exactly 180°. (See Fig 2.22 for a schematic.)

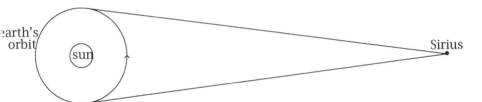

Figure 2.22. Lobachevsky's big triangle.

Lobachevsky's talk did not have much impact, nor did papers, entitled "On the Principles of Geometry," which he published in a math journal in 1829 and 1835. Most readers were just not ready for such revolutionary new ideas. One reviewer wrote that the paper was mistitled, and that the title should have been "A Satire on Geometry." In 1840 Lobachevsky wrote a book in German describing his work. Gauss was very impressed with this volume, and even taught himself Russian so that he could read Lobachevsky's other papers. In a letter written in 1846, Gauss notes that "I have found in Lobachevsky's work nothing that is new to me, but the development is made in a way different from that which I have followed, and certainly by Lobachevsky in a skillful way and in truly geometrical spirit."[13] As a result of Gauss's suggestion, Lobachevsky was elected a corresponding member of the Göttingen Scientific Society. (Gauss taught at the University of Göttingen, and ever since that time it has been the center of German mathematics.)

Despite this modicum of recognition from Gauss, Lobachevsky's work did not receive the prominence it deserved. It wasn't until the publication of the German book *Elements of Mathematics* by R. Baltzer in 1867 that the work of Bolyai and Lobachevsky became widely known. This book, which described their work in some detail, came after the death of both of these discoverers.

13. Faber 1983a, p. 164.

Before we proceed with the mathematics, we will summarize the great achievement of Bolyai and Lobachevsky. They showed that there is another kind of geometry that begins with assumptions dramatically different than Euclid's and develops a body of consequences of those assumptions. These theorems, although they seem strange at first glance, do not imply any obvious contradictions, and so they are apparently just as consistent as the theorems of Euclidean geometry. (Neither Bolyai nor Lobachevsky actually proved the consistency of their geometry; such a proof required another forty years.) Moreover, this new body of theorems could in fact be the true geometry of space.

### SOME THEOREMS OF HYPERBOLIC GEOMETRY

The form of non-Euclidean geometry developed by Bolyai and Lobachevsky has since become known as *hyperbolic geometry*. We shall list many of the most striking results of hyperbolic geometry, and sketch proofs of many of them. There are, of course, many theorems of hyperbolic geometry that we will not discuss. For example, one of the main thrusts of Lobachevsky's work was trigonometric formulas in this new system; we shall ignore that completely.

We begin by assuming as postulates the first four postulates of Euclid, together with:

POSTULATE 5H. *Given a point not on a line, there exists more than one line passing through the point and not meeting the given line, no matter how far extended.*

Since *point* and *line* are being thought of as undefined terms, there really should be more postulates, similar to Hilbert's treatment of Euclidean geometry, but we won't worry about such technicalities. Also, we won't distinguish between the words "line" and "straight line," as Euclid did. The word "degree" has the same meaning as in Euclidean geometry, basically that a straight angle has 180 degrees. Because Euclid's first four postulates are assumed, all the theorems of absolute geometry, in particular the first twenty-eight propositions of Euclid, are theorems in this geometry.

We will indicate theorems of hyperbolic geometry with a letter "H" following the number of the theorem. We will not follow the usual practice of listing the most elementary theorems first, because we want to emphasize as quickly as possible the fundamental differences between hyperbolic and Euclidean geometry. In our discussion of precursors to non-Euclidean geometry in the preceding section, we gave disguised proofs of some theorems of hyperbolic geometry. We begin with two of the most important ones.

THEOREM 2.3.1H. *The sum of the angles of a triangle is less than* 180°.

*Proof.* Theorem 2.2.4 said that the sum cannot be greater than 180°. Theorem 2.2.5 said that either all triangles have exactly 180° or else all triangles have less than 180°. Theorem 2.2.3 said they cannot all have 180° in hyperbolic geometry, since Playfair's version of Euclid's fifth postulate is assumed to be false. From these statements, the conclusion of this theorem is immediate. Q.E.D.

THEOREM 2.3.2H. *Two triangles whose corresponding angles are equal are congruent.*

*Remark.* This is what we might call an Angle-Angle-Angle Theorem. No such result is true in Euclidean geometry. It can also be stated as saying that there are no similar triangles in hyperbolic geometry (unless they are congruent), since triangles with the same angles must be the same size. It is one of the most marked differences between hyperbolic and Euclidean geometry.

*Proof.* This was proved as Exercise 4 of the preceding section. We summarize the proof by saying that if one triangle is smaller, then it can be placed inside the other at one of the agreeing angles, and the portion of the larger triangle outside the smaller one will be a quadrilateral with exactly 360°, violating an immediate consequence of Theorem 2.3.1H. Q.E.D.

One can deduce from Postulate 5H the following elaboration of the properties of parallel lines.

THEOREM 2.3.3H. *Given a point P not on a line $\ell$, there are two lines, m and n, passing through P which do not meet $\ell$ and satisfy*
  i. *All lines between m and n do not meet $\ell$.*
  ii. *All lines not between m and n which pass through P do meet $\ell$.*

In Figure 2.23, line *a* illustrates a line of type (*i*), and line *b* illustrates a line of type (*ii*) in Theorem 2.3.3H.

We shall say that the lines *m* and *n* are the *parallels* to $\ell$ through *P*, and that lines such as *a*, which do not meet $\ell$ but are not the lowest such line, are *divergent* to $\ell$. We shall show in the next subsection why the word "divergent" is appropriate. Thus there are two parallels to $\ell$ through *P* but infinitely many lines through *P* that do not intersect $\ell$.

Intuitively, you can think of rotating the lines through *P* until they finally meet $\ell$. By Postulate 5H, there must be more than one line that does not meet $\ell$. If *m* is a line that separates those that meet $\ell$ from those that do not, then *m* cannot meet $\ell$ because it is easily

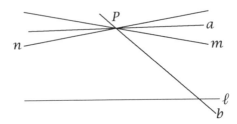

Figure 2.23. Illustration of Theorem 2.3.3H.

proved that a slight rotation of intersecting lines leaves them still intersecting. Although this argument seems quite convincing, it is really relying on our intuition, and Hilbert's rigorous treatment of hyperbolic geometry, which he performed along with his rigorous treatment of Euclidean geometry that we discussed in Section 2.1, uses the entire content of our 2.3.3H as his parallel postulate.

### PROPERTIES OF PARALLELS AND DIVERGENT LINES

Next we will investigate some properties of parallel lines and divergent lines in hyperbolic geometry. We begin with a preliminary theorem that says that it makes sense to use the term "parallel lines."

THEOREM 2.3.4H.
   *i. If $m$ is a parallel to $\ell$ through $P$, and $P'$ is another point on $m$, then $m$ is a parallel to $\ell$ through $P'$. That is, $m$ is the lowest line (in one direction) through $P'$ which does not intersect $\ell$.*
   *ii. If $m$ is a parallel to $\ell$, then $\ell$ is a parallel to $m$.*

This theorem, which may seem self-evident or overly pedantic, is quite difficult to prove, and we shall omit the proof. It says that we can talk about lines being *parallel to each other*. We can also talk about the *direction of parallelism* of parallel lines. It is the direction (either right or left in our drawings) in which all lines that lie between them and intersect one of them also intersect the other.

Now we can state, in the next three results, the principal relationships regarding the distance between parallel lines and divergent lines.

THEOREM 2.3.5H. *Parallel lines approach each other asymptotically in their direction of parallelism. Moreover, the distance between them always decreases as one moves in the direction of parallelism.*

This means that if $m$ and $\ell$ are parallel, and $P$ and $P'$ are points on $m$ with the direction from $P$ to $P'$ being the direction of parallelism,

then the length of the perpendicular line from $P'$ to $\ell$ is less than
the length of the perpendicular line from $P$ to $\ell$, and by moving
$P'$ far enough in the direction of parallelism, this distance can be
made less than any preassigned positive distance. The reader may
have seen the word "asymptote" in a precalculus course; asymptotes
are curves that become arbitrarily close to each other without ever
touching (see Fig. 2.26, where lines $m$ and $n$ are asymptotes to $\ell$).
This asymptotic nature of straight lines was particularly bothersome
to Saccheri.

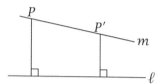

Figure 2.24. Illustration of Theorem 2.3.5H.

THEOREM 2.3.6H. *Any two lines that have a common perpendicular are
divergent, and divergent lines have a unique common perpendicular.*

In Euclidean geometry, parallel lines have infinitely many common perpendiculars, while in hyperbolic geometry lines that are
(strictly) parallel do not have any, and divergent lines have one ($PQ$
in Fig. 2.25).

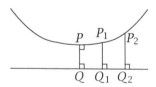

Figure 2.25. Illustration for Theorems 2.3.6H and 2.3.7H.

THEOREM 2.3.7H. *The distance between divergent lines increases as you
move away from the common perpendicular.*

This theorem says that the distance $P_2Q_2$ is greater than the distance $P_1Q_1$ in Figure 2.25. One can in fact prove that the distance
between divergent lines becomes infinitely large (i.e., greater than
any prescribed value) in either direction, but we shall not prove it.
This is the reason for the name "divergent." Because they lie above
the divergent lines, it follows that the distance between parallel lines
becomes infinitely large in the direction opposite to the direction of
parallelism, since here they lie above the divergent lines. Thus the sit-

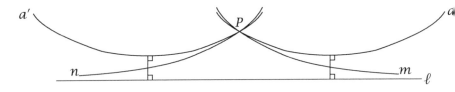

Figure 2.26. Parallel and divergent lines.

uation for distance between lines may be depicted as in Figure 2.26, where $m$ and $n$ are parallel to $\ell$, while $a$ and $a'$ are divergent to $\ell$, with their common perpendiculars drawn in. If you object to the apparent "nonstraightness" of the lines, we remind you of our admonishments in the previous section:

1. You don't know what "straight" means; it is undefined.
2. The picture may be greatly exaggerated.

We will sketch the proofs of Theorems 2.3.5H, 2.3.6H, and 2.3.7H. Readers who are interested more in the results than in the proofs may skip to the next subsection, "Area and Defect." We begin with the easy parts, the first of which is the first part of 2.3.6H. If lines $\ell$ and $a$ have a common perpendicular that intersects $a$ at $P$, then the two parallels to $\ell$ at $P$ must certainly lie below $a$, essentially because of 2.3.3H. Hence $a$ is a divergent to $\ell$. (See Fig. 2.27.)

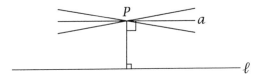

Figure 2.27. Lines with a common perpendicular are divergent.

The following easy lemma of absolute geometry, whose proof we relegate to Exercise 2, will be useful. Note that, unlike most results of this section, this lemma is not followed by a letter H, because it does not require Postulate 5H in its proof.

Lemma 2.3.8. *Assume only absolute geometry. In a quadrilateral ABCD with right angles at A and B as in Figure 2.28, then $\angle C < \angle D$ if and only if $AC > BD$.*

This statement says that the longer line is the one coming down from the smaller angle.

You can now easily prove 2.3.7H, following the hints of Exercise 3.

The part of 2.3.5H that says the distance decreases in the direction of parallelism is proved similarly, and is even easier. (See Exercise 4.) Here you will use a fact that has been implicit in our drawings but has not been explicitly stated—if $m$ and $\ell$ are parallel, the angle which $m$ in the direction of parallelism makes with the perpendicular from any point of $m$ to $\ell$ is acute. We expand this in the following important result, which should really precede Theorems 2.3.5H, 2.3.6H, and 2.3.7H since it is used in some of their proofs.

Figure 2.28. Illustration for Lemma 2.3.8.

THEOREM 2.3.9H. *If $m$ and $n$ are the parallels to $\ell$ through $P$, then the angles they make with the perpendicular from $P$ to $\ell$ are equal acute angles.*

In other words, $\alpha = \beta < 90°$ in Figure 2.29.

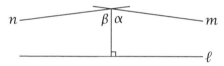

Figure 2.29. Illustration of Theorem 2.3.9H.

This proof is fun to perform, and is relegated, with hints, to Exercise 5. This result can be strengthened considerably to the following result, which we will not prove or use. This theorem is a statement about uniformity of space, closely related to congruence theorems.

THEOREM 2.3.10H. *If $m$ is parallel to $\ell$ at $P$, and $PQ$ is the perpendicular from $P$ to $\ell$, then the angle which $m$ makes with $PQ$ (called the* **angle of parallelism***) depends only on the length $PQ$.*

This theorem says that anywhere there are parallel lines separated by the same perpendicular distance, the angle $\theta$ in Figure 2.30 will be the same.

Next we prove the "asymptotic" part of 2.3.5H. Suppose in Figure 2.31 that $m$ and $\ell$ are parallels, $PQ$ is perpendicular to $\ell$, and $QR$ is any tiny little distance marked off along $PQ$. (We have drawn it as large as we have in order to illuminate our argument, but it could

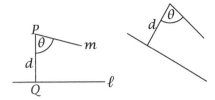

Figure 2.30. Illustration of Theorem 2.3.10H.

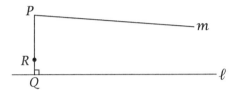

Figure 2.31. We will show that the distance from $m$ to $\ell$ becomes smaller than $RQ$.

really be extremely small.) We want to show that for some point far enough to the right along $m$, the perpendicular distance to $\ell$ will be equal to $RQ$.

Let $n$ be the left parallel to $\ell$ passing through $R$, as in Figure 2.32. Extend $n$ to the right until it meets $m$. (A bit of argument, which we omit, is required to show that it actually does meet $m$.) Let $O$ be the point where $n$ and $m$ meet, and let $OT$ be perpendicular to $\ell$. Draw $OS$ along $m$ so that $OS = OR$, and let $SU$ be perpendicular to $\ell$. In Exercise 6, you are given hints to help prove that $SU = RQ$, as desired.

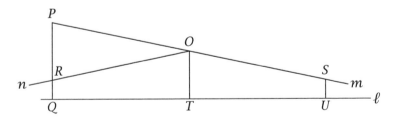

Figure 2.32. Used in proof of Theorem 2.3.5H.

In Exercise 7, the proof of the uniqueness part of 2.3.6H is outlined. The proof that there is a common perpendicular is much more difficult and is omitted.

AREA AND DEFECT

By now you should have a fairly good feeling for the vagaries of parallel lines in hyperbolic geometry, and the strange pictures should no longer bother you so much. In this subsection, we will show that there is a very simple relationship between the area of a triangle and its angle sum. It is most easily expressed in terms of defect.

DEFINITION 2.3.11. *The **defect** of a triangle is 180° minus its angle sum.*

In Euclidean geometry all triangles have defect 0, but by 2.3.1H all triangles in hyperbolic geometry have defect greater than 0. The following result, whose proof you will discuss in Exercise 8, is the reason that defect is nicer to work with than angle sum.

PROPOSITION 2.3.12H. *If a triangle is composed of smaller triangles, then the defect of the big triangle equals the sum of the defects of the smaller triangles.*

For example, if in Figure 2.39a the angle sums of triangles $ABD$ and $ADC$ are 175° and 177°, respectively, then the angle sum of triangle $ABC$ must be 172°, so that its defect equals 5 + 3.

The two principal results relating area and defect are the following.

THEOREM 2.3.13H. *Triangles have the same area if and only if they have the same defect, or, equivalently, if and only if they have the same angle sum.*

THEOREM 2.3.14H. *The ratio of the areas of two triangles equals the ratio of their defects.*

The latter result says, for example, that if one triangle has twice the area of another, then it has twice the defect. This result allows us to work many simple problems.

**Problem.** Suppose one triangle has an area of 10 square miles and an angle sum of 178°. What is the angle sum of a triangle whose area is 100 square miles?

*Solution.* Since the second triangle has an area ten times as large as the first, its defect must also be ten times as large. Since the first triangle has a defect of 2°, the second must have a defect of 20°, and hence an angle sum of 160°.

Both area and defect are additive. This means that if a triangle is decomposed into a number of smaller triangles, then the area of the big one is the sum of the areas of the smaller ones, and the defect of the big one is the sum of the defects of the smaller ones (by 2.3.12H).

This is the main observation underlying the proofs of 2.3.13H and 2.3.14H, but we shall postpone a formal proof until the end of this subsection. As a corollary of 2.3.14H, we have

COROLLARY 2.3.15H. *The ratio area/defect is the same for all triangles in hyperbolic geometry.*

*Proof.* Suppose one triangle has area $A_1$ and defect $d_1$, and another triangle has area $A_2$ and defect $d_2$. Then 2.3.14H says that

$$\frac{A_1}{A_2} = \frac{d_1}{d_2}.$$

Multiplying both sides by $A_2/d_1$ yields the desired result. Q.E.D.

This result leads one to wonder, "What is the value of this ratio?" The answer is, "It could have any value." This should become clearer in later sections when we talk about models of hyperbolic geometry. For now, let it suffice to say that you can have hyperbolic geometry in which the area/defect ratio for all triangles equals one square inch or one million square miles. Abstractly, the value of this ratio is not a matter of consequence; we can just call it $k$ if we want. But if one believes that the geometry of the universe is hyperbolic, then the value of this constant for the universe is of the utmost importance.

If the universe is hyperbolic, then this constant will have an extremely large value, since a huge triangle will be required in order for it to have any noticeable defect. This result is why Gauss and Lobachevsky knew that they should be looking at huge triangles in order to observe any deviation of the angle sum from 180°. Gauss was very taken with this idea that hyperbolic geometry has an intrinsic constant, given to us solely by the geometry. There is no such thing in Euclidean geometry. The existence of an intrinsic constant in hyperbolic geometry had been noticed earlier (in 1766) by J. H. Lambert, who, as we noted in Section 1.1, was also the first to prove that $\pi$ is an irrational number.

In Gauss's letter of 1824 from which we quoted at the end of the subsection of 2.2 entitled "Gauss's Work," we omitted his remarks about the constant, which we now quote:

> I can solve every problem in it with the exception of the determination of a constant, which cannot be designated *a priori*. The greater one takes this constant, the nearer one comes to Euclidean Geometry, and when it is chosen infinitely large the two coincide. The theorems of this geometry appear to be paradoxical and, to the uninitiated, absurd; but calm, steady reflection

reveals that they contain nothing at all impossible. For example, the three angles of a triangle become as small as one wishes, if only the sides are taken large enough; yet the area of the triangle can never exceed a definite limit, regardless of how great the sides are taken, nor indeed can it ever reach it. All my efforts to discover a contradiction, an inconsistency, in this Non-Euclidean Geometry have been without success, and the one thing in it which is opposed to our conceptions is that, if it were true, there must exist in space a linear magnitude, determined for itself (but unknown to us). . . . If this Non-Euclidean Geometry were true, and it were possible to compare that constant with such magnitudes as we encounter in our measurements on the earth and in the heavens, it could then be determined *a posteriori.* Consequently in jest I have sometimes expressed the wish that the Euclidean Geometry were not true, since then we would have *a priori* an absolute standard of measure."[14]

We note, with Gauss, that one consequence of 2.3.15H is the existence of an upper bound on the areas of triangles. If $R$ is the constant area/defect ratio, then, since defect is less than 180° for all triangles, we must have area less than $R \cdot 180°$ for all triangles. There will be triangles, which look something like Figure 2.33, whose angles are very close to 0° and whose area is very close to $R \cdot 180°$, but that area will not be achieved by any triangle.

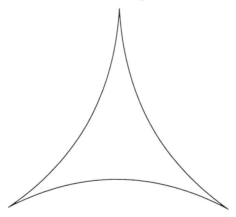

Figure 2.33. A triangle with large area and defect.

Gauss's note talks about the constant as a linear magnitude, whereas the constant $R$ we have described has units of area. There are many ways of choosing an intrinsic constant with units of length.

14. Greenberg 1980, p. 144.

One example would be the length of the side of an equilateral triangle, each of whose angles is 45°. Another would be the distance of a point from a line for which the angle of parallelism (of Theorem 2.3.10H) is 60°. Once one of these has been selected as the natural unit of length, all the others can be determined in terms of it using pure mathematics. Most of these formulas involve trigonometry and were worked out by both Bolyai and Lobachevsky. See Exercise 17 for one example. As a simpler example, suppose that we decide to define 1 Lob to be the length of side of a 45-45-45 triangle. Then we can work out a formula for the number of Lobs in an 80-80-80 triangle (Will it be more or less than 1?) and a formula for how many square Lobs equal $R$, the area/defect ratio. We note that since hyperbolic geometry has no squares, some care is required in defining area, but we shall not worry about this.

The point is that there is nothing like this intrinsic distance in Euclidean geometry. A meter or a foot is an artificially defined distance. In Euclidean geometry there is no need for the National Bureau of Standards to keep a standard degree, because that is defined from geometry as $1/180$ of a straight angle. In a world of hyperbolic geometry, there is no need for the National Bureau of Standards to maintain a standard of length, either.

Now that we have discussed the important consequences of Theorems 2.3.13H and 2.3.14H, we proceed to their proofs. Again, many readers may prefer to skip these proofs, which will not be used in subsequent reading. Note that 2.3.14H is actually a stronger result than 2.3.13H (for let the ratio = 1), but 2.3.13H must be proved first, since it is the primary tool in the proof of 2.3.14H. Our proof will utilize the idea of equivalent polygons.

DEFINITION 2.3.16. *Two polygons are equivalent if they can be decomposed into congruent triangles.*

For example, the two polygons in Figure 2.34 are equivalent.

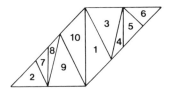 

Figure 2.34. Equivalent polygons.

By the additivity of area and defect, one immediately deduces that equivalent polygons have the same area and the same defect. Thus

Theorem 2.3.13H is an easy consequence of the following result, which is proved in Exercise 15.

THEOREM 2.3.17H. *Two triangles with the same defect are equivalent.*

Now that 2.3.13H has been proved, it is not too difficult to deduce 2.3.14H.

*Proof of 2.3.14H.* For any triangle and any positive integer $n$, by sweeping out from a vertex as indicated in Figure 2.35, the triangle can be divided into $n$ subtriangles of equal area. By 2.3.13H and the additivity of defect, the defect of each of these subtriangles will equal $\frac{1}{n}$ times the defect of the whole triangle.

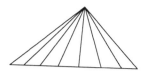

Figure 2.35. Dividing a triangle into smaller triangles of equal area.

Now suppose triangle $T_1$ has area $A_1$ and defect $d_1$, while triangle $T_2$ has area $A_2$ and defect $d_2$. If $A_1/A_2$ is a rational number $p/q$, then divide $T_1$ into $p$ triangles of area $A_1/p$ and defect $d_1/p$, and divide $T_2$ into $q$ triangles of area $A_2/q$ and defect $d_2/q$. Since $A_1/p = A_2/q$, Theorem 2.3.13H implies that these little triangles also have equal defects, that is, $d_1/p = d_2/q$. Hence, $d_1/d_2 = p/q = A_1/A_2$, as desired.

We omit the modification in the argument required if $A_1/A_2$ is not a rational number.

### ASIDE: KANT AND GEOMETRY

Immanuel Kant (1724–1804) was the preeminent philosopher during the century preceding the discovery of non-Euclidean geometry, and his ideas about space and geometry were influential in delaying the acceptance of non-Euclidean geometry. In fact, there are some who would say that reluctance to contradict Kant was a major factor in Gauss's decision not to publish his ideas about non-Euclidean geometry. The ultimate acceptance of non-Euclidean geometry was a major blow to some of the central ideas of Kant.

In his *Critique of Pure Reason*, Kant wrote that geometry is the study of space, and our knowledge of space is not empirical, but rather it is a consequence of the way in which our mind is structured. Moreover, the mind is structured in such a way that Euclidean geometry is the only conceivable geometry. Thus the reason that one is so naturally inclined to accept Euclid's fifth postulate is not due to experimental observations, but rather to the structure of the brain.

This theory made the statements of geometry (both postulates and propositions) a rare and prized sort of statement—a priori synthetic statements. "A priori" means that they are prior to experience, that their verification does not require experience (which is good, because experience can be fallible). "Synthetic" means that their verification requires something more than just the meaning of the words (and hence is nontrivial). In the case of geometry, that "something" is the structure of the brain, which forces it to see geometry as Euclidean.

Of course, non-Euclidean geometry shows that Kant was wrong about this. The modern position is that pure mathematics (which includes geometry) consists of formal deductions from abstract postulates, and as such is not giving necessary truths about the real world. The many applications of pure mathematics to the real world are due to the fact that many real-world phenomena can be modeled by various systems of pure mathematics, and the deductions of pure mathematics, which have been more readily obtained in the more abstract scenario, can then be applied with the knowledge that the applicability of the deductions is limited by the validity of the modeling.

### Exercises

1. Prove that if a line $a$ is a divergent to $\ell$ passing through a point $P$, then it is a divergent to $\ell$ passing through any point on it. (Hint: This is immediate from the definitions and 2.3.4H.)

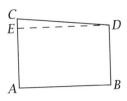

Figure 2.36. For Exercise 2.

2. Prove Lemma 2.3.8, using the following hints. If $AC = BD$, we have already proved that $\angle C = \angle D$. (When?) Assume $AC > BD$ as in Figure 2.36, and draw $AE = BD$. Compare angle $AED$ with angle $C$ and with the whole angle at $D$. For one of these, use Euclid's Proposition 16, and for the other use the first part of

this hint. Resist the temptation to say that these are right angles below the dashed line. Now you should have proved that $\angle C <$ $\angle D$. Similarly, if $BD > AC$, then $\angle D < \angle C$. Why is this enough to prove "if and only if" in Lemma 2.3.8?

3. Prove Theorem 2.3.7H, using Lemma 2.3.8 and the following hint. In Figure 2.25, tell whether the following angles are acute or ob-tuse: $\angle PP_1Q_1$, then $\angle Q_1P_1P_2$, then $\angle P_1P_2Q_2$. Give reasons.

4. If $m$ and $\ell$ are parallel to the right, prove that $P_2Q_2 < P_1Q_1$ in Figure 2.37. This is part of the proof of Theorem 2.3.5H. Use Lemma 2.3.8 and the acuteness in 2.3.9H.

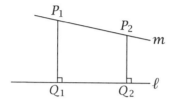

Figure 2.37. For Exercise 4.

5. Prove part of Theorem 2.3.9H by showing that $\angle \alpha = \angle \beta$ in Figure 2.29, where $m$ and $n$ are parallels to $\ell$. Follow these hints, and refer to Figure 2.38. Assume $\alpha \neq \beta$, and let $\beta$ be the smaller angle. Draw a line from $P$ making angle $\beta$ with $PQ$ on the other side of $PQ$ from the given angle $\beta$. This line must meet $\ell$ (Why?) at some point $R$. Draw $QS = QR$ on the other side. Show that triangles $PQR$ and $PQS$ are congruent. Conclude that $PS$ and $n$ coincide. Why is this a contradiction? Conclude that the assumption $\alpha \neq \beta$ must have been false.

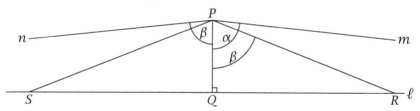

Figure 2.38. For Exercise 5.

6. Finish the proof of 2.3.5H by showing that $SU = RQ$ in Figure 2.32. (Hint: First show that triangles $ORT$ and $OST$ are congru-ent, and then show that triangles $RTQ$ and $STU$ are congruent. You will have to use 2.3.9H once.)

7. Show that in hyperbolic geometry two lines cannot have more than one common perpendicular by considering the angle sum of the quadrilateral that would be formed.

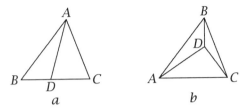

Figure 2.39. For Exercises 8 and 9.

8. Prove Proposition 2.3.12H in the two cases pictured in Figure 2.39. (Hint: The angles at $D$ count in the little triangles, but not in the big triangle.)
9. If the angle sums of triangles $ABD$, $ACD$, and $BCD$ in Figure 2.39b are $173°$, $175°$, and $178°$, respectively, then what is the sum of the angles of triangle $ABC$?
10. Suppose one triangle has an area of 10 square miles, and angles of $90°$, $65°$, and $20°$. Another triangle has an area of 2 square miles, and two of its angles are also $90°$ and $65°$. What is its third angle?
11. Formulate a notion of defect for quadrilaterals, and prove that the obvious analogue of 2.3.14H holds. (Hint: Use Corollary 2.3.15H.)

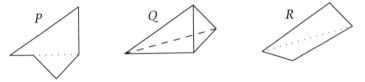

Figure 2.40. For Exercise 12.

12. In order to prove Theorem 2.3.17H, one needs the fact that two polygons that are equivalent to the same polygon are equivalent to each other. Illustrate this by showing that $P$ is equivalent to $R$ in Figure 2.40. Note that the solid line in $Q$ shows that $Q$ is equivalent to $P$, and the dashed line in $Q$ shows that $Q$ is equivalent to $R$. Use the dashed line to decompose $P$ farther, and the solid line to decompose $R$ farther.
13. We define the $V$-edge of a Saccheri quadrilateral to be the side connecting the vertices that have the acute angles. Prove that any triangle $ABC$ is equivalent to a Saccheri quadrilateral whose $V$-edge equals one of the sides of the triangle, following these suggestions. Let $M$ and $N$ be the midpoints of $AB$ and $AC$, respectively, and drop perpendiculars from $A$, $B$, and $C$ to the ex-

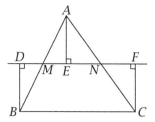

Figure 2.41. For Exercise 13.

tension of the line through $M$ and $N$, as in Figure 2.41. Prove certain triangles are congruent, and deduce the desired result.

14. Prove that two triangles with the same defect and a side of one equal to a side of the other are equivalent, following these suggestions. By Exercise 13, the two triangles are each equivalent to Saccheri quadrilaterals with the same V-edge and the same vertex angles. If we knew that such quadrilaterals are congruent, then we would be done by Exercise 12. To show that they are congruent, explain why a picture as in Figure 2.42 cannot occur in hyperbolic geometry.

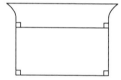

Figure 2.42. For Exercise 14.

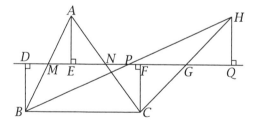

Figure 2.43. For Exercise 15.

15.* Prove 2.3.17H, following these suggestions. Suppose that triangles $ABC$ and $A'B'C'$ have the same defect. We may suppose that triangle $A'B'C'$ has $B'C' > BC$, for if they have two equal sides we are done by Exercise 14. Form the diagram of Figure 2.43, which includes Figure 2.41 and has $CG = GH = \frac{1}{2}B'C'$. Show that the defects of $BCH$ and $ABC$ are equal. (This requires some work. First show that triangles $GHQ$ and $GCF$ are congruent. Then, using also the congruences established in Exercise 13, show that $DBP$ and $QHP$ are congruent. Now you can match up angles in triangles $BCH$ and $ABC$.) Hence, $BCH$ is equivalent to $ABC$ by Exercise 13 (since they agree in $BC$). Also, $BCH$ is equivalent to $A'B'C'$, since $CH = B'C'$. Thus we are done by Exercise 12.

16. In Lobachevsky's axiom system, which of the following statements are equivalent to Postulate 5H? (Saying that a statement is equivalent to Postulate 5H is the same as saying that it is true in hyperbolic geometry but not true in Euclidean geometry.)

   a. The sum of the angles of a triangle is less than $180°$.
   b. The vertex angles of a Saccheri quadrilateral are equal.
   c. The vertex angles of a Saccheri quadrilateral are acute.
   d. Triangles whose corresponding angles are equal are congruent.
   e. Triangles whose corresponding sides are equal are congruent.
   f. Triangles with the same angle sum have the same area.
   g. Triangles with the same area have the same angle sum.

17.** A basic formula relating distance and angle in hyperbolic geometry involves the angle of parallelism (Theorem 2.3.10H). If $\ell$ is any length, let $\alpha(\ell)$ denote the associated angle of parallelism. It was proved by both Bolyai and Lobachevsky that there is a certain length $k$ (one choice of the intrinsic distance) such that for any number $\ell$

$$\tan\left(\frac{\alpha(\ell)}{2}\right) = e^{-\ell/k}.$$

   a. Show that $k$ equals the length whose angle of parallelism is $40.38°$.
   b. Let $\ell_1$ denote the length whose angle of parallelism is 45 degrees, and $\ell_2$ the length whose angle of parallelism is 89 degrees. What is the ratio $\ell_2/\ell_1$?

## 2.4   Spherical Geometry

There is another form of non-Euclidean geometry, which is more intuitive than hyperbolic geometry in terms of a model, but less intuitive in terms of its theorems. This is the geometry on a sphere, such as the surface of the earth (which we will assume to be a perfect sphere). The great German mathematician Bernhard Riemann (1826–1866) is usually credited with realizing that this familiar world can be taken as an alternative to Euclidean geometry.

We want to think that the analogue of plane geometry is done on (the surface of) a large sphere. The word "sphere" in mathematics usually means just the surface, not the solid region enclosed, and that is the way we will use it. The notion of "point" on a sphere poses no problem, but how about "straight line"? Think of two points on a large spherical balloon. The shortest distance in space between those points does not lie on the sphere; it cuts through the region enclosed by the sphere. But we are not allowing that to occur. We are restricting everything to lie *on* the sphere. There is a shortest distance on the sphere, and this is a portion of what is called a "great circle." One way of visualizing a great circle is to think of a point-source of light at the center of the sphere, and take the shadow on the sphere of the straight line in space that cuts through the region enclosed by the sphere. (See Fig. 2.44.)

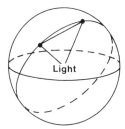

Figure 2.44. Great circle as a shadow.

A *great circle* can be defined as a circle on a sphere that divides the sphere into two portions of equal area. The equator is the most familiar example of a great circle, but there are lots of other great circles on the earth. Lines of longitude on a globe are great circles, while lines of latitude are circles that are *not* great circles. One can prove that a great circle will be the shortest path on the sphere between points on it. For two cities such as New York and Rome, which are at roughly equal latitudes, the shortest distance between them will be on the great circle that goes quite far north of them, rather than the circle of latitude. (See Figure 2.45, which also includes Anchorage and St. Petersburg as a more extreme example of the advantage of flying north to go east or west.)

We want to think of geometry on the sphere, and use the word "straight line" to refer to great circles. This actually should have been the most important form of geometry to the pre-Greek civi-

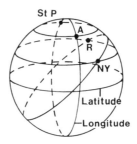

Figure 2.45. Latitude and longitude.

lizations, who were concerned with practical measurement of land on the earth's surface. As we shall see, for the relatively small figures they were measuring, the deviation of the actual geometry of the figures on the sphere from Euclidean geometry was small enough to be negligible. However, for figures that cover a significant portion of the sphere, geometry of the sphere deviates drastically from Euclidean geometry.

FLATLAND

If you object to calling these great circles "straight lines" because your 3-dimensional insight allows you to connect points by shorter paths that are not confined to the sphere, then imagine yourself as a tiny being that lives on the sphere and does not project outside the sphere at all. Before trying to imagine this, you might think about Flatland, a 2-dimensional world described in the classic book, *Flatland, a Romance of Many Dimensions*, written by Edwin Abbott Abbott in 1884. Abbott was one of the leading scholars and theologians of the Victorian era, and is best known for his *Shakespearian Grammar* and his biography of Francis Bacon.

The world of Flatland is a plane inhabited by figures such as squares, triangles, and circles, whose motion and sight is completely restricted to this plane. The book is in part a social satire of Victorian Britain: status is determined by shape, ranging from circular priests down to straight line women. The narrator, whose name is A Square, describes how people could discern the size of angles because an everpresent fog would make nearby points seem brighter, and how coloring of one's body was outlawed because it could be used to incorrectly render one's angles. In Figure 2.46a, which along with the adjacent figures is taken from *Flatland*, the shading on *DE* shows how the triangle *ABC* appears to the viewer.

Midway through the book, A Square is shocked when a tiny circle

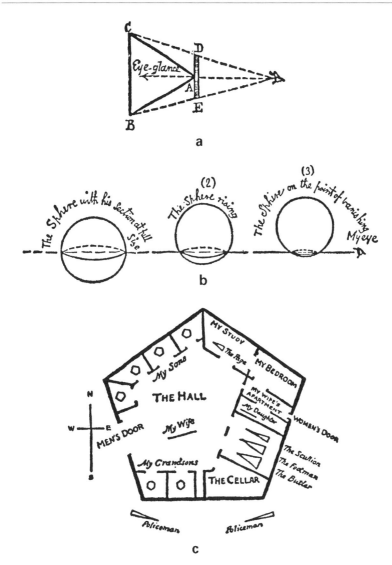

Figure 2.46. Scenes from *Flatland*. (From Edwin A. Abbot, *Flatland: A Romance of Many Dimensions* [Princeton, N.J.: Princeton University Press]. Reprinted with permission

that gradually grows larger suddenly appears inside his house. It is a sphere from the 3-dimensional world, which is penetrating the plane of Flatland (see Fig. 2.46b). The sphere tries to explain 3-dimensionality to A Square, and even resorts to deeds such as removing items from a cupboard without opening the cupboard, which it can do because the insides of all flatland objects are exposed to it (see depiction in Fig. 2.47). A Square doesn't understand 3-dimensionality until the sphere lifts him above the plane, so that he can now see the insides of all of the rooms of his house at once (Fig. 2.46c), and can even see the insides of Flatlander bodies. When the sphere leaves and A Square tries to explain the third dimension to the Council of Flatland, using the expression "Upward, not Northward," he is jailed for life.

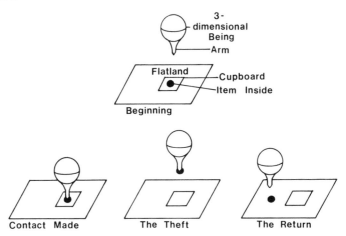

Figure 2.47. A theft from Flatland.

The book's aim is to help us envision the fourth dimension in a way that is analogous to the way a Flatlander must struggle to envision the third dimension, and we shall return to this idea several times later in this chapter. For now, however, use the idea to help put yourself in the place of tiny 2-dimensional creatures whose entire being and sensing is confined to a sphere. This world was described in *Sphereland*, written in 1965 by Dionys Burger.

Burger's thesis is that Flatland was actually part of a huge sphere, but the Flatlanders were living on such a small portion of it that for a long time everyone thought it was flat. The first bit of evidence that something was awry occurred when two explorers ventured in opposite directions and eventually met up with each other. It is easy for us to see that they had gone to the opposite side of their sphere, but Flatlander scientists came up with an alternative explanation—that they had just walked in a big circle in Flatland. Then a surveyor started observing that angle sums of large triangles were consistently greater than 180°. He and the grandson of A

Square, with some help from the spherical visitor from Flatland, deduced that their world must be a sphere, but they couldn't convince anyone else.

So think of Flatlanders (squares, octagons, straight lines, etc.) who actually live on a huge sphere. In their world, light travels along shortest paths on the sphere, and these they naturally call "straight lines." For example, the shortest path from a (male) triangle to a (female) straight line is the solid path in Figure 2.48. It makes no sense to say that the dashed path is shorter because it is not part of their world.

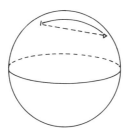

Fig. 2.48. Shortest path in Sphereland.

This spherical geometry is a very close approximation of Euclidean geometry for small figures, but behaves very differently when the figures begin to take on a significant fraction of the circumference or area of the sphere. Spherelanders living on a sphere the size of the earth (radius 4000 miles), but so small and slow that their entire lives and experience were restricted to a football field, would probably adopt Euclidean geometry as the geometry of their world, unless they were capable of extremely accurate measurements. As we shall see later in this section, the sums of angles of all their triangles is greater than 180°, but the largest angle sum they would ever see is less than 180.00000001°. They would have no parallel lines, although they would probably think they did. The lines they think of as parallel and everywhere equidistant would not actually be equidistant. Their "parallel" lines are always getting closer together or farther apart, depending on which way you look at it. But for the lines that are the longest and farthest apart that they would ever see, the sides of the football field, the amount by which they get closer together is about .0000001 inch. See Figure 2.49, which is a greatly exaggerated picture of a football field on the earth, but illustrates the way in which it is impossible to draw a perfect rectangle on a sphere. Of course, for an actual football field this is of no practical significance,

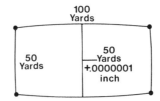

Fig. 2.49. A football field on the earth.

since the width of a blade of grass (or a particle of artificial turf) is much greater than the amount by which the field fails to be a perfect rectangle. However, if a Spherelander could follow those sidelines in a perfectly straight path for about 6250 miles, he would find that they would actually meet.

### POSTULATES

It is quite difficult to set up this spherical geometry as a formal axiom system. Of Euclid's five postulates, two (numbers 2 and 5) must be changed drastically, two (the first and third) must be changed slightly, and one (the fourth) remains the same. The substitute for #1 is: "Two points lie on a unique straight line unless those points are antipodal, in which case they lie on many straight lines." Here we have introduced the word *antipodal* for points on a sphere that are directly opposite from one another. Each point has a single antipodal point. Spherelanders would not have the perspective to see points as directly opposite, and so they might just note that for every point there seems to be exactly one other point to which it is connected by more than one straight line, and give that point a name such as "antipodal." In Figure 2.50a, points $P$ and $P'$ are antipodal, and we

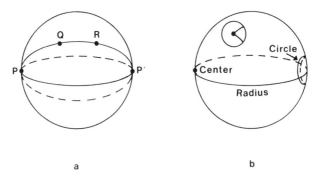

Figure 2.50. Antipodal points and circles on a sphere.

have pictured three straight lines passing through them; one is the equator, one is the circle that defines our drawing, and one is the great circle passing through $Q$ and $R$. Only one straight line passes through $Q$ and $R$, the indicated great circle. If we want to talk about line segments between $Q$ and $R$, we should consider the short portion of the great circle that connects them, not the long portion that goes around the back of the sphere. This is one of the messy parts of axiomatizing.

The substitute for Euclid's Postulate 2 is: "There is a certain distance $C$ such that if a straight line is extended to a distance $C$, it meets itself." This distance $C$ is, of course, the circumference of the sphere, but Spherelanders would not be able to visualize it that way. This fact that a straight line is unbounded but still finite is the most significant difference between spherical geometry on the one hand and Euclidean and hyperbolic on the other. Mathematicians since the time of the Greeks had thought about doing geometry on a sphere. That great circles were the shortest distance between points was well known. The formula (Theorem 2.4.1 below) relating angle sums of triangles to their area had been known at least since 1629. But no one considered this as an alternative to Euclidean geometry until Riemann gave his famous lecture, "On the Hypotheses Which Lie at the Foundation of Geometry," in 1854, a talk that we shall discuss in more detail in Section 2.6. Here is a paraphrase of the relevant part:

> In the application of geometry to space in the large, we must distinguish between unboundedness in extent and infinitude in measure. That space is without boundary and three-dimensional is an empirical certainty that is continually reaffirming itself. (No matter how far we travel, we never encounter a boundary marking the end of space.) This, however, by no means implies that space is infinite (in total volume). From the assumed possibility of motion of bodies without distortion (i.e., rigid motions), it follows only that space has constant curvature. However, this curvature could conceivably have a positive value, and though this value might be exceedingly small, space would nevertheless be curved and closed on itself—like a three-dimensional analogue of the surface of a sphere—and therefore finite.[15]

The main thrust of these remarks deals with the possible geometry of the universe; we will return to this subject in Section 2.6. For our purposes in this section, the important aspect of the talk is the idea that geometry on a sphere should be considered as an alternative

15. Faber 1983a, p. 113.

Georg Friedrich Bernhard Riemann

to Euclidean geometry in the plane, with the implication that little Spherelanders might think they were Flatlanders.

The substitute for Euclid's Postulate 3 is: "A circle can be drawn with any point as center and any radius, provided the radius is less than $C/2$." Figure 2.50b shows two circles on a sphere, one with a small radius and one with a large radius, yet the circles themselves have roughly the same circumference. Note how the circumference of circles increases with the radius until the radius equals $C/4$, while for radii greater than $C/4$, circumference decreases with radius, with the circle shrinking to a point as the radius approaches $C/2$. In Exercise 1, you are asked to investigate a circle of radius $C/4$.

Euclid's Postulate 4 says, "All right angles are equal," which is also true in spherical geometry. It is a statement about uniformity of space, and a sphere has this uniformity property, which is what Riemann was talking about when he mentioned the "possibility of motion of bodies without distortion." The statement that right angles are equal is a weak form of rigid motion. For example, on the surface of a football, where "straight lines" are the curves that are shortest distances between points on them (called *geodesics*), it is still true that all right angles are equal, but you cannot cut out a piece of one part of the ball and paste it to another part without some distortion. The Side-Angle-Side Theorem (SAS) is a stronger statement about uniformity of space. Remember that in Hilbert's rigorous form of Euclidean geometry, the SAS theorem had to be taken as an axiom. SAS is true in spherical geometry, but not in the geometry on a football. See Figure 2.51, in which the two triangles have right angles at $A$ and $A'$ and $AB = A'B'$ and $AC = A'C'$, but $BC > B'C'$.

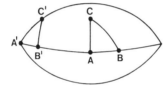

Fig. 2.51. SAS fails on a football.

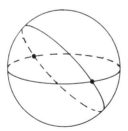

Fig. 2.52. All lines meet in two points.

Finally, Euclid's fifth postulate should be replaced by "All lines meet in two points." See Figure 2.52, which shows the two points of intersection of two great circles, and that there are no parallel lines in this geometry. Thus Euclidean, hyperbolic, and spherical geometries encompass the three possibilities regarding parallel lines—one, many, and none. Hyperbolic is considered to be more important than spherical from the point of view of pure mathematics, because it agrees with Euclid's first four postulates, and so it can be used to show that Euclid's fifth postulate cannot possibly be deduced from his first four. Spherical geometry cannot be used to show this impossibility because Euclid's first four postulates are not all true in it.

Note the occurrence of the intrinsic distance $C$ that appeared in Postulates 2 and 3. You should have no trouble understanding that all the formulas of spherical geometry can be based on this number $C$. Equivalently, we could use the radius $R$ of the sphere as the underlying constant for our geometry. These constants are related by $C = 2\pi R$. There is a different sphere for every value of $R$, but qualitatively the theorems of spherical geometry do not depend on the value of $R$. As we shall see below, if our two-dimensional beings restricted to a football field on the earth were smart enough and could measure accurately enough, they could determine the value of $R$ that is relevant for their geometry. Note that this is different than the determination of the circumference of the earth, which was accomplished by a Greek, Eratosthenes, around 200 B.C. He did this by using 3-dimensional data, namely, the length of shadows cast by the sun at different points on the earth.[16] In Sphereland, there is no such thing as 3-dimensional data; all the Spherelanders have is measurements along their sphere.

The constant for hyperbolic geometry, which was discussed in the previous section, is completely analogous to the $C$ or $R$ that can be used for spherical geometry. In the next section, we will discuss models for hyperbolic geometry, similar to the sphere as a model for the five strange axioms of spherical geometry we just discussed. When you think of the constant as a parameter that determines the size of the model, it becomes much less mysterious. But the existence of such a constant can be deduced abstractly from the axioms, and there exists no such constant for Euclidean geometry.

*PROPOSITIONS*

We have seen that some of the propositions of Euclidean geometry, such as 180° in triangles, and the existence of lines that are everywhere equidistant, are almost true in spherical geometry when viewed on a small enough scale. Some other propositions of Euclidean geometry are actually true for small figures but false for large ones. For example, Proposition 1 should be replaced by something

16. See Section 3.1 for a more detailed discussion.

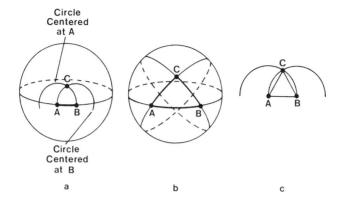

Figure 2.53. Constructing an equilateral triangle on a sphere.

like: "For lines less than a certain length, there is an equilateral triangle with that line as one side, but this is not true for lines longer
than that length." You haven't been told what that length is, because
you are asked to determine it in Exercise 2.

In Figure 2.53a, we are given a line connecting points $A$ and $B$. Draw
circles (not great circles) with one of the points $A$ or $B$ as center and
passing through the other, and let $C$ be a point of intersection of
these circles. The distance from $A$ to $C$ and from $B$ to $C$ equals the
distance from $A$ to $B$. The equilateral triangle $ABC$ is pictured in Figure 2.53b, slightly enlarged, using arcs of great circles (the straight
lines in this geometry) to connect the points. If we try the same construction with a big line $AB$, as in Figure 2.54a, the circles centered
at $A$ and $B$ will not meet, and so there is no point $C$ whose distances
from both $A$ and $B$ equal the distance from $A$ to $B$. Thus there is no
equilateral triangle with $AB$ as side.

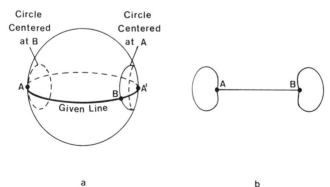

Figure 2.54. Equilateral triangle cannot be constructed on a long line.

The construction for a small line is exactly like the construction in Euclidean geometry, which is recalled in Figure 2.53c. Note that the "straight lines" in Figure 2.53b lie inside the circles of Figure 2.53a, similar to the Euclidean situation pictured in Figure 2.53c. For a large line, if we try to depict in Euclidean geometry what has happened in spherical geometry, it is as if the circles stay completely apart, as shown in Figure 2.54b. We commented on this possibility in Section 1.2 when we discussed the construction illustrated here in Figure 2.54a. When we asked there, "How do we know that the second circle contains points both inside and outside the first circle?" it may have seemed like crazy question, but now it seems reasonable.

Another proposition and proof of Euclidean geometry that works nicely for small figures of spherical geometry, but fails for large figures, is Proposition 16: "An exterior angle of a triangle is greater than either opposite interior angle." Review the proof given in Section 1.2 that angle $A$ was less than the exterior angle at $C$. Figure 2.55a shows how this proof works verbatim for a small triangle in spherical geometry; but in Figure 2.55b, where the triangle is large, the line $EF$ drops below the line $CD$, so we cannot say that angle $ECF$ is less than the exterior angle $ECD$.

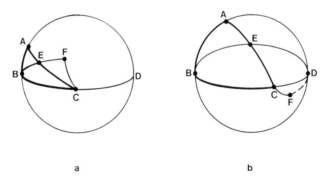

a                                   b

Figure 2.55. Proposition 16 works for small triangles, but not for large ones, in spherical geometry.

*AREA AND EXCESS*

The most quantitative result measuring the deviation of spherical geometry from Euclidean is the following.

THEOREM 2.4.1. *On a sphere of radius R, if a triangle has area A and angle sum S, then*

$$A = \frac{\pi}{180}R^2(S - 180).$$

This formula appeared in a book by Albert Girard in 1629, which makes it somewhat surprising that the prediscoverers of hyperbolic geometry found the analogous results so strange.

The amount $S - 180$ by which the angle sum of a triangle exceeds 180 is called the *excess* of the triangle. Thus, analogously to Corollary 2.3.15H, Theorem 2.4.1 says that the ratio area/excess is the same for all triangles, but, more than that, it says that its value is $\pi R^2 / 180$. This formula makes it clear how the ratio is the same for all triangles on a sphere, but the ratio can have any value, depending on the size of the sphere. Corollary 2.3.15H should now seem a little less mysterious.

Note that a consequence of Theorem 2.4.1 is that triangles have the same angle sum if and only if they have the same area. In particular, there are no "similar triangles," triangles that have the same angles but the sides of one are a multiple ($\neq 1$) of the other. Thus, as in hyperbolic geometry (Theorem 2.3.2H), an Angle-Angle-Angle Theorem exists in spherical geometry.

Theorem 2.4.1 can be used by Spherelanders to determine the radius of their sphere (provided they know that it is a sphere). If they measure a triangle to have area 10 square miles and angle sum 180.1°, they can find the radius of their sphere by solving the equation

$$10 = \frac{\pi}{180} R^2 (180.1 - 180)$$

for $R$, yielding

$$R = \sqrt{\frac{10}{.1} \frac{180}{\pi}} = 75.7 \text{ miles.}$$

A nice proof of Theorem 2.4.1, involving lunes, is given in Weeks (1985). We will be content to show that Theorem 2.4.1 is true for triangles that have two right angles. We can rotate the sphere so that it appears that the side of the triangle connecting the two right angles is the equator. The other two sides are then what we generally view as lines of longitude, and the triangle can be viewed as $ABC$ in Figure 2.56.

Note that the area of the triangle is proportional to the angle $D$. This should be clear; for example, doubling $D$ doubles the area. We compare the pictured triangle with the triangle having $D = 180$. This latter triangle covers the front upper portion of the sphere, and so its area is one-fourth the area of the sphere. Since the area of a sphere of radius $R$ is $4\pi R^2$, the area of the triangle with $D = 180$ is $\pi R^2$. By the stated proportionality, the area of the triangle with arbitrary angle $D$ is $D/_{180}\pi R^2$. Noting that the angle sum $S$ of our triangle is

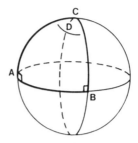

Figure 2.56. A triangle with two right angles.

$90 + 90 + D$, we can replace $D$ in our formula by $S - 180$, obtaining the desired formula,

$$A = \frac{S - 180}{180} \pi R^2.$$

This simple proof works only for triangles with two right angles, but it accomplishes two things. First, it emphasizes how there can be triangles with two, or even three, right angles; and second, it establishes the plausibility of Theorem 2.4.1.

CONSISTENCY

Spherical geometry is not considered to be as important in pure mathematics as hyperbolic geometry because not all the theorems of absolute geometry are valid in it, and because it is quite difficult to treat spherical geometry as a formal axiom system. Nevertheless, because of its intuitive model, it gives a very nice illustration of how some very strange-sounding axioms can actually yield a consistent body of mathematics.

There is a fine point here. We should say that spherical geometry is consistent *if* Euclidean geometry is consistent. This is true because our model of spherical geometry has been constructed within 3-dimensional Euclidean geometry. But how do we know that Euclidean geometry is consistent? No one has ever seen an infinite Euclidean plane. The answer is that Euclidean geometry can easily be reduced to *real analysis*, which means just the standard properties of the real numbers, using coordinate geometry. All the theorems of Euclidean geometry can be reduced to statements about relationships between numbers, by considering the equations of the geometric figures. So everything has been reduced to the consistency of real analysis. Moreover, the consistency of real analysis can be reduced to that of *arithmetic*, which means the standard properties of the positive integers. Whether or not arithmetic has been proved to be consistent depends on what techniques we allow ourselves to use in our proof. Indeed, the consistency of arithmetic cannot be proved using just arithmetic. The famous twentieth-century mathematician Hermann Weyl remarked that God exists because arithmetic is consistent, but the devil exists because we cannot *prove* that it is consistent. Almost all mathematicians are happy to accept the consistency of arithmetic, and hence that of Euclidean and spherical geometry.

Another very significant factor in the importance of spherical geometry will be discussed in Section 2.6. This is the possibility that the geometry of the universe is the 3-dimensional analogue of spherical geometry. We saw that Riemann considered this possibility in 1854. The idea that Euclidean geometry is not necessarily the geometry of the universe, but only one of several candidates, and that the determination of the geometry of the universe belongs to physics and not to mathematics, is one of the most important contributions of non-Euclidean geometry to human thought. Perhaps an even more important consequence of non-Euclidean geometry was the removal of belief in absolute truth. The realization that Euclidean geometry was not the only possible type of geometry had ramifications in such diverse areas as religion and physics.

### ASIDE: A FLAT MODEL OF SPHERICAL GEOMETRY

Spherical geometry is considered to be a 2-dimensional geometry, even though we like to use our 3-dimensional insight to picture the whole sphere. It is considered to be 2-dimensional because locally it can be flattened out to look like a 2-dimensional plane. A coordinate system with two coordinates can be used to parametrize a region around any of the points on the sphere.

We can actually give a flat model of spherical geometry, which clinches its 2-dimensionality and will be useful for our discussions in the next two sections. This model was discussed by the principal figures we will encounter in the next two sections, Henri Poincaré, in his 1902 essay, *Science and Hypothesis*, and Albert Einstein, in a 1921 lecture entitled "Geometry and Experience."

Think of a sphere resting on a plane with a light at the top of the sphere casting projections or shadows of all figures on the sphere onto the plane. Readers familiar with Plato's *Allegory of the Cave* may note some similarity of the two ideas. Figures that have the same size on the sphere will have larger projections on the plane if they are near the top of the sphere than if they are near the bottom. See Figure 2.57, which illustrates this situation for three little circles on the sphere.

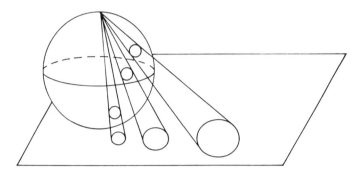

Figure 2.57. Projecting spherical figures onto the plane.

The flat world of the projections of the spherical figures contains all of the information of their round world. From our external vantage point, figures seem to grow larger as they go out from the center of the plane. But beings in this projection world wouldn't notice this growth. Think of the three circles in the plane in Figure 2.57 as being along the path of the same projected circle moving out from the center of the plane. If the circle carries a measuring stick with it, the projections of the measuring stick will (from our perspective) grow at the same rate as the circle, and so if it measures itself at any point along the way it will always obtain the same value.

If we want to try to think of this flat model of spherical geometry without using projections from the sphere, that is, without using any 3-dimensional intuition, we could think of a flat Euclidean world that has a temperature gradient or funny gas that causes everything to expand at a certain rate as it moves out from the center. Again, the objects in this world will not know they are expanding because their measuring sticks expand along with them. It is only we, from our external viewpoint, who think they are expanding.

Regardless of whether you think of it as projected from a sphere or having an expanding gas, the measurement of distances in this flat world will differ from the way they look to an external viewer. It is easier to see this difference through projections. The geodesics (shortest paths between points) in the plane will be the projections of great circles. Two of these are pictured in Figure 2.58.

It turns out that these are actually circles in the plane. (See Exercise 10.) It seems surprising at first that these circles in the plane are the shortest paths between points on them. The reason is that the measuring sticks will be longer at points farther from the center of the plane, and so fewer of them will be required than would have been to follow what appears to be a shorter straight line path through

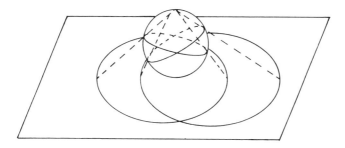

Figure 2.58. Projecting great circles onto the plane.

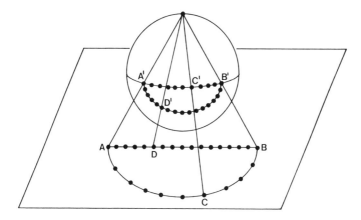

Figure 2.59. $ACB$ is shorter than $ADB$ in the flat model of spherical geometry.

the interior of the circle, but using smaller measuring sticks. We hope that Figure 2.59 will clarify this situation. Here the path $ACB$, which is the projection of the path $A'C'B'$, is 12 sticks long, while the path $ADB$ is 16 sticks long. Notice that all sticks have the same length on the sphere; it is their projections which differ in size.

The biggest problem with this flat model of the sphere is what to do for a point in the plane corresponding to the North Pole. You have to adjoin an artificial "point at infinity," and then give complicated rules for projections of figures that contain the point at infinity. We will not worry about this.

Like the round sphere, this flat model of spherical geometry is a finite world. Only a finite number of disks of the "same" size can be placed on the plane, because the disks appear to become much larger as they move out from the center. It is totally analogous to the finite

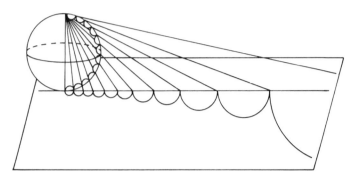

Figure 2.60. Finite number of disks of same size cover the flat model of spherical geometry.

number of disks of the same size that one can place on a sphere. Figure 2.60 shows a sequence of half-disks of the same size both on the sphere and its projection. The projection of the top half-disk will be infinite.

### *Exercises*

1. What are the circumference and area of a circle of radius $C/4$ on a sphere whose great circles have circumference $C$? (Hint: Draw a picture of the circle, whose radius goes one-fourth of the way around the sphere. You will need to use the formula that the area of a sphere of radius $R$ is $4\pi R^2$, but your answer for the area of the circle on the sphere should be expressed in terms of $C$, since that is the length which is apparent to the beings of spherical geometry.) Compare these formulas with the formulas relating circumference and area of circles in Euclidean geometry to their radius. In other words, are the standard formulas of Euclidean geometry that relate the circumference and area of a circle to its radius valid for the circle of spherical geometry we just studied? How badly do they fail?

2.* In terms of the circumference $C$, what is the largest size side of an equilateral triangle in spherical geometry? Explain why, with pictures. Draw this largest equilateral triangle. What is strange about it? Draw one slightly smaller. (Hint: If, in Figure 2.54a, $A'$ is the point antipodal to $A$, then each of the little circles can be thought of as having radius equal to the distance from $A'$ to $B$. One of these circles is entirely on the back half of the sphere, while the other is half in front and half in back. It is the disjointness of these circles that causes the proposition to fail.)

3. Explain exactly how the change in one of Euclid's postulates in spherical geometry causes a very careful proof of Euclid's Proposition 16 to go wrong for a large figure in spherical geometry. (See Fig. 2.55b.) Tell which postulate is involved and how it was used in Euclid's proof. Draw a picture in the Euclidean plane of the failure of the proof of Proposition 16 for large triangles in spherical geometry. It should be just a flattened-out version of Figure 2.55b. Something will have to be distorted, of course.

4. Which of the following of Euclid's propositions are true in spherical geometry?

    a. Prop. 5. Base angles in an isosceles triangle are equal.

    b. Prop. 12. A perpendicular can be constructed from a point to a line.

    c. Prop. 17. The sum of two angles of a triangle is always less than 180°.

    d. Prop. 26. The Angle-Side-Angle congruence theorem.

    e. Prop. 27. If lines have alternate interior angles equal, then they are parallel.

    f. Prop. 48. In a right triangle, the square of the hypotenuse equals the sum of the squares of the other sides.

5. Assume the earth is a sphere whose radius is 4000 miles. What is the sum of angles of a triangle whose area is .001 square mile? (Remark: This is relevant to an example discussed in the text of beings living on a football field. The largest triangle that can be drawn on a football field of 100 by 50 yards has an area about .001 square mile.)

6. Some 2-dimensional beings know that they live on a sphere. They measure the angles of a triangle to be 50.1°, 60.2°, and 70.3°, and the area of the triangle to be 10,000 square feet. What is the radius of their sphere?

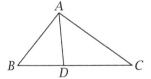

Figure 2.61. For Exercise 7.

7. In Figure 2.61 in spherical geometry, triangle $ABD$ has angle sum 181, and triangle $ACD$ has angle sum 182. What is the angle sum of triangle $ABC$? Explain why.

8. Derive a formula relating the angle sum of a quadrilateral on a sphere of radius $R$ to its area.

9. Some 2-dimensional beings living on a sphere define a *Rie* to be the length of side of a 90-90-90 triangle. One of these beings walks at a speed of $1/2$ Rie/hour in a straight line. How long does it take until it returns to its starting position?

10.[**]Use analytic geometry to show that, in the flat representation of spherical geometry, the projection of a great circle that does not pass through the North Pole is a circle. Write out all the steps in the following outline of a proof. Refer to Figure 2.62. Choose the scale so that the equation of the sphere is $x^2 + y^2 + (z - 1/2)^2 = 1/4$.

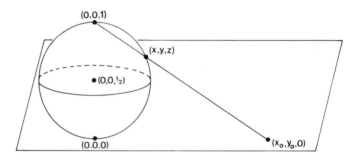

Figure 2.62. For Exercise 10.

The great circle will be the intersection of this with the plane $Ax + By + (z - 1/2) = 0$, for some numbers $A$ and $B$. Let $(x_0, y_0, 0)$ be the point where the line from $(0, 0, 1)$ through $(x, y, z)$ meets the $xy$-plane. Use similar triangles to show that

$$(2.4.2) \qquad \frac{x}{x_0} = \frac{y}{y_0} = \frac{1 - z}{1} = \frac{\sqrt{x^2 + y^2 + (z - 1)^2}}{\sqrt{x_0^2 + y_0^2 + 1}}.$$

Use the equation of the sphere to replace $x^2 + y^2$ in (2.4.2) by $z - z^2$, and then use the last equation of (2.4.2) to deduce $1 - z = 1/{x_0^2 + y_0^2 + 1}$. Now replace $x$, $y$, and $z$ in the equation of the plane by expressions in $x_0$ and $y_0$, and deduce, after simplifying, that $x_0^2 + A x_0 + y_0^2 + B y_0 = 1$, which is the equation of a circle.

11. A photographer goes bear hunting. From camp, he walks one mile due south, then one mile due east, then shoots a bear, and, after walking one more mile, is back at camp. What color was the bear?

12. On a sphere, triangle $ABC$ has three right angles, while triangle $A'B'C'$ has angles $90°, 90°$, and $30°$. Which is greater, the sides of $ABC$ or the equal sides of $A'B'C'$? Explain. Generalize.

13. Recall that the area of a sphere of radius $R$ is $4\pi R^2$. What is the angle sum of a triangle that covers one-tenth of the sphere?

14. Two-dimensional beings who live on a sphere define a *Rie* to be the length of side of a 90-90-90 triangle. What is the perimeter (in Ries) of a 90-90-30 triangle?

15. A Mercator projection map of the earth uses parallel vertical lines to represent lines of longitude. Explain, with drawings, why this makes regions such as Greenland, that are very far north or south appear larger on these maps than they actually are.

## 2.5  Models of Hyperbolic Geometry

In Section 2.3 we discussed how the axiom system for hyperbolic geometry, in which Euclid's fifth postulate was replaced by one which said that there is more than one line parallel to a given line through a given point, seemed to be consistent. Even though its theorems seemed counter to one's intuition about space, they did not seem to imply any mathematical contradictions. But neither Lobachevsky nor Bolyai could prove that hyperbolic geometry was consistent. It was conceivable at that time that if one worked hard enough, eventually a contradiction might be obtained, that is, something could be proved to be both true and false. Recall that the way we showed that axiom systems were consistent in Section 2.1 was to construct a model of the axioms, which is a way of interpreting the undefined terms in such a way that all of the axioms are true of the interpretation. This implies consistency because then every statement that can be logically deduced from the axioms will necessarily be true of the model, and a statement about a model cannot be simultaneously true and false.

It was not until at least forty years after the work of Bolyai and Lobachevsky that a model of hyperbolic geometry was discovered. This long wait was primarily due to the fact that most mathematicians did not want to deal with such strange ideas as non-Euclidean geometry.

We will discuss several models of hyperbolic geometry discovered between 1868 and 1881. These models are defined in terms of Euclidean geometry; hence they are consistent if Euclidean geometry is consistent. This situation is totally analogous to that of spherical geometry discussed in the preceding section. The model of hyperbolic or spherical geometry is constructed within 2- or 3-dimensional

Euclidean space. Using standard methods of Euclidean geometry, it is shown that the axioms of hyperbolic or spherical geometry are true in the model. Thus if hyperbolic or spherical geometry implies a contradiction, so does 3-dimensional Euclidean geometry. As we mentioned in the last section, most mathematicians would say that Euclidean geometry has not been proved to be consistent, but they are happy to believe that it *is* consistent. They must be equally happy to say that hyperbolic and spherical geometry are consistent.

### BELTRAMI'S PSEUDOSPHERE

The first model of hyperbolic geometry was proposed by the Italian mathematician Eugenio Beltrami (1835–1900) in 1868. This model is obtained by revolving a certain curve called a *tractrix* around its axis to obtain a figure that you might imagine as an infinite horn, or two infinite horns stuck together at their bells, as in Figure 2.63a. It is called a *pseudosphere* because its role for hyperbolic geometry is similar to that of the sphere for spherical geometry.

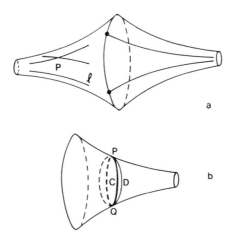

Figure 2.63. Beltrami's pseudosphere.

The interpretation of the word "point" in hyperbolic geometry is a point on the pseudosphere, while the interpretation of "straight line" is a geodesic on the pseudosphere. The word *geodesic*, as introduced in the previous section, is a curve that is the shortest distance between points on it. Some geodesics are drawn on the pseudosphere in Figure 2.63a, in a way that illustrates that Postulate 5H is satisfied: there are many lines through the point $P$ that do not meet the line $\ell$.

Beltrami didn't just pick this surface because it looked like a horn. It was already known (by H. F. Minding in 1839) that this surface, obtained by revolving a tractrix, was a surface of constant negative curvature. We shall say more about curvature at the end of this subsection. The sphere is a surface of constant positive curvature, and it had also been proved by Minding that on a surface of constant curvature geometric objects can be moved from one part to another without changing such things as angles. The ability to do this is of course an important part of geometry, as in the Side-Angle-Side Proposition. (Recall that in Figure 2.51 we illustrated how on a surface of nonconstant curvature, such as a football, the SAS Proposition is not valid.) Beltrami pointed out that this property of constant curvature should make the pseudosphere a model for hyperbolic geometry, and he noted such things as the fact that triangles on the pseudosphere have less than 180°, and that Postulate 5H is satisfied.

The pseudosphere does not serve as a complete model of hyperbolic geometry. One problem is due to the place where the two horns are attached. Not all curves can be continued smoothly over this ridge. Another is the circular symmetry it has. It seems as if the closed curve $C$ in Figure 2.63b, which is the intersection of the pseudosphere with a plane perpendicular to its axis, should be a geodesic that intersects itself after a finite length, similarly to the geodesics on a sphere. However, this curve is not a geodesic, just as circles on spheres that are not great circles are not geodesics. Also in Figure 2.63b, we show how a curve, labeled $D$, which veers slightly into the narrower part of the horn is a shorter path between $P$ and $Q$ than the closed curve $C$. But this is a problem, too, since there is a similar geodesic on the back of the pseudosphere, and hence there are two straight lines passing through the points $P$ and $Q$. The way to get around these problems is to say that the pseudosphere cut along one of its tractrices is a model for the portion of the hyperbolic plane between two parallel lines and bounded on one side.

Finally, in 1901 the great German mathematician David Hilbert proved that there can be no surface in 3-dimensional space that has constant negative curvature and provides a complete model of hyperbolic geometry. So Beltrami's pseudosphere didn't quite prove the consistency of hyperbolic geometry, but it had a great influence on people's acceptance of and interest in it.

The tractrix, the curve that is revolved to give the pseudosphere, is an interesting curve. It is pictured in Figure 2.64, in a position so that revolving around the $x$-axis gives the pseudosphere. It has a complicated equation, which we will not list, but it has an interesting geometric property, which involves the tangent lines to the curve at

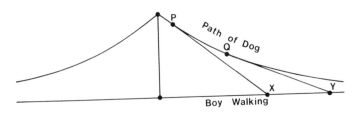

Figure 2.64. A tractrix.

points on it. These are lines that make the highest degree of contact with the curve. The tangent lines to the tractrix at points $P$ and $Q$ are indicated in Figure 2.64. The property of the tractrix is that these lines $PX$ and $QY$, and others like them, have the same length.

The basic geometric property of the tractrix says that if at each point on the curve you draw the portion of the tangent line connecting the point on the curve and the point where the tangent line meets the $x$-axis, then all of these lines have the same length. This curve is sometimes called the "path of the obstinate dog," because it is the curve that a dog would follow if it were attached by a leash to its master who was walking along the $x$-axis. The resisting dog would always draw the leash taut so that it always has the same length, but the dog would be pulled so that its motion would have its tangent line in the direction of the leash.

We close this subsection with a brief discussion of curvature of surfaces, which was mentioned earlier in the subsection. Around 1760, Euler, whom we met in Section 1.3, defined the principal curvatures of a surface, which are two numbers that can be positive or negative and can vary from point to point on the surface. Euler's definition required a 3-dimensional external perspective on the surface, such as we have when we picture a sphere or pseudosphere.

In 1828 Gauss published what he called his *Theorema Egregium* (Extraordinary Theorem), which dealt with curvature. He noted that the product of the two principal curvatures at a point had certain nice properties. This product is now called the Gaussian curvature. The most important property is that the Gaussian curvature can be determined by measurements within the surface. Two-dimensional beings living within the surface would not have the perspective required to find Euler's principal curvatures. But there are various ways in which they could do measurements that would determine the Gaussian curvature of their surface at any of its points, even for a surface in which the curvature varies from point to point. One of the ways in which they could determine this curvature is by comparing the angle sum and area of tiny triangles surrounding the point. The formula is that the Gaussian curvature at $P$ equals the limit of the ratios $(\pi/180)(S-180/A)$, where $A$ is the area and $S$ the angle sum of small triangles around $P$. The limit is taken as the triangles get smaller and smaller. Note that this curvature will be positive if $S > 180$, for example, on a sphere, and negative if $S < 180$, for example, on a pseudosphere. The formula is a little bit nicer if we measure angles in radians instead of degrees, but we have been avoiding that. This idea of 2-dimensional beings being able to determine the curvature of their surface is extremely important, as we will see in the next section.

### THE KLEIN-BELTRAMI DISK MODEL

We have seen that Beltrami's pseudosphere did not quite prove the consistency of hyperbolic geometry, because it was only a model of part of the hyperbolic plane. However, it was believed for quite a few years that this was good enough. Hence the first actual model, which was given independently by Beltrami and Felix Klein in 1871, did not receive quite the attention it deserved. We shall discuss it briefly, and then in the next subsection we will discuss in more detail an improved version of it due to Henri Poincaré .

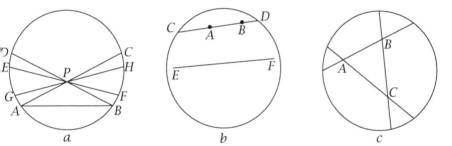

Figure 2.65. Klein-Beltrami disk.

Draw a disk in a plane. The word "point" will be used to refer to a point inside the disk, not including the boundary. The words "straight line" will be used to refer to a straight line inside the disk, again not including points on the boundary. Postulate 5H is satisfied in this model. This is illustrated in Figure 2.65a, which shows four lines passing through $P$ not intersecting the line $AB$. Remember that points on the boundary of the disk are not part of the model, so when we write $AB$ we mean all the points on the line between $A$ and $B$, not including $A$ and $B$ themselves. The lowest lines that do not meet $AB$

are $AC$ and $BD$. These appear to meet $AB$ at $A$ and $B$, respectively, but, as we keep repeating, $A$ and $B$ are not part of the model. Thus $AC$ and $BD$ are the parallels to $AB$ passing through $P$ in the strict sense of hyperbolic geometry, while $EF$ and $GH$ are what we called divergents in Section 2.3. They do not meet $AB$, but they are not the lowest such lines.

However, there are some aspects of this model that should be bothering you. How about Euclid's Postulate 2, that a line can be extended to any length? It appears that there is a bound on the length of lines, namely, they can only go out as far as the boundary. That is correct, of course, but length is defined differently in this model. A complicated formula[17] is given for the length of a line segment $AB$ in Figure 2.65b, in terms of the apparent Euclidean lengths $AC$ and $BD$ as well as $AB$. Under this definition, as $A$ gets very close to $C$, the defined length of the the segment $AB$ becomes arbitrarily large. (See Exercise 4.)

Another bothersome point is the angle sum of a triangle. It certainly looks like the angle sum of triangle $ABC$ in Figure 2.65c is $180°$. The explanation is similar to the preceding one. Angles in this model are defined in a way that makes them differ from their Euclidean value. We will not worry about the definition of angle in the Klein-Beltrami model because this unpleasant aspect of the model was fixed in Poincaré's modification of the model.

In a careful treatment of geometry, size of lines and angles are undefined terms, and so the fact that this model's definition of them differs from the Euclidean definition is not a flaw, axiomatically. In fact, you don't even have to talk about an actual measure of length or angle size—all you have to do is give a rule for when two lines are equal (or congruent), and for when two angles are congruent. Then Postulate 2 says that a line can be extended so that it is longer than any given line. In Figure 2.65b, under the Euclidean definition of equality of lines, $AB$ cannot be extended to be longer than $EF$, but under the Klein-Beltrami definition of equality of length, it can. One can prove that under the Klein-Beltrami interpretations of the words "point," "straight line," "congruence of lines," and "congruence of angles," all the postulates of a rigorous treatment of hyperbolic geometry are satisfied, and this gives the first proof that hyperbolic geometry is consistent.

### THE POINCARÉ DISK MODEL

In 1881 the great French mathematician Henri Poincaré modified the Klein-Beltrami model in such a way that angles are measured in the

---

17. $\log(^{BC \cdot AD}/_{AC \cdot BD})$.

Euclidean way. The primary advantage of this method is intuitive—
figures in this model really look like they are illustrating hyperbolic
geometry. Poincaré's model still has the property that distance (or
congruence of lines) has to be defined in a non-Euclidean way, but
he presented his model in a manner that made this new notion of
distance seem quite reasonable.

In the Poincaré model, "points" again are points in the interior (not
including boundary) of a disk. The "straight lines" are arcs of circles
that meet the boundary at right angles. In Figure 2.66a, five of these
straight lines are indicated. Note that a Euclidean straight line that
passes through the center of the disk will meet the boundary at right
angles and is considered to be a "straight line" in the Poincaré model.
It can be thought of as an arc of a circle of infinite radius, or as a
limit of arcs of actual circles that meet the boundary at right angles.
Note that a Euclidean straight line that does not pass through the
center of the disk will not meet the boundary in right angles on both
sides, and hence is not a "straight line."

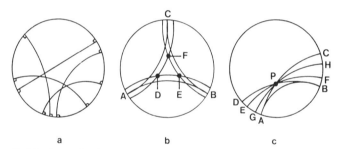

a                          b                          c

Figure 2.66. Straight lines, triangles, and parallels in the Poincaré
disk.

Angles are defined in this model to be the same as their Euclidean
value. Of course, when we talk about the Euclidean angle between
two curves, we mean the angle between their tangent lines. In Figure 2.66b, triangle $DEF$ quite clearly has angle sum less than $180°$,
while "triangle" $ABC$ is the limiting triangle of largest possible area
and smallest possible angle sum, namely $0°$. This is not actually a
triangle, because points $A$, $B$, and $C$ are not part of the model, but
can be approximated as closely as desired by an actual triangle in
the model, whose angle sum is as close as you want to $0°$, and whose
area is as close as you want to that of $ABC$. What is meant by area
depends on what is meant by length, and we haven't come to that yet.
This limiting triangle is this model's version of the one that Gauss
talked about in Section 2.3. (See Fig. 2.33.)

In Figure 2.66c, we have illustrated that Postulate 5H is satisfied in this model. The situation here is totally analogous to that for the Klein-Beltrami model. The two parallels to the line $AB$ are $AC$ and $BD$, while $EF$ and $GH$ are divergent to $AB$ through $P$, that is, lines through $P$ that do not meet $AB$, but not the lowest such lines. Remember that boundary points are not part of the model, and so $AC$ and $AB$ do not meet, since $A$ is not part of the model.

In order to know that this is a model for hyperbolic geometry, we must verify that it also satisfies the first four postulates of Euclidean geometry. The only one that is obvious is #4, since right angles here are the same as those of Euclidean geometry. The verification of the first postulate requires a construction in Euclidean geometry: given two points inside a circle, there is a unique circle that passes through the two points and intersects the given circle perpendicularly in two points. This is a difficult exercise in Euclidean geometry, but can be proved. Make sure that you understand why this is an interpretation of Euclid's first postulate in this model.

The second postulate does not appear to be satisfied at first glance, because it seems that straight lines can be extended only to a certain distance. But, with our discussion of the Klein-Beltrami disk model, we have prepared you for this occurrence. The answer is that distance is measured differently in this model than it appears to an external Euclidean viewer. A definition similar to that given for the Klein-Beltrami model in a footnote can be given. However, we shall present a less complicated formula that works well for small figures.

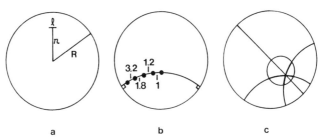

a                              b                              c

Figure 2.67. Length and a circle in the Poincaré disk.

Suppose the radius of the disk is $R$, and we have a tiny straight line of Euclidean length $\ell$ at distance $r$ from the center of the disk, as in Figure 2.67a. The length of this tiny segment in Poincaré's model is defined to be $\ell \cdot R^2 / (R^2 - r^2)$. For example, suppose the radius of the disk is 10 feet. A straight-line segment of Euclidean length .01 foot is considered to have length .0101 foot if it is 1 foot out from the center, length .0526 foot if it is 9 feet out from the center, and length .5025 foot if it is 9.9 feet from the center. We illustrate the calculation in the second case:

$$\text{length} = .01\frac{100}{100 - 81} = \frac{1}{19} = .0526.$$

Note how the Poincaré length becomes much greater than the Euclidean length for a figure that is close to the boundary, while for a figure close to the center, the two distances are practically the same.

An arc of a Euclidean circle (which is a "straight line" in this geometry) can be thought of as being approximately composed of lots of tiny segments, and the length of a Poincaré straight line is defined to be approximately the sum of the Poincaré lengths of these tiny segments. The more finely you divide the arc into segments, the better is the approximation. Calculus is required to make it precise. In Figure 2.67b, a portion of a circle is approximated by five line segments, which look equal to us but have lengths approximately 1, 1.2, 1.8, and 3.2 in this model. Thus the Poincaré length of that portion of a Poincaré straight line is approximately 7.2. As you get even closer to the boundary, the hyperbolic lengths become such large multiples of the apparent lengths that each of the straight lines of hyperbolic geometry has infinite length; this means that by extending it close enough to the boundary, its length will exceed any prescribed number. Thus Postulate 2 (the extendability of lines) is satisfied.

Another way of thinking about the distance formula is that figures which have the same size in the Poincaré model look to us as if they have different sizes. Their apparent sizes become smaller as they get farther from the center, that is, closer to the boundary. The formula relating these sizes is obtained by inverting the previous formula:

$$\text{apparent length} = \frac{R^2 - r^2}{R^2} \cdot \text{Poincaré length.}$$

Note that the apparent length equals the Poincaré length if the figure is at the center of the disk.

A brilliant illustration of Poincaré's disk was given by the famous Dutch artist M. C. Escher in 1958. He depicted a disk full of flying fish, which are congruent to one another in hyperbolic geometry. It appears to us as if they are getting smaller as they approach the boundary, but in the hyperbolic way of measuring distance they all are the same size! You can see the straight lines of this geometry by following the fishes' backbones; and, by counting fish along one of

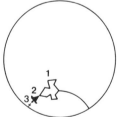

Figure 2.68. Escher's depiction of the Poincaré disk. (B-26983 Circle Limit I, 1958, M. C. Escher, 1898–1972, National Gallery of Art, Washington, D.C., Cornelius Van S. Roosevelt Collection. c 1958 M. C. Escher/Cordon Art-Baarn Holland.)

these curves, you can see how it will take infinitely many of these fish to get to the boundary, up to the limitations of the physical model. Escher made several more graceful depictions of Poincaré's disk in *Circle Limit III* and *Circle Limit IV,* but we have chosen *Circle Limit I* (Fig. 2.68) because it shows the straight lines of the geometry most clearly. We have also depicted how one can count fishes along a straight line.

In order to establish that Poincaré's disk satisfies all of the postulates of hyperbolic geometry, we have yet to consider Postulate 3, the existence of circles. Choose a point as center, and a distance as radius. You can move from this center point along a "straight line" in any direction for exactly the "distance" that was specified. When you mark off the end points of all these "radii," you will have a circle in this geometry. The radii won't look straight or equal to us external Euclideans, but the figure is certainly a circle in this geometry. Somewhat surprising is the fact that the figure actually is a Euclidean circle. Its center will be farther to the outside than where we think the center should be. One of these circles is depicted in Figure 2.67c.

Thus Poincaré's model satisfies Euclid's first four postulates and Postulate 5H. It shows that hyperbolic geometry is just as consistent as Euclidean. And it proves that Euclid's fifth postulate is independent of the other four, because in one model of the other four (the Euclidean plane) it is true, while in another (the Poincaré disk) it is false.

Poincaré offered an ingenious interpretation of this disk world that makes it seem quite real. Think of it as being a Euclidean Flatland filled with a funny gas or temperature differential that causes everything in it to shrink as it moves away from the center according to our earlier formulas. Thus if this Euclidean disk has radius $R$, an object that moves from the center to a distance $r$ from the center has its length multiplied by $(R^2 - r^2)/R^2$. That could explain why all those flying fish are the same size, even though they look different to us, looking in from the outside.

Figure 2.69a shows how a little Flatland triangle carrying a ruler with it will shrink as it moves toward the boundary. It will not know that it is shrinking, because its ruler shrinks right along with it. If it was 2 sticks long at the start of its trip, it will be 2 sticks long at the end. Everything else (clothes, friends, birds, etc.) shrinks as it moves along with it. Figure 2.69b shows how the shortest path between two points in this Euclidean disk with shrinkage is not the Euclidean straight line, but rather the arc of a circle that meets the boundary at right angles. The rulers along the straight line path are farther from the center of the disk than are those along the circular path;

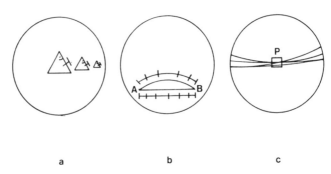

a                          b                          c

Figure 2.69. Equal measuring sticks and an almost Euclidean portion of the Poincaré disk.

hence they appear shorter to us, and so more of them are required to measure the distance. What looks to us to be the longer path is 5 sticks long, while the Euclidean straight line path is 6 sticks long. If we assume that whatever it is that causes things to change in size also affects light rays, then light will travel along these circular arcs. If our interpretation of the words "straight line" is the path that light follows, we will naturally call the circular arcs "straight lines." Of course, all of our earlier discussion of the geometry of the Poincaré disk applies to the world we have just described. In particular, as in Figure 2.66c, there are many lines through a point that do not meet a given line.

This interpretation of the Poincaré disk in terms of shrinking figures can be thought of in two ways, depending on our use of the term "straight line." If we use the term to mean the path that light travels, which is the shortest number of shrinking sticks, then hyperbolic geometry holds. This is the interpretation with nice laws of physics (light follows the shortest path), but the slightly strange geometry of many parallels. On the other hand, if we use the term "straight line" to refer to the paths that are straight to the external Euclidean viewer, then Euclidean geometry holds, but we have strange laws of physics, such as light does not follow the path we are calling a straight line. Poincaré's point of view was that these two ways of viewing this world were equally good, and so asking whether Euclidean or hyperbolic geometry is true in the universe is a meaningless question. Most thinkers today would reject the second (external) viewpoint, and say that "straight line" should be used to describe the shortest distance between points, and so the hyperbolic description of Poincaré's disk is the preferable one.

If the inhabitants of Poincaré's world are very tiny and limited in motion compared to the entire disk, they will probably not be able to measure deviations from Euclidean geometry. In Figure 2.69c, suppose that the widest expanse they can ever hope to see is the little square. To them, the two parallels and one divergent to $\ell$ through $P$ probably all look like the same line, unless they are capable of measuring very tiny deviations. So they think they live in a Euclidean world, until their science advances enough so that they can see outside this little rectangle or make accurate enough measurements to distinguish the lines inside the square of Figure 2.69c.

You should have no trouble envisioning a 3-dimensional analogue of Poincaré's disk—the inside of a ball, complete with funny gas or temperature gradient that makes things shrink as they approach the boundary. Three-dimensional hyperbolic geometry will be valid here. You should already have been led to the conclusion that we could be the tiny people, thinking that our geometry is Euclidean because we cannot see enough of the universe or make accurate enough measurements, but actually we could be living in a world of hyperbolic geometry.

Any of the models we have discussed allow us to comprehend the constant or intrinsic distance of hyperbolic geometry that was so bothersome to Gauss. Analogous to the radius of the sphere for spherical geometry, you could take it to be the radius of the wide part of the pseudosphere (where the two horns are attached), or the radius of the disk in either of the disk models. These are constants that are not readily observable to inhabitants of the model. So, as suggested in Section 2.3, they might take something like the length of side in a 45-45-45 triangle. Formulas could be derived relating this distance to either of the global radii just mentioned, or relating them to other choices of the intrinsic distance. If you just call 1 Lob the length of side of a 45-45-45 triangle, then all your formulas in terms of number of Lobs will be the same, regardless of whether 1 Lob is 1 inch or 1 billion miles.

*ASIDE: HYPERBOLIC PAPER*

You can make your own partial model of hyperbolic geometry, called hyperbolic paper, following an idea in Weeks (1985). Tape together a lot of equilateral triangles so that seven of their 60° angles meet at each vertex. This will cause a high degree of floppiness, and the more triangles you use, the floppier it will get. Figure 2.70 shows a photograph of some hyperbolic paper with some lines drawn suggesting that there will be more than one line through a point not intersecting a given line. This is suggested by a pair of lines that seem to be con-

Figure 2.70. Hyperbolic paper.

verging but will begin to diverge when one of the vertices with the 420° gets in the way. In the actual hyperbolic plane, this tendency of lines to spread out more than you would expect in Euclidean geometry takes place continually, whereas in our imperfect model it just happens at the vertices of our grid of triangles. The smaller we make our triangles, the better will be our approximation to the hyperbolic plane.

Think back to your first encounters with diagrams of hyperbolic geometry and our admonition that maybe space spreads out at great distances. This hyperbolic paper should help to confirm your growing feeling of comfort with this idea. As usual, you should try to imagine 2-dimensional beings who can only experience phenomena within this surface.

In order to make some hyperbolic paper, you will need a lot of equilateral triangles in a grid. Make about four photocopies of the next page to serve as a source of triangles. Wherever six triangles come together at a vertex, cut along one of the edges and tape another triangle into the gap. You might as well make the triangle that you tape in have some other triangles already attached to it, so as to

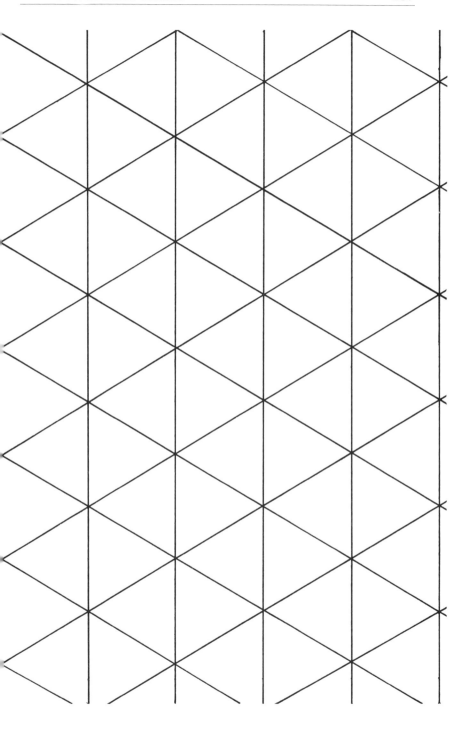

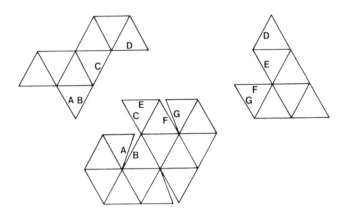

Figure 2.71. An aid for constructing hyperbolic paper.

minimize taping. You won't be able to keep those four photocopied
pages intact. To get you started, you might try cutting out from your
photocopies pieces analogous to the three figures in Figure 2.71, and
taping together edges that have the same letter.

### Exercises

1. Draw a triangle and Saccheri quadrilateral on a sphere and on a
   pseudosphere. Make them large enough relative to the size of the
   sphere and pseudosphere so that the distinguishing properties
   of angle size are apparent. What are these properties?
2. Write a short essay about why Euclid's fifth postulate is independent of the first four. Include some historical comments.
3. Discuss Theorems 2.3.4H and 2.3.5H in terms of the Klein-
   Beltrami model.
4. Show that the Klein-Beltrami distance formula given in footnote
   17 becomes arbitrarily large as $A$ approaches $C$.
5. Show that the Klein-Beltrami distance formula has the additivity
   property that any measure of distance ought to have, namely
   that in Figure 2.72 the distance from $A$ to $B$ plus the distance
   from $B$ to $E$ equals the distance from $A$ to $E$. (Hint: All you need
   to know about log is that $\log xy = \log x + \log y$.)
6. Draw a disk to represent Poincaré's disk model, and inside it
   draw a large Saccheri quadrilateral. Point out the property of
   angles that distinguishes it from a Saccheri quadrilateral in Euclidean geometry.

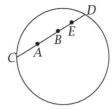

Figure 2.72. For Exercise 5.

7. Illustrate Theorems 2.3.6H and 2.3.7H in the Poincaré disk. Explain how the model shows that the distance between divergent lines becomes arbitrarily large.

8. What fact in Euclidean geometry is required to prove Postulate 5H in the Poincaré model? (Hint: This is analogous to the Euclidean fact necessary to prove that Euclid's first postulate is true in the Poincaré disk, which was discussed in the text.)

9. In a Poincaré disk of radius 100 miles, there are sticks that look to us Euclideans to be 1 foot long. One is at the center, one is 50 miles out from the center, and the third is 75 miles out from the center. What is the length of each defined to be in Poincaré's model? (You need not worry about the use of both feet and miles in this problem. The units of square miles over square miles will cancel out in the fraction.)

10. At the center of a Poincaré disk of radius 100 miles, there is a stick that has length 1 foot. It moves out to radius 50 miles. How long does it look to us now? (It is still considered to have length 1 foot in their system.) Now it moves out to 75 miles. How long does it look to us now?

11. Show that the ratio of Poincaré length divided by Euclidean length only depends on the ratio $r/R$, which gives the fraction of the way out to the boundary of the object. What will be the ratio of Poincaré length over Euclidean length when $r/R$ has each of the following values: .5, .9, .99, .9999?

12. In the copy in Figure 2.73 of Escher's depiction of hyperbolic geometry, compare the two paths from $P$ to $Q$ in terms of number of flying fish.

13. Suppose that the Klein-Beltrami disk included the boundary. In other words, you have a model in which the points and lines are points and lines inside a disk including the boundary, and you want to write some axioms that describe it. Write the analogue of Theorem 2.3.3H, which states the properties of divergent and parallel lines.

P                                    Q

Figure 2.73. Counting fish. (Detail of B-26983 Circle Limit I, 1958, M. C. Escher, 1898–1972, National Gallery of Art, Washington, D.C., Cornelius Van S. Roosevelt Collection. © 1958 M. C. Escher/Cordon Art-Baarn-Holland.)

## 2.6    The Geometry of the Universe

In this section, we will discuss how the ideas of non-Euclidean geometry, particularly those of Riemann, were influential on Einstein's General Theory of Relativity, which deals with the curvature of space-time. We will close with an overview of contemporary thoughts on the large-scale geometry of the universe.

### HIGHER-DIMENSIONAL SPACES

The space around us is quite clearly 3-dimensional. That means that three numbers are required to specify the position of an object. This is usually expressed in mathematics using an $(x, y, z)$-coordinate system, as pictured in Figure 2.74. For example, the point (2,3,4) is obtained by starting at the point where the three axes meet, and from there moving 2 units in the $x$-direction, 3 units in the $y$-direction, and 4 in the $z$-direction.

Surfaces in 3-dimensional space can be expressed by equations in-

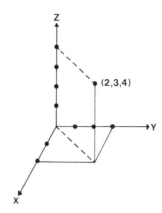

Figure 2.74. $(x, y, z)$-coordinates.

volving $x$, $y$, and $z$. For example, the set of points $(x, y, z)$ satisfying the equation $x^2 + y^2 + z^2 = 1$ forms a sphere, since this equation describes exactly the points whose distance from the point $(0,0,0)$ is 1. We remind you that the sphere is considered to be 2-dimensional, because, locally at least, points on it can be specified by two numbers, such as latitude and longitude. (See Fig. 2.75.)

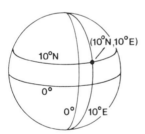

Figure 2.75. Latitude and longitude.

Algebraically, one can consider $n$-dimensional space defined by ordered lists of $n$ numbers, for any number $n$. In order to avoid running out of letters, we usually name these variables using subscripts. Thus we write $x_1$, $x_2$, and $x_3$ instead of $x$, $y$, and $z$, and we could just as easily talk about the points $(x_1, x_2, x_3, x_4, x_5, x_6)$ in 6-dimensional space. Such ideas could actually come up in real-world considerations such as manufacturing, where $x_6$ is your profit, and $x_1$ through $x_5$ might be various parameters that affect your profit, such as $x_1$ = cost of energy, $x_2$ = number of items manufactured, $x_3$ = price you charge for them, $x_4$ = number of employees, and

$x_5$ = cost of a certain ingredient. Then you might be interested in equations relating $x_6$ to $x_1$ through $x_5$, and geometrical properties of the 5-dimensional surface consisting of solutions of these equations might be helpful to you in designing a production strategy. In fact, linear programming is a field of mathematics that considers computerized methods of solving such optimization problems, sometimes involving hundreds of variables.

But geometrically it is very difficult for us to comprehend more than three dimensions. We made this point in Section 2.4 in our discussion of the book *Flatland*. These 2-dimensional beings were, for the most part, unable to visualize a third dimension. The point of that book was that in order to envision 4-dimensional space we should do whatever was required for them to imagine the third dimension. So just as a 3-dimensional ball passing through Flatland looks to them like a series of disks of increasing size followed by a series of shrinking disks, we might imagine a 4-dimensional ball passing through our world as a sequence of growing balls followed by a sequence of shrinking balls.

You can get some feeling for the fourth dimension by thinking about the boxes of zero, one, two, three, and four dimensions pictured in Figure 2.76. Each one is obtained from that of one less dimension by pushing it in a direction perpendicular to itself. On the sheet of paper we cannot even draw the third perpendicular dimension, and yet the drawing helps us to imagine the cube. The 4-dimensional hypercube, or tesseract, doesn't look too much more complicated.

However, it takes a lot of training and concentration to gain any intuition about where the fourth dimension in which the cube is being pushed to form the hypercube is. Thomas Banchoff, professor

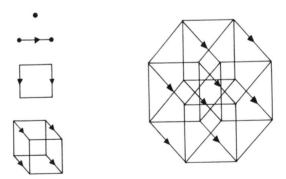

Figure 2.76. Building a 4-dimensional hypercube.

of mathematics at Brown University, has made a number of movies that are quite useful in developing the ability to imagine the fourth dimension, and especially to get a feeling for the hypercube. Charles Hinton (1853–1907) was a mathematician who wrote extensively on methods of visualizing the fourth dimension, at approximately the time of Abbott's *Flatland.* One of his favorite ways was to imagine coloring each of the eighty-one parts (vertices, edges, faces, cubes, and hypercube) of the tesseract a different color, and then to memorize which colors were attached to which.

In the late part of the nineteenth century, there was a lot of interest in the theory that various unexplained occurrences were caused by 4-dimensional beings interfering with life in our 3-dimensional universe, similarly to the way in which the 3-dimensional Sphere had such an impact in Flatland. These phenomena are best imagined by analogy with Flatland. One example is that ghosts could be 3-dimensional cross sections of 4-dimensional beings.

A leading proponent of this theory was Johann Zöllner, professor of physical astronomy at the University of Leipsic, whose book, *Transcendental Physics*, published in 1888, describes his many observations of unusual phenomena, for which he says the only explanation is intervention from the fourth dimension. Most of these were occasioned by his medium, "Mr. Henry Slade, the American," and could not be repeated when more objective observers were present. One of his favorites was begun by placing a solid ring on the floor next to a table, as in Figure 2.77a. A blanket was placed on top of these so that the visitor from the fourth dimension could do its business in private. When the blanket was removed, the ring was looped around the stem of the table, as in Figure 2.77b. This trick could be performed if the ring was considered as an object in 4-dimensional space, with

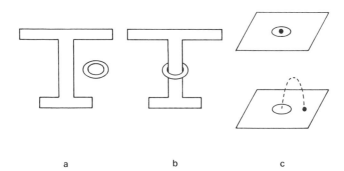

a                    b                    c

Figure 2.77. A trick from the fourth dimension and an analogy one dimension lower.

the fourth dimension effectively allowing the ring to pass through the stem of the table. To understand this, it is useful to consider a similar phenomenon one dimension lower. In Figure 2.77c, the dot cannot be moved outside the circle without crossing through it in 2-dimensional space, but it can in 3-dimensional space by moving over the circle in the third dimension and then back into the plane. Similarly, the ring can be moved "up" into the fourth dimension. While its fourth coordinate is different from that of the table, it will not intersect the table in 4-dimensional space, and so at this different level it can be moved so that its first three coordinates are passing through those of the table. Once its first three coordinates are such that they encircle the table stem, the ring can be moved back "down" so that it intersects our 3-dimensional space. Because Slade could not repeat this feat for observers other than Zöllner, it was not taken seriously by many people.

Another of Zöllner's favorite ways of seeing evidence of a 4-dimensional being was in "slate writing." A blank slate, after being held under a table for a while, would emerge with writing on it. His explanation here is again best explained by analogy from Flatland, as in Figure 2.78a, where the Sphere writes on a Flatlander's slate with a piece of chalk that is perpendicular to Flatland.

One experiment which Zöllner's 4-dimensional friend was unable to accomplish was to turn some 3-dimensional object, such as a seashell, into its mirror image. In Figure 2.78b, we show how, by ro-

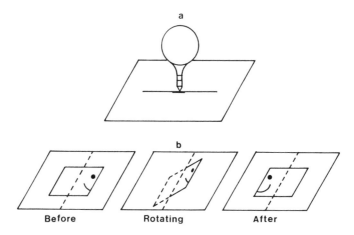

Figure 2.78. Analogues of slate writing and turning an object into its mirror image.

tating through the third dimension, a Flatlander can be turned into its own mirror image. It was noted in 1827 by the well-known mathematician A. Möbius that this should be possible for 3-dimensional objects if there is a 4-dimensional reality.[18]

The easiest way to think of a fourth dimension added onto the three spatial dimensions is to think of time as the fourth dimension. As we shall see in the next subsection, there are some subtleties in doing this for objects moving with a speed that is not negligible when compared with the speed of light. Ignoring these for the moment, we can think of an event as having three spatial coordinates and one time coordinate, which tell where and when it occurred.

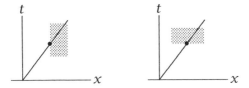

Figure 2.79. Space-time graphs of a car driving into a rainstorm.

For example, if you are driving along and it starts to rain, it is not always clear whether the start of the rain was due to the change of position (you just moved into a rainy spot) or a change in time (all of a sudden it started raining over the entire region). Simplifying by overlooking two spatial coordinates, we can graph one dimension of position as the horizontal axis and time as the vertical axis. Then the path of the car in the previous paragraph can be shown as in either of the graphs in Figure 2.79, which starts at $x = 0$ when $t = 0$ and proceeds at constant speed. This is called the *worldline* of the car, consisting of all values of $(x, t)$ during the relevant portion of the life of the car. Each point $(x, t)$ is called an *event*, which means the specification of a place and a time. If the shaded region indicates rain, then the left graph represents driving into a rainy spot, since for the value of $x$ where you first encountered the rain, it was already raining there at times $t$ before you arrived. The right-hand graph depicts a rainstorm that just began over the entire region, for the space-time graph shows that the rain over the whole region began at the same time—the time you first encountered it.

18. See Exercise 2 for his most famous example, the Möbius strip. This is an important example in topology. See the "Aside" to Section 1.2 for another topic in topology.

## RIEMANN'S LECTURE

In Section 2.5, we mentioned Gauss's *Theorema Egregium*, which showed that curvature of a surface can be determined intrinsically by measurements within the surface. We will now explain that the important thrust of Riemann's famous 1854 lecture was to discuss higher-dimensional analogues of this theorem.

A 2-dimensional surface, such as a sphere or pseudosphere, can be defined without reference to its embedding in 3-dimensional space. Gauss had noted how this can be done by setting up a system of coordinates using any grid of intersecting curves on the surface. The geometry of the surface can then be determined by measuring the way in which distance between nearby points is determined by changes in the values of the two coordinates.

Riemann had to give his famous 1854 lecture in order to obtain the position of lecturer at the University of Göttingen, a position that entitled him to no salary but only to fees obtained from whatever students chose to attend his classes. He was to give Gauss and the other examiners three topics on which he was prepared to lecture, and they would choose one. His third topic, "On the Hypotheses That Lie at the Foundations of Geometry," was the one on which he had done the least work, but it was the one of most interest to Gauss.

In this lecture, Riemann said that we should consider manifolds[19] of $n$ dimensions for any number $n$, and showed how to study distance and define curvature on them, in a manner that generalized Gauss's *Theorema Egregium*. Other mathematicians had talked about $n$-dimensional Euclidean space before, but the idea of an $n$-dimensional manifold was new. This idea meant that a space can be defined using $n$ coordinates to specify the position of a point, but that these coordinates need not be related to distance in the usual Euclidean way. A manifold as defined today only requires that these coordinate systems exist locally, with additional information needed to show how to patch them together. For many important examples, this added technicality is not required.

Examples of manifolds include obvious $n$-dimensional analogues of the sphere and the Poincaré disk. One does not have to think of the manifold as being embedded in a higher dimensional Euclidean space in order to understand curvature; it is just a measure of how its geometry deviates from Euclidean geometry. In general, this curvature might vary from point to point, but if you want the manifold

19. The word "manifold," introduced in the "Aside" to Section 1.2, generalizes the idea of "surface."

to have a homogeneity property that says that figures can be moved around on the manifold without changing sizes or angles, then the curvature should be constant.

Riemann was very concerned with the geometry of the universe. He said that clearly the universe is 3-dimensional, and apparently it is homogeneous (i.e., objects can move from one point to another without distortion). But it is not clear that it is Euclidean. Determining what 3-dimensional manifold it is should primarily be the job of physicists. However, mathematicians can help by showing physicists what the possible forms of 3-dimensional manifolds are. This work is still in progress. The American topologist William Thurston has made major strides in the classification of all possible 3-dimensional manifolds. He won the Fields Medal, the mathematical equivalent of the Nobel Prize, for this work in 1982.

We close this subsection by returning to the idea that the universe might possibly be a 3-dimensional sphere. One profound difference between the 3-dimensional sphere and the 3-dimensional Euclidean space is that the former is bounded. Just as in the 2-dimensional sphere we studied in Section 2.4, straight lines in a 3-dimensional sphere eventually return to their starting point. It would then seem to be possible that an astronomer who is looking at a galaxy that is 4 billion light years away from us is actually looking at our galaxy 4 billion years ago. This would be the case if the universe were spherical of circumference 4 billion light years. In a later subsection, "The Large-Scale Geometry of the Universe," we will say more about the possibility that the universe is bounded, a possibility that is taken seriously by many esteemed scientists. We will also discuss why light that has gone completely around the universe could not appear in a telescope.

### SPECIAL THEORY OF RELATIVITY

In the next two subsections, we give a brief sketch of Einstein's Theory of Relativity, and show how non-Euclidean geometry is an essential ingredient of it. There are two parts to this theory: the Special Theory, published in 1905, which deals with the relationship between space and time; and the General Theory, published in 1916, which deals with gravity and the curvature of space-time. Although the general theory is the more direct outgrowth of the ideas we have been studying, the special one is necessary to understanding it properly. The special theory also involves ideas related to our topics from geometry.

Albert Einstein (1879–1955) was a physicist who was born in Ulm, Germany, and worked in Switzerland and Germany before moving to the United States in 1933. (See the brief biography in the "Focus" at the end of this section.) His Special Theory of Relativity deals with the consequences of two postulates which he put forward to explain the results of some experiments, performed by others during the preceding twenty-four years, which did not quite jibe with the prevailing ideas in physics, most of which dated back to Newton in the 1600s. The postulates are the following:

1. All physical laws valid in one frame of reference are equally valid in any other frame moving at constant velocity with respect to the first.
2. The speed of light in a vacuum is the same regardless of the motion of its source.

The first law says that if we are considering two systems moving with constant velocity relative to one another, we may think of either one as being at rest and the other as moving. For example, a boy in a moving train may think of the outside as moving relative to the train in the same way that a girl on the outside considers the boy in the train to be moving. The motion of a ball tossed up by the boy in the train is, in his frame of reference, the same as the motion of a ball tossed similarly by the girl outside the train, relative to her frame of reference.

The second law is more surprising. A pitcher who can throw a baseball 90 miles per hour on the ground could throw it 150 miles per hour if he threw it forward from a car moving 60 miles per hour. But a ray of light projected from the moving car travels no faster than a ray of light projected from a stationary source on the ground. It is an easy consequence of these two laws that the speed of light also does not depend on the speed of the observer. Light from the headlight of an oncoming car approaches you at the same speed regardless of whether you are traveling 1 mile per hour or 100 miles per hour. This is very different than the situation for ordinary moving particles, but it was consistent with the troubling experiments that had been performed in the decades prior to Einstein's hypothesis, and its consequences have been verified in many experiments since that time.

One of the most shocking consequences of Einstein's postulates is that, for observers who are moving with respect to one another, events that are simultaneous according to one observer need not be simultaneous from the point of view of the other. This is the *relativity of time*—that each observer should be considered to have a personal clock, and comparisons with the clock of another observer are limited, at best. There is also a *relativity of distance*—that comparisons of distance can be relative only to the observer. This is true because in order to measure the distance between two points, you should be observing those points at what you think is the same time.

Another strange aspect about relative time is that a clock moving with constant velocity with respect to an observer will measure time more slowly than a clock that is at rest with the observer. The formula is that if a clock is moving with velocity $v$, then it will measure time

$$(2.6.1) \qquad t' = t\sqrt{1 - (\tfrac{v}{c})^2}$$

while the observer's clock measures time $t$. Here, and elsewhere from now on, $c$ is the speed of light, which is roughly 186,000 miles per second or, more roughly, 1 billion miles per hour. If $v$ is small compared to the speed of light, this difference will be minuscule, but for speeds a significant fraction of the speed of light, $t'$ will be noticeably smaller than $t$. (See Exercise 5.) Human clocks (e.g., heartbeats) will run more slowly, too, on a moving object, and so someone traveling on a spaceship at a speed that is a significant fraction of the speed of light will age more slowly than someone on the ground.

The most bothersome aspect of this is that we could alternatively consider the spaceship to be at rest, and the earth moving with respect to it. If we think of it this way, then it appears to someone on the spaceship that the clocks on earth are running slowly. But from the first point of view it seemed to someone on the earth as if the clocks on the spaceship are running slowly. Isn't this a paradox? The explanation is that they won't be able to compare clocks without the spaceship having to decelerate to turn around, and then to decelerate again in order to stop. The above formula doesn't explain what happens to clocks undergoing such changes, and we won't go into it here.

In 1908 Hermann Minkowski (1864–1909), who had once been Einstein's mathematics teacher, put Einstein's Special Theory of Relativity into a mathematical context that became important in the later extension to the General Theory. Minkowski's idea was to consider space-time as a 4-dimensional space with coordinates $x$, $y$, $z$, and $ct$, but with distance (called *interval*) given by

$$(2.6.2) \qquad \sqrt{(\Delta ct)^2 - (\Delta x)^2 - (\Delta y)^2 - (\Delta z)^2}.$$

Here $t$ is multiplied by $c$ in order to give it units of distance, and the $\Delta$ in front of each of the letters refers to change in the value of that variable. It can be shown, using (2.6.1), that the quantity in (2.6.2) will appear the same to any observer. Minkowski's space was in the context originally proposed by Riemann, which had been developed considerably by mathematicians in the fifty intervening years. Minkowski began his lecture "Space and Time" with the following famous paragraph:

> The views of space and time which I wish to lay before you have sprung from the soil of experimental physics, and therein lies their strength. They are radical. Henceforth space by itself, and time by itself, are doomed to fade away into mere shadows, and only a kind of union of the two will preserve an independent reality.

### GENERAL THEORY OF RELATIVITY

The General Theory of Relativity, as published by Einstein in 1916, can be summarized in the brief statement, "Gravity is curvature in space-time due to matter." This is primarily a statement in physics, but it has a very significant contribution from geometry, and that is why it is included in this book. We wish to point out one last time that the work of Gauss and Riemann in geometry, which was derived ultimately from the study by Gauss and others of Euclid's fifth postulate, gave the mathematical basis required by Einstein in order to thoroughly elucidate his ideas about time, space, and gravity. Einstein spent much of 1912 to 1914 learning the requisite mathematics from his friend Marcel Grossmann, who was a professor of mathematics in Zurich, where Einstein was a professor of theoretical physics. His respect for the mathematics is clear in the following excerpt from a 1912 letter to a colleague:

> I am now exclusively occupied with the problem of gravitation and hope, with the help of a local mathematician friend, to overcome all the difficulties. One thing is certain, however, that never in life have I been quite so tormented. A great respect for mathematics has been instilled in me, the subtler aspects of which, in my stupidity, I regarded until now as pure luxury. Against this problem the original problem of the theory of relativity is child's play.[20]

And in a 1915 letter, Einstein wrote: "When all my confidence in the old theory vanished, I saw clearly that a satisfactory solution could only be reached by linking it with the Riemann variations."

We have seen how the Special Theory could be viewed using 4-dimensional space-time with the Minkowski distance given in (2.6.2).

---

20. Schwinger 1986, p. 252.

This is called *flat* space-time. Einstein's idea was that wherever matter is present, it causes space-time to become curved near it. This curvature does not have to be envisioned by looking from some 5-dimensional perspective. It just means that the geometry is changed, that is, distances between points are modified somewhat from what they would have been in the flat model with no matter. The world-lines of objects will be geodesics in this curved space-time; this means that they will be paths that are the straightest paths between points on them.[21] If you just look at the 3-dimensional spatial projection of one of these geodesics, it will not look straight. For example, reducing the number of dimensions by 1, we picture in Figure 2.80a the orbit of a planet around the sun as a worldline in space-time. You can see its revolution around the sun as time changes. The diagram is greatly exaggerated; it would have to be elongated very much in order to be drawn to scale. However, the point is that the presence of the sun causes space-time to have its distance function modified so that this is the straightest path between points on it.

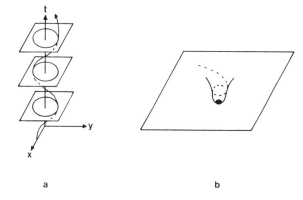

Figure 2.80. Two interpretations of the General Theory of Relativity.

A heavy dose of differential geometry is required in order to really understand the General Theory. We will present two analogies to try to convey its flavor. For the first, imagine an elastic rubber sheet, such as a trampoline, and think of it as a 2-dimensional world. If we place an extremely heavy disk on the trampoline, it causes the trampoline to curve. This is the curvature of the space caused by the existence of matter. If we propel a small disk in the general direction of the heavy one, but not straight toward it, the path of the small disk will be bent in the direction of the heavy one, and if it

21. It turns out that they are the longest intervals between the points, not the shortest, as in ordinary geometry.

has sufficient speed, it will orbit the heavy disk. It appears as if the heavy disk is attracting the little one, which Newton's theory would have said is due to gravity, but Einstein's theory says that it is due to the curvature of the space (i.e., the trampoline). (See Fig 2.80b.)

There are several flaws in this analogy. The main one is that in this model it is really the gravitational attraction of the earth,[22] acting on both the heavy disk and on the little one, that is causing this to happen, whereas in Einstein's theory the curving of the space is caused by the heavy object within the system. A heavy disk on a trampoline in outer space would not cause the trampoline to become curved. The other flaw is that this model does not include the temporal variable. Einstein's theory refers to curvature of space-time, not pure space.

In order to demonstrate the other analogy, we begin by raising the question that ultimately led Einstein to the General Theory. An object has two kinds of mass. One is sometimes called *inertial mass*; it is the resistance to motion of a body at rest when a force is applied to it. It is the $m$ in Newton's famous formula $F = ma$, where $F$ is the force on the object, and $a$ the acceleration caused by the force. The other is the *gravitational mass*, sometimes called weight. It is the mass which Newton's theory said would attract other bodies near it. These two masses turn out to be proportional for all bodies, and so we might as well choose a scale so that they are equal. Einstein wondered why they can be made equal just by a change of units, because they are really measuring different things. He concluded that it must be because gravity and acceleration are intimately related.

One way to picture this relationship is to imagine being in an elevator at the top of a tall building, and to suppose that the elevator's cables are cut, so that it goes into free fall. Since the elevator and all objects in it fall with the same acceleration relative to the ground, they do not have any acceleration relative to one another. If one of the passengers had been holding a ball and lets it loose while the elevator is falling, the ball will seem to float right along beside the passenger. If the ball is pushed away, it will seem to move away in a straight line until it hits the wall of the elevator. Thus the elevator in free fall is indistinguishable from what it would have been like if it were floating in outer space. Since it is impossible to tell whether your weightless state is produced by accelerated motion in a gravitational field or by the absence of a gravitational field, we conclude that gravity and acceleration must be the same notion.

Upon closer inspection, however, we realize that there will be some small differences between the two systems. Two balls falling side

---

22. Or, more precisely, the curvature of space-time due to the presence of the earth.

by side in the elevator will be falling directly toward the center of the earth and thus will get slightly closer together as they fall. The effect is very minute for the situation we have described, but it is nevertheless worth our consideration. (See Fig. 2.81a and Exercise 9.)

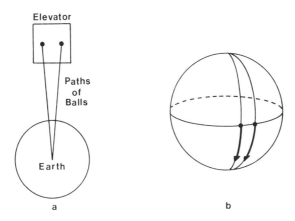

Figure 2.81. An analogy between gravity and geometry.

Now consider something different—two objects moving south from the equator along lines of longitude. Although both are moving due south, they get closer together because of the curvature of the earth. (See Fig. 2.81b.) This is the same phenomenon we noted when we talked about the Mercator projection of a map in Section 2.4. There is a strong similarity between the motion toward each other of the objects moving on the earth, which is due to geometry, and that of the two balls in the freely falling elevator, which is due to gravity. Einstein used this analogy to help him toward the conclusion that gravity and geometry are the same thing.

There were several experimental verifications of Einstein's General Theory, but by far the most important was the observation of the bending of light rays due to the gravitational attraction of the sun. Einstein's theory said that light follows a geodesic in space-time, and calculated that path for a ray passing very close to the sun. (See Fig. 2.82.) It turns out that the angle of deflection predicted by Einstein's theory was twice as great as that which Newton's theory would predict, and in 1915 Einstein suggested this difference as a way of putting his theory to the test. You cannot ordinarily see stars when their light is passing the sun, because the sun is too bright, but you can see them during a total eclipse of the sun. In 1919 a total eclipse was visible from Africa, and a group of distinguished scientists, led

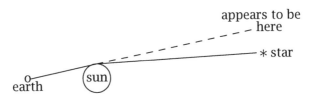

Figure 2.82. Bending of light from a star due to sun's gravity.

by the British astronomer Sir Arthur Eddington, went to Africa to determine the apparent position of a certain star whose light was passing very close to the sun during the eclipse. This new position was compared with the usual position of the star, and the scientists found that the deflection due to the sun agreed with Einstein's prediction. Einstein's fame among the general public skyrocketed after this discovery.

### THE LARGE-SCALE GEOMETRY OF THE UNIVERSE

We have seen that near matter the geometry of the universe is certainly not Euclidean, as the matter causes curvature of space-time. But the question remains: What is the (large-scale) geometry of the universe? Is it Euclidean in regions far removed from all matter? The answer to this question is still not known, and is more in the domain of physics, astronomy, or cosmology than of mathematics.

The universe is expanding. All galaxies are moving away from all other galaxies. This expansion was observed by the American astronomer Edwin Hubble in 1929. It had been pointed out in 1922 by a Russian, Alexander Friedmann, that expansion was a consequence of the equations of Einstein's General Theory of Relativity, along with the assumption that the universe looks essentially identical in all directions and from any position. This expansion is most easily visualized by thinking of the universe as a balloon that is being blown up. This model of the universe is of course one dimension lower than the actual 3-dimensional universe. Most of the surface of the balloon represents empty space, but on the balloon are drawn some marks, which represent stars or galaxies. As the balloon is being blown up, the galaxies are moving farther apart. One minor inaccuracy in this model is that the galaxies themselves do not get larger, so perhaps you should think of them as being drawn on the balloon with some kind of nonexpanding ink.

If the universe is like this balloon, then it is a 3-dimensional sphere. The idea of the universe as being finite, but unbounded, was championed by Einstein. Much earlier, Riemann had mentioned this possibility for the geometry of the universe, and much more recently it has been presented by Stephen Hawking, probably the most famous contemporary theoretical physicist. Hawking, who holds Newton's chair at Cambridge University, is author of the book *A Brief History of Time: From the Big Bang to Black Holes*, which was on the best-seller lists for virtually all of 1988, 1989, and 1990. This book presents modern theories of cosmology "for those who prefer words to equations." In 1963, while still a student, Hawking was stricken by the debilitating motor-neuron disease known as Lou Gehrig's disease;[23] but despite great difficulty in communicating, he has produced many of the most important ideas about the evolution of the universe.

One of the big ideas associated with Hawking is that of the "big bang." The fact that the universe is expanding suggests that long ago it must have been very small. The hypothesis of the big bang is that there was a beginning of time, when the universe was infinitely small and infinitely dense, and it has been expanding ever since then. It makes no sense to ask what came before the big bang, because *there was no time* before the big bang. The big bang theory was opposed by many, often on religious grounds, until 1970, when Hawking and the British mathematician and physicist Roger Penrose proved that the existence of a big bang is a consequence of Einstein's equations.

We now return to the idea of the spherical universe. We can draw a nice picture of space-time under this hypothesis, two dimensions lower than the real thing. Think of an ordinary 2-dimensional sphere in 3-dimensional space as representing space-time. Think of the vertical axis as time, and the horizontal circles at each level as the universe at the appropriate time. This is like Lineland, a 1-dimensional universe. At the South Pole, we have the big bang, then as time increases moving up, we have an expanding universe up to a point (the equator), after which time the universe begins to contract, ending in a "big crunch." (See Fig. 2.83.)

The spherical universe is one of three conceivable forms for the universe. In the 1920s Friedmann showed that a consequence of Einstein's equations plus his (Friedmann's) homogeneity assumptions mentioned above was that there are three possibilities for the large-scale geometry of the universe. One of these is the spherical universe just described. Another is the Newtonian-Euclidean model of the universe. And the third is the 3-dimensional analogue of hyperbolic geometry. This latter can be considered as analogous to Poincaré's disk, except inside a ball.

You may be bothered that we are considering space without con-

---

23. Named after the famous baseball player of the 1930s, who suffered from the same disease.

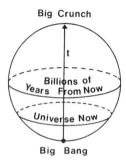

Figure 2.83. Space-time chart for a 1-dimensional spherical universe.

sidering time, which seems to go against Minkowski's dictum that neither one should be considered alone. But a *spatial section* of space-time is closely related to the structure of the entire space-time; it shows what the universe looks like to some observer at some time. It is like a horizontal cross-section of Figure 2.83.

We know that the density of matter in the universe determines which of the three forms for the geometry of the universe—spherical, Euclidean, or hyperbolic—is valid. *Density* means the total mass of matter in the universe per volume of space; it does not mean the density of individual pieces of matter. There is a certain critical value, $\rho_0$, roughly equal to $4 \cdot 10^{-27}$ kilograms per cubic meter. If the density of matter in the universe exceeds $\rho_0$, then the universe has spherical geometry; if the density is less than $\rho_0$, the universe has hyperbolic geometry; if the density is precisely equal to $\rho_0$, then the universe is Euclidean. The density of matter observed to date is only 10% of $\rho_0$, so many scientists believe that the geometry of the universe is probably hyperbolic. Many think, however, that there may be "dark matter" in the universe that has not been observed, and could appreciably change our knowledge of the density of matter in the universe. Finding this dark matter is one of the principal goals of the Hubble space telescope, launched in 1990.

The determination of the amount of matter in the universe is of more than theoretical interest. It is also known that if the density of matter exceeds $\rho_0$, then the universe will eventually collapse, as pictured in Figure 2.83; if it is less than $\rho_0$, the universe will expand forever. If the density is precisely equal to $\rho_0$, then the universe will expand forever, but at a much slower rate. This seems reasonable, because the matter in the universe is countering the tendency of the universe to expand by pulling it together. It also seems reasonable that having more matter in the universe would tend to make

Figure 2.84. A large amount of mass causes the universe to become spherical.

the universe have spherical geometry. Think of the curvature of the trampoline due to matter. Many of these effects would seem to be trying to cause the surface to close up on itself. (See Fig. 2.84 for a sketch for a 1-dimensional universe.) The proof that these consequences of the density of matter are as stated requires analysis using Einstein's equations.

The fate of the universe is thus linked to its geometry. This geometry could, in principle, be determined using measurements such as angle sums of triangles, but huge distances and extreme accuracy would be required. Consequently, the analysis using density of matter is the one preferred by today's scientists.

We summarize these possibilities in Figure 2.85. We close by pointing out that if the universe is spherical, then a ray of light will eventually return to its starting point, but it has been proven that the universe will collapse before that can happen. We also know that the time until the collapse of the universe is greater than the present age of the universe, which is roughly 10 billion years, so we need not yet worry about its demise.

| density | geometry | fate |
|---------|----------|------|
| $> \rho_0$ | spherical | collapse |
| $= \rho_0$ | Euclidean | slow expansion |
| $< \rho_0$ | hyperbolic | expansion |

Figure 2.85. Possibilities for geometry of universe.

*FOCUS: ALBERT EINSTEIN*

Albert Einstein was probably the most celebrated scientist in history, and he was also a very interesting person. He was not a child genius like Gauss. In fact he was extremely slow in learning to talk, and his parents feared for some time that he might be retarded. He did show some scientific inclination, though; a gift of a pocket compass at age five and Euclid's *Elements* when he was twelve had a great

Albert Einstein

impact on his developing mind. He failed the entrance exam to the Polytechnic School in Zurich on his first attempt, but passed it on his second one—after a year of remedial work.

After graduation, he took some part-time teaching jobs, and finally obtained a position as clerk in the Swiss Patent Office in Bern. He spent his evenings studying physics, and in 1905, while still a patent clerk, he developed his Special Theory of Relativity. The famous equation $E = mc^2$, which shows that mass can be converted to energy in a quantifiable way, was also published in 1905. Einstein, however, was not involved in the practical application of this formula to the development of the atom bomb forty years later. His name is nevertheless linked with the bomb for two reasons: his early work provided the theoretical underpinnings for it, and in 1939 he sent a famous letter about the bomb to President Roosevelt. The letter was drafted by Leo Szilard, who, like Einstein, had emigrated from Europe to the United States in the 1930s but was working on atomic research at Columbia University. In it, the scientists urged Roosevelt to commit more federal monies to American atomic research and to have the government play a major role in it, speculating that the Germans were already developing an atom bomb. The letter was signed by Einstein because of his enormous prestige, which was sure to get the president's attention.

The letter had the desired effect, but Einstein was not personally involved in the research. In March 1945, when it was clear that a bomb had been built, he again wrote to Roosevelt on behalf of Szilard and other scientists who now felt that the bomb need not and should not be used for military purposes. Roosevelt died shortly after receiving the letter, so it had no effect, and, as we know, it was used over Hiroshima and Nagasaki. After the war, Einstein was a prominent proponent of several ideas of world government. One was "that the United Nations be transformed from a league of sovereign states into a government deriving its specific powers from the peoples of the world." He was afraid of letting atomic secrets get into the hands of small countries. In the United States he was active as chairman of the Emergency Committee of Atomic Scientists, which dealt with policy issues concerning atomic energy.

Another favorite cause was Zionism. Although he rejected the biblical idea of God, he was a strong spokesman for the Jewish people and the idea that they have a homeland. He went on a fund-raising tour of the United States in 1921 for the Zionist cause. Prior to that (1919–1920), he was the target of numerous anti-Semitic diatribes in Germany. He was visiting California in 1933 when Hitler came to power, and never returned to Germany. In 1952 he was offered the presidency of Israel, but declined, pointing out that he lacked "both the natural aptitude and the experience to deal properly with people and to exercise official functions."[24]

He settled in the United States in 1933 as the first professor of the newly founded Institute for Advanced Study in Princeton, New Jersey. By this time, he was at odds with the prevailing trend in physics—*quantum mechanics.* According to this theory, at the atomic level things happen statistically, not deterministically. Einstein's favorite comment on this theory was that he could not believe that "God plays dice with the universe."

Two of Einstein's favorite pastimes were playing the violin and sailing. He was married twice and had two sons and two stepdaughters.

## Exercises

1. Fill in the rest of the table below.

|  | Vertices | Edges | Faces | Cubes | Hypercubes |
|---|---|---|---|---|---|
| Point | 1 | | | | |
| Line | 2 | 1 | | | |
| Square | | | | | |
| Cube | | | | | |
| Tesseract | | | | | |

Each entry in the table should equal twice the number above it plus the number to the left of that. Explain why, using the way in which one of these figures is formed from the previous one.

2. Make a Möbius strip by cutting out a strip of paper, twisting it once, and taping the short ends together. Show that it has only one side, and is nonorientable. This means that an object that traverses the length of the Möbius strip will return as its own mirror image. What happens when you cut it along the middle?

3. Draw a graph similar to those in Figure 2.79 representing the worldline in $(x, t)$-space of a car that starts from rest, accelerates to a maximum speed, which it holds briefly, and then decelerates to a stop, all the time moving in the $x$-direction.

4.[*] Take the usual $x$- and $y$-coordinates in the plane and stereographically project them onto the sphere. Sketch the resulting coordinate system on the sphere. (Hint: Under stereographic projection, straight lines in the plane correspond to circles on the sphere that pass through the North Pole.)

24. French 1979, p. 206.

5. The speed of light is roughly 1 billion ($10^9$) miles per hour. What is the approximate difference in time measured by a clock on the following moving objects compared to a clock on the earth? [Hint: In (a) and (b), you may use the fact that if $x$ is very small, then $\sqrt{1-x}$ is approximately equal to $1 - \frac{1}{2}x$.]
   a. A satellite traveling 18,000 miles per hour for one year.
   b. An airplane traveling 600 miles per hour for six hours.
   c. A spaceship of the future traveling at half the speed of light for ten years.

6. One can prove that the length $d$ in the direction of motion of an object moving with velocity $v$ is related to its length $d_0$ when at rest by the formula

$$d = d_0\sqrt{1 - (v/c)^2}.$$

   A spaceship that is 100 feet long when measured at rest has a length of 50 feet when moving away from the earth. How fast is it moving?

7. Show that the interval between two events in Minkowski's space-time is zero if and only if a ray of light could pass from one event to the other. In other words, the interval is zero between a light ray at one place and time and the same light ray at another place and time.

8. Draw the worldlines of two objects moving toward each other until they collide and start moving apart.

9. Suppose two balls are 4 feet apart in an elevator at the top of a 2640-foot ($= \frac{1}{2}$ mile) building. How much closer together will they get if the elevator goes into free fall to the surface of the earth? Assume the radius of the earth is 4000 (or, if you prefer, $3999\frac{1}{2}$) miles. Refer to Figure 2.81.

# 3. Number Theory

## 3.1 Prime Numbers

THE NOTION of prime number is very simple—one with which we become familiar in grade school. The study of properties of prime numbers has engaged countless mathematicians, both professional and amateur, since the time of the Greeks. This fascination is due in part to their regular/irregular behavior, and in part to the many simply stated unsolved questions about prime numbers. For example, as we will see, there is a formula (the Prime Number Theorem) that tells the approximate number of primes through any range of numbers, yet sometimes they occur very far apart and sometimes very close together. The latter case is illustrated most vividly by twin primes, pairs of prime numbers such as 17 and 19 or 41 and 43, which differ by only 2. No one knows for sure whether there are infinitely many twin primes. In this section, we will glimpse a number of famous theorems about prime numbers that have been proved, such as the Prime Number Theorem, and also a number of unresolved questions about them, such as whether there are infinitely many twin primes.

DEFINITION. *A prime number is an integer greater than 1 that cannot be evenly divided by any positive integers except itself and 1. An integer greater than 1 that is not prime is called a* composite number.

Thus a number is composite if it can be written as $a \cdot b$, where $a$ and $b$ are both integers greater than 1; it is prime if this cannot be done. Note that 1 is not considered to be either prime or composite, and you must resist the temptation to think of it as being prime. One reason that 1 is not considered to be prime is that, as we shall see below, the decomposition of integers into prime factors is a very important idea in number theory, and writing 1 as a factor of a number is not a very interesting thing to do. The first few primes are 2, 3, 5, 7, 11, 13, 17, and 19. You will soon learn a systematic method of tabulating prime numbers, called the Sieve of Eratosthenes. For now, you should at least be able to verify by yourself that the above list is correct as far as it goes.

UNIQUE FACTORIZATION

By repeated factoring, it is clear that any composite number can be written as a product of prime factors. For example,

$$168 = 2 \cdot 84 = 2 \cdot 2 \cdot 42 = 2 \cdot 2 \cdot 2 \cdot 21 = 2^3 \cdot 3 \cdot 7.$$

This procedure always ends with prime factors, because if the factors in a decomposition are not prime, they can be factored into smaller numbers, and this procedure must eventually terminate. What is not so obvious is that the resulting factorization into primes is unique. It seems conceivable that if you started by factoring 168 as $7 \cdot 24$, you might eventually end up with prime factors that are different than the ones you got by starting as above. If it seems obvious to you that the factorization into primes can end up only in one way, you were probably taught this fact many years ago and were never challenged to question it. This unique factorization property of integers is called the Fundamental Theorem of Arithmetic.

THEOREM 3.1.1. *(Fundamental Theorem of Arithmetic). Every composite number can be written uniquely as a product of primes.*

The uniqueness here means "up to reordering"; for example, $2 \cdot 3$ and $3 \cdot 2$ are considered to be the same factorization of 6. This theorem was proved by the Greeks; it appeared in Euclid's *Elements*, which included a lot of number theory. We will discuss the proof, which utilizes the Euclidean algorithm, in Section 3.2.

The Fundamental Theorem of Arithmetic gives one method of finding the *greatest common divisor* (GCD) of a set of integers. To find the largest number that is a divisor of a set of numbers, write the factorization of each into primes, and then take as the GCD the product of all primes that occur as factors of each of the numbers. If a prime is repeated a certain number of times as a factor of all of the numbers, then it should be repeated that many times as a factor of the GCD. For example, the GCD of 168 and 180 is $2^2 \cdot 3 = 12$ since $168 = 2^3 \cdot 3 \cdot 7$, as we saw above, and $180 = 2^2 \cdot 3^2 \cdot 5$. The $2^2$ that occurs as a factor of the GCD is present because both 168 and 180 have $2^2$ as part of their factorization, but 180 does not have $2^3$ as part of its factorization.

To understand why this method gives the GCD, note how the factorization of a number into prime powers gives a systematic way of listing *all* positive divisors of the number. If $p$ is a prime number and $p^e$ is part of the factorization of the number, then $p^0$, $p^1$, $\ldots, p^e$ are all divisors of the number. We follow the usual convention that the 0th power of any positive number is 1. All positive divisors of the number are obtained by choosing, for each $p^e$ in the factorization of the number, one of these $p^i$'s with $i \leq e$, and taking the product of these. For example, the divisors of $168 = 2^3 \cdot 3 \cdot 7$ are obtained by multiplying together 0, 1, 2, or 3 of the 2's, 0 or 1 of the 3's, and 0 or 1 of the 7's. We list these in Table 3.1.

TABLE 3.1. The Divisors of 168.

| | | | |
|---|---|---|---|
| $2^0 3^0 7^0 = 1$ | $2^0 3^1 7^0 = 3$ | $2^0 3^0 7^1 = 7$ | $2^0 3^1 7^1 = 21$ |
| $2^1 3^0 7^0 = 2$ | $2^1 3^1 7^0 = 6$ | $2^1 3^0 7^1 = 14$ | $2^1 3^1 7^1 = 42$ |
| $2^2 3^0 7^0 = 4$ | $2^2 3^1 7^0 = 12$ | $2^2 3^0 7^1 = 28$ | $2^2 3^1 7^1 = 84$ |
| $2^3 3^0 7^0 = 8$ | $2^3 3^1 7^0 = 24$ | $2^3 3^0 7^1 = 56$ | $2^3 3^1 7^1 = 168$ |

Clearly the greatest divisor common to a set of numbers will be the largest product of prime factors common to all the numbers. In Exercise 3, you are asked to carry out for 180 the procedure used above for 168. A comparison of the two lists may help you understand why 12 is their GCD.

Note that if two numbers have no common prime divisors, then their GCD is 1 (not 0). Certainly 1 is a common divisor of the two numbers, and if they have no common prime divisors, then 1 is the *greatest* common divisor. Such numbers are said to be *relatively prime*. For example, 35 and 72 are relatively prime.

Although the unique factorization of integers may seem obvious to you, there are systems of numbers that occur naturally in number theory and have many of the properties of the integers, but do not admit a unique factorization into primes. You have the opportunity to learn about these in Exercises 4 through 8. The failure of the numbers discussed in those exercises to have unique factorization was important in the history of mathematics. You are encouraged to read now the "Aside" at the end of this section, which deals with Fermat's Last Theorem, probably the most famous unsolved question in mathematics. It shows there how certain mathematicians in the eighteenth and nineteenth centuries thought they could prove Fermat's Last Theorem, but made the mistake of assuming unique factorization into primes in a situation where it didn't exist.

### DIVISIBILITY AND PRIMES

The Fundamental Theorem of Arithmetic is the basis of the following useful fact.

PROPOSITION 3.1.2. *If $n$ is not divisible by any prime number less than or equal to $\sqrt{n}$, then $n$ is prime.*

This is the same as saying that if $n$ is not prime, then $n$ must be divisible by some prime less than or equal to $\sqrt{n}$. To prove this restatement, we note that if $n$ is not prime, and $p$ is the smallest prime divisor of $n$, then $n$ must have at least two prime factors (possibly both the same) at least as large as $p$, so that $n \geq p^2$; that is, $p \leq \sqrt{n}$.

Proposition 3.1.2 is very useful because it greatly reduces the number of trial divisors that you must consider in order to be sure that a number is prime. For example, to verify that 239 is prime, it is enough to verify that it is not divisible by 2, 3, 5, 7, 11, or 13, since $\sqrt{239} \approx 15.45$.

The ancient Greeks had a cumbersome notation for numbers that made division rather difficult. The Sieve of Eratosthenes was a method of generating the prime numbers through any range without actually doing any division. You write down all the numbers in the desired range starting with 2, circle the number 2 as a prime, and then cross out every second number after it. These numbers will all be multiples of 2, and hence not prime. Note that by mechanically counting off by 2's, you are not actually doing any division. Now the first number neither circled nor crossed out, in this case 3, must be prime, since it is has been tested for divisibility by all smaller numbers. Circle the 3, and now cross out every third number after it. These will be multiples of 3, and hence not prime. Some numbers will get crossed out for the second time. When you count off every third number, you must consider all numbers, not just those that haven't been crossed out. The procedure is now repeated for the next unmarked number, 5. After circling the 5 and crossing out every fifth number after it, you will have crossed out every number $\neq 2, 3,$ or 5 that is divisible by 2, 3, or 5. You continue this procedure until you have considered multiples of numbers that are less than or equal to the square root of the range of your table. For example, to find the primes through 100, write the numbers from 2 to 100, and cross out multiples of 2, 3, 5, and 7. (Why is that enough?) All remaining numbers are guaranteed to be prime. This is done in Table 3.2 below, in which numbers divisible by 2 are crossed out by \, those divisible by 3 by /, those divisible by 5 by |, and those divisible by 7 by −.

Eratosthenes, who invented this sieving procedure, lived roughly 276 B.C. to 194 B.C., about the same time as Archimedes and slightly after Euclid. He is most famous for his accurate estimate of the circumference of the earth. This is particularly impressive when you consider that seventeen centuries later many Europeans still thought the earth was flat. Eratosthenes' method was to consider the shadows cast by the sun at two cities that were about 500 miles apart. At

TABLE 3.2. The Sieve of Eratosthenes.

| ② | ③ | ~~4~~ | ⑤ | ~~6~~ | ⑦ | ~~8~~ | ~~9~~ | 10 |
|---|---|---|---|---|---|---|---|---|
| 11 | ~~12~~ | 13 | ~~14~~ | ~~15~~ | ~~16~~ | 17 | ~~18~~ | 19 | ~~20~~ |
| ~~21~~ | ~~22~~ | 23 | ~~24~~ | ~~25~~ | ~~26~~ | ~~27~~ | ~~28~~ | 29 | ~~30~~ |
| 31 | ~~32~~ | ~~33~~ | ~~34~~ | ~~35~~ | ~~36~~ | 37 | ~~38~~ | ~~39~~ | ~~40~~ |
| 41 | ~~42~~ | 43 | ~~44~~ | ~~45~~ | ~~46~~ | 47 | ~~48~~ | ~~49~~ | ~~50~~ |
| ~~51~~ | ~~52~~ | 53 | ~~54~~ | ~~55~~ | ~~56~~ | ~~57~~ | ~~58~~ | 59 | ~~60~~ |
| 61 | ~~62~~ | ~~63~~ | ~~64~~ | ~~65~~ | ~~66~~ | 67 | ~~68~~ | ~~69~~ | ~~70~~ |
| 71 | ~~72~~ | 73 | ~~74~~ | ~~75~~ | ~~76~~ | ~~77~~ | ~~78~~ | 79 | ~~80~~ |
| ~~81~~ | ~~82~~ | 83 | ~~84~~ | ~~85~~ | ~~86~~ | ~~87~~ | ~~88~~ | 89 | ~~90~~ |
| ~~91~~ | ~~92~~ | ~~93~~ | ~~94~~ | ~~95~~ | ~~96~~ | 97 | ~~98~~ | ~~99~~ | ~~100~~ |

the same moment, the sun cast no shadow at one of them and made an angle of about 7° with a post at the other. This is the marked angle at *P* in Figure 3.1, which is not drawn to scale. Assuming the sun to be far enough from the earth that its rays may be assumed to be parallel at the two points, he deduced that the angle from the center of the earth to the two points is also about 7°, or roughly $1/50$ of 360. Thus 500 miles is $1/50$ of the circumference of the earth, and so the circumference is 25,000 miles. His estimate was accurate within 1%, and was not improved upon for a thousand years.

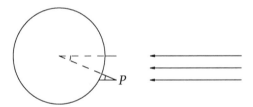

Figure 3.1. How Eratosthenes determined the circumference of the earth.

There are easy rules for testing for divisibility by 2, 3, and 5, but for larger primes the testing should be done by ordinary division.

PROPOSITION 3.1.3.

   *i. A number is divisible by 2 if and only if its last digit is 0, 2, 4, 6, or 8.*

   *ii. A number is divisible by 5 if and only if its last digit is 0 or 5.*

   *iii. A number is divisible by 3 if and only if the sum of its digits is divisible by 3.*

   *iv. A number is divisible by 9 if and only if the sum of its digits is divisible by 9.*

For example, 24567 is divisible by 3 because $2 + 4 + 5 + 6 + 7 = 24$ is divisible by 3, but 41765 is not divisible by 3 because the sum of its digits is 23. The rule for divisibility by 9 is not so important as the others for primality testing, since to test whether a number is prime you only have to check whether it is divisible by prime numbers, and 9 isn't prime. A similar, but slightly more complicated, rule for divisibility by 11 is discussed in Exercise 11.

*Proof of Proposition 3.1.3.* A number $n$ can be written as $10m + d$, where $d$ is the last digit of the number, and $m$ is the number obtained by removing the last digit. The equation,

$$n = 2 \cdot 5 \cdot m + d,$$

shows that $n$ is divisible by 2 or 5 if and only if $d$ is, and parts (i) and (ii) of the proposition are just a restatement of this. Note that the last digit determines divisibility by 2 or 5 because $10 = 2 \cdot 5$. Also note that in this argument we are using a fact we will use over and over again:

if $A$ is divisible by $K$, then $A + B$ is divisible by $K$ if and only if $B$ is.

This is really just a restatement of the distributive law $K\alpha \pm K\beta = K(\alpha \pm \beta)$, which shows that the sum or difference of multiples of $K$ is a multiple of $K$.

We illustrate the proof for divisibility by 3 or 9 with the three-digit number $n$ whose digits from left to right are $a$, $b$, and $c$. The reader should be able to modify this proof to a number with any number of digits. Thus $n = 100a + 10b + c$, while the sum of the digits $S$ satisfies $S = a + b + c$. Subtracting and rearranging yields

$$n = (99a + 9b) + S.$$

Since $99a + 9b$ is divisible by 3 and by 9, this shows that $n$ is divisible by 3 or 9 if and only if $S$ is. Here we have used the general principle about divisibility that was observed at the end of the previous paragraph. Q.E.D.

## THE INFINITUDE OF THE PRIMES

Because large numbers have more potential divisors, primes become sparser as one looks at larger numbers. One might even guess that the primes would stop eventually, that is, it would seem to be conceivable that beyond some point all numbers are composite. This is not the case, as was proved by Euclid more than twenty-two centuries ago. His proof of this fact is considered a paradigm of elegance. In fact, in a poll of mathematicians taken in 1990, this theorem and proof were voted the third most beautiful in all of mathematics.[1] We have already seen in Section 1.3 the theorems ranked numbers 2 and 4, and will see the top-ranked theorem in Exercise 2c of Section 5.3. Numbers 2 and 4 were, respectively, Euler's formula $v - e + f = 2$ (Theorem 1.3.4), and the Greek theorem about the existence of exactly five regular polyhedra (Theorem 1.3.3).

THEOREM 3.1.4. *(Euclid). There are infinitely many primes.*

*Proof.* Suppose that there exist only a finite number of primes, so that a complete list can be given as $p_1, p_2, \ldots, p_k$. This notation with subscripts means that $p_1$ is the first prime, hence $p_1 = 2$, $p_2$ is the second prime, so that $p_2 = 3$, etc., and that there are exactly $k$ primes altogether, with $p_k$ being the last one. It doesn't matter what the precise value of $k$ is, just that some such number exists. Either such a number $k$ exists or else there are infinitely many primes. That is what "infinitely many" means: they never stop.

Now consider the integer obtained by multiplying all those prime numbers together and adding 1 to the product. Call this huge new number $N$. Thus

$$N = (p_1 p_2 \cdots p_k) + 1.$$

Such a number can be formed under our supposition that the number of primes is finite. For each of the primes $p_i$, this big number $N$ is 1 greater than a multiple of $p_i$. Thus $N$ is not a multiple of $p_i$, since it has remainder of 1 upon division by $p_i$. Here $p_i$ refers to any of the primes in our list, which is assumed to be a complete list of all prime numbers. Thus $N$ is not divisible by any prime, and hence $N$ itself must be prime. Since $N$ is (much) larger than any of the primes in our list, this contradicts the assumption that the list was complete. Q.E.D.

In case you are bothered by the indirect nature of this elegant proof, we note that a slight modification of the argument produces an algorithm that can be repeated indefinitely, always yielding new primes. Start with just the prime 2, and at each step let $N = P + 1$, where $P$ is the product of all primes produced up to that point. This $N$ either will be itself a new prime, or else can be factored as a product of new primes. The primes in this factorization will be new, because $N$ is not divisible by any of the primes generated up to that point. We show the first three steps, and ask you to do the fourth in Exercise 12.

   1. $N = 2 + 1 = 3$, yielding the new prime 3.

---

1. D. Wells, "Are these the most beautiful?" *Mathematical Intelligencer* 12 (1990): 37–41.

   2. $N = 2 \cdot 3 + 1 = 7$, yielding the new prime 7.

   3. $N = 2 \cdot 3 \cdot 7 + 1 = 43$, yielding the new prime 43.

Note that this method makes no claim to generate all the primes, just that it keeps generating new ones at each step.

   Note that all primes except 2 are odd numbers, and that every odd number can be written as either $4k + 1$ or $4k + 3$ for some number $k$. We show now how a slight modification of Euclid's proof can be used to show that there are infinitely many primes of the form $4k + 3$. There are also infinitely many primes of the form $4k + 1$, but this is a little bit harder to prove. In Exercise 13, you are asked to perform a direct modification of this argument to show that there are infinitely many primes of the form $6k + 5$.

PROPOSITION 3.1.5. *There are infinitely many primes of the form* $4k + 3$.

   *Proof.* Suppose that there are only finitely many primes of the form $4k + 3$. Let $N$ be the number obtained by multiplying them all together, multiplying the product by 4, and then subtracting 1. Thus

$$N = 3 \cdot 7 \cdots (4L + 3) \cdot 4 - 1,$$

where $4L + 3$ is the hypothetical largest prime of the form $4k + 3$. First note that $N$ is 1 less than a multiple of 4, and hence it is also of the form $4k + 3$, since the number 1 less than a multiple of 4 is 3 greater than the previous multiple of 4.

   Next note that $N$ cannot be divided by 2, since $N$ is an odd number, and $N$ is not divisible by any of the primes of the form $4k + 3$ that were in our list, because it is 1 less than a multiple of each of them. But our list was assumed to contain all primes of the form $4k + 3$, and so $N$ is not divisible by 2 or *any* prime of the form $4k + 3$. Thus, either (i) $N$ itself is prime, or else (ii) it is a product of primes of the form $4k + 1$. However, (i) is ruled out by the fact that $N$ is of the form $4k + 3$ and is much larger than any of the numbers on our list, which was assumed to contain all primes of the form $4k + 3$. Also, (ii) is ruled out because the product of numbers of the form $4k + 1$ is 1 greater than a multiple of 4, while our $N$ is 3 greater than a multiple of 4. We prove this fact about products of numbers of the form $4k + 1$ by considering the product of two of them:

$$(4k + 1)(4\ell + 1) = 4(k\ell + k + \ell) + 1.$$

A similar argument works for a product of more than two of them. Thus both possibilities (i) and (ii) have been eliminated, and we arrive at a contradiction of our hypothesis that the number of primes of the form $4k + 3$ is finite. Q.E.D.

## THE PRIME NUMBER THEOREM

One is naturally inclined to wonder how many primes there are through any range of numbers. For example, how many primes are $\leq 1{,}000{,}000$? The name $\pi(n)$ is traditionally used for the number of primes less than or equal to $n$. This has nothing to do with the number $\pi$. The letter $\pi$ is used for the name of this function because both $\pi$ and "prime" start with the letter "p". This is a function just like the function $f(x) = x^2$. We remind the reader that a function is a rule that associates to each element of a certain set a unique number. For example, $\pi(20) = 8$ because 2, 3, 5, 7, 11, 13, 17, and 19 are the eight primes $\leq 20$. Here $\pi$ is the name of a rule that associates to the number 20 the number 8.

It seems inconceivable that a *useful* formula for $\pi(n)$ that works for all $n$ will ever be obtained, because that is tantamount to being able to list all the primes. In Exercise 15 a formula that yields the $n$th prime when $n$ is plugged into it is discussed. This formula is not very useful because computing it requires more work than performing the Sieve of Eratosthenes.

In 1792, at the age of fifteen, Carl Friedrich Gauss made a guess of an approximate formula for $\pi(n)$. His guess was based on extensive calculation. He once wrote in a letter that he "very often used an idle quarter of an hour to count through another chiliad here and there" until finally he had listed all the prime numbers up to 3,000,000. A chiliad is an interval of one thousand numbers, and by "counting" he meant determining which of the numbers were prime. In order to understand his guess, we need a brief review of logarithms.

The classical definition of $\log_b x$, read "logarithm (or log) to the base $b$ of $x$," is that it is the number $y$ such that $b^y = x$. For example, $\log_{10}(100) = 2$ because $10^2 = 100$. Prior to the invention of electronic calculators, tables of logarithms were used frequently for calculations involving multiplication, division, and exponentiation. The reason for this usefulness is that

$$(3.1.6) \qquad \log_b(w \cdot x) = \log_b(w) + \log_b(x).$$

Thus to multiply 3.1416 by 2.3456, you could look up the logarithms of these two numbers, add the logarithms (addition being easier than multiplication), and then use the table again to find the number whose logarithm equaled that sum. Of course, pushing buttons on a calculator is much easier, and somewhat more reliable. Students of calculus learn a different approach to logarithms. In calculus and most other forms of higher math, $\log X$ is defined to be the area

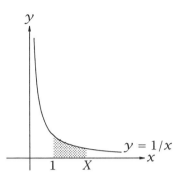

Figure 3.2. The definition of log $X$.

bounded by the $x$-axis, the curve $y = 1/x$, and the lines $x = 1$ and $x = X$. (See Fig. 3.2.)

This is compatible with the previous definition if the base $b$ equals the number $e$ (named for Euler), defined by

$$e = \lim_{n \to \infty} (1 + \tfrac{1}{n})^n \approx 2.71828.$$

The number $e$ runs a close race with $\pi$ for the most important number in mathematics. In this section, when we use log without a subscript, we mean this log to the base $e$. It comes up in number theory because of (3.1.6), and because of the ability of areas to approximate certain complicated sums. On many calculators, the LOG key is for $\log_{10}$, while the LN key is for $\log_e$.

Gauss's guess was that the approximate value of $\pi(n)$ is $n/\log n$. This says that of the numbers from 1 to $n$, roughly 1 of every $\log n$ of them is prime. Since $\log(10^k)$ is approximately $2.3k$, this says that of the first million ($10^6$) numbers, roughly 1 out of every $13.8 (= 2.3 \times 6)$ is prime, while out of the first billion ($10^9$) numbers, roughly 1 out of every 20.7 is prime. This illustrates how the primes become sparser as you look at larger and larger numbers.

Saying that $\pi(n)$ is approximately equal to $n/\log(n)$ is not a precise mathematical statement, and therefore not capable of being proved, unless we say what we mean by "approximately equal." This was made precise by Gauss, but he was unable to prove that what he had observed to be apparently true was necessarily always true for any range of numbers. In fact, it took over one hundred years before Gauss's conjecture about the approximate value of $\pi(n)$ was proved. It was finally proved in 1896, independently by Jacques Hadamard of France and Charles de la Vallée Poussin of Belgium. Their life spans, 1865–1963 and 1866–1962, respectively, would suggest that proving the Prime Number Theorem, as it was called, was good for one's health.

PRIME NUMBER THEOREM 3.1.7. $\pi(n)$ *is approximately equal to* $n/\log n$. *More precisely, the ratio*

$$\frac{\pi(n)}{n/\log n}$$

*approaches 1 as n approaches* $\infty$.

Even this statement may not sound very precise to someone who has not studied a little bit of calculus, but statements about limits such as this are made very precise in calculus. It means that by considering large enough values of $n$, the ratio can be made to be as close to 1 as desired. Extremely large values of $n$ are required before the approximation becomes very good. (See Table 3.3.) When $n = 10,000,000,000,000,000(= 10^{16})$, the ratio of $\pi(n)$ to $n/\log n$ has still only dropped to 1.028. Nevertheless, the theorem guarantees that it will eventually get as close to 1 as you want.

TABLE 3.3. Data for the Prime Number Theorem.

| $n$ | $\pi(n)$ | $n/\log(n)$ | $\pi(n)/n/\log n$ |
|---|---|---|---|
| 1,000 | 168 | 145 | 1.158 |
| 10,000 | 1,229 | 1,089 | 1.128 |
| 100,000 | 9,592 | 8,696 | 1.103 |
| 1,000,000 | 78,498 | 72,464 | 1.083 |
| 10,000,000 | 664,579 | 621,118 | 1.070 |
| 100,000,000 | 5,761,455 | 5,434,783 | 1.060 |
| 1,000,000,000 | 50,847,478 | 48,309,185 | 1.052 |
| 10,000,000,000 | 455,052,511 | 434,782,650 | 1.046 |

Refinements of the formula in the Prime Number Theorem are known that give better approximations, but they lack the simplicity of the original formula. There is a surprising overlap between the Prime Number Theorem and *complex analysis*, the field of mathematics that deals with the calculus of complex numbers. (We will learn a little bit about complex numbers in Section 5.3.) The great German mathematician of the mid-nineteenth century, G.F.B. Riemann, whose contributions to geometry we have already discussed at length, also made the most important contributions here. The Riemann Hypothesis is a conjecture he made around 1859, which, if proved, would have tremendous implications in number theory. It says that a certain function[2] in complex analysis can be zero only when its variable lies on a certain line. It is arguably the most important unsolved question in mathematics, although it doesn't have the popular appeal of Fermat's Last Theorem because its statement is more technical.

Typical of the mystique of the Riemann Hypothesis is a story concerning the great British analyst G. H. Hardy (1877–1947), who had made one of the greatest advances toward proving the Riemann hypothesis by showing that Riemann's function has infinitely many zeros along the critical line. Hardy was taking a ferry back to England across the North Sea on a stormy day. He sent a postcard to his colleague J. E. Littlewood, claiming that he had proved the Riemann Hypothesis. His thinking was that God would not allow him to receive undeserved credit for proving such an important result, and therefore his ship would not sink. His trip was successful, and he was able to explain his joke.

Prime numbers are irregularly distributed, even though the Prime Number Theorem says that there is some regularity to their approximate frequency of occurrence. We close this subsection by discussing how sometimes they come close together, and sometimes they occur far apart.

*Twin primes* are primes that differ by 2. For example, 17 and 19 are twin primes, as are 41 and 43. It is widely believed, but not yet proved, that there are infinitely many twin primes. This is another important open question in number theory, whose statement can be understood by anyone but whose proof has resisted attempts by many of the best mathematicians in the world.

On the other hand, there are arbitrarily large gaps between primes. This means that for any number you can think of, there is a string of at least that many consecutive composite numbers. This can be done systematically using factorials. The reader is reminded that $n!$, read "$n$ factorial," is the product of all positive integers less than or equal to $n$. For any value of $n$, the numbers $n! + 2$ through $n! + n$ are all composite. This is true because $n!$ is divisible by all numbers from 2 through $n$. Hence $n! + 2$ is the sum of two multiples of 2, and hence is a multiple of 2, $n! + 3$ is a sum of two multiples of 3 and hence is a multiple of 3, etc. Thus we have given an explicit string of $n - 1$ composite numbers for any number $n$, no matter how large. The string of composite numbers $n! + 2$ through $n! + n$ can often be extended to a longer string of composite numbers. For example, if $n = 25$, the numbers $25! + 26$, $25! + 27$, and $25! + 28$ are all composite,

---

2. The Riemann zeta function $\zeta(s) = 1/1^s + 1/2^s + 1/3^s + \cdots$, where $s$ is a complex number.

the first being divisible by 2 and 13 (since both 25! and 26 are), the second by 3, and the third by 4 and 7. Thus we have a string here of at least 27 consecutive composite numbers, beginning with $25! + 2$. It is not easy to tell whether or not $25! + 1$ and $25! + 29$ are prime. In fact, both $25! + 1$ and $25! + 29$ can be written as the product of two primes, as follows:

$$25! + 1 = 401 \times 38681321803817920159601$$
$$25! + 29 = 48511549 \times 319742625479367521$$

*MERSENNE PRIMES AND PERFECT NUMBERS*

Even though Theorem 3.1.4 says that there is definitely no largest prime number, there is a largest *known* prime number, that is, a largest specific number known to be prime. As of mid-1992, the largest known prime was $2^{756,839} - 1$, a number with 227,832 digits.

This prime is one of a family of prime numbers called *Mersenne primes*, named after a French monk who worked in the early 1600s. He stated that $2^n - 1$ is prime for $n = 2, 3, 5, 7, 13, 17, 19, 31, 67, 127$, and 257, and is composite for all other values of $n$ less than 257. His statement was correct for $n < 61$, but was wrong for $n = 61, 89$, and 109, for which $2^n - 1$ is prime, and for $n = 67$ and 257, for which $2^n - 1$ is not prime. Lest one think that this record hardly justifies having one's name live in perpetuity, one should keep in mind that he correctly classified as prime or not all numbers of the form $2^n - 1$ with fewer than nineteen digits, no mean feat considering that there were no calculating devices in those days. It was not until 250 years later that any mistakes were found in his guesses.

The following proposition is basic to the study of which Mersenne numbers are prime. This result was known to the ancient Greeks.

PROPOSITION 3.1.8. *If $n$ is composite, then so is $2^n - 1$.*

*Proof.* We assume that $n = ab$ with $a > 1$ and $b > 1$, and will obtain from this an explicit factorization of $2^n - 1$.

First we note the following important algebraic identity.

$$(3.1.9) \qquad x^b - 1 = (x - 1)(x^{b-1} + x^{b-2} + \cdots + x + 1).$$

This is valid for any positive integer $b$. It is proved by simply multiplying out the right-hand side, obtaining

$$(x^b + x^{b-1} + \cdots + x^2 + x) - (x^{b-1} + x^{b-2} + \cdots + x + 1) = x^b - 1,$$

since all intermediate terms cancel out.

Since (3.1.9) is an algebraic identity, it remains valid when any number is substituted for $x$. We let $x = 2^a$, obtaining

$$2^n - 1 = 2^{ab} - 1 = (2^a)^b - 1$$

$$= (2^a - 1)\Big((2^a)^{b-1} + (2^a)^{b-2} + \cdots + 2^a + 1\Big).$$

Since $a > 1$, the first factor on the right hand side, $2^a - 1$, is greater than 1, and since $b > 1$, the second factor is also greater than 1. Thus we have a nontrivial factorization of $2^n - 1$. Q.E.D.

Primes of the form $2^n - 1$ were avidly studied by the Greeks, in part because of a connection with perfect numbers, which we will discuss below. The Greeks knew that $2^n - 1$ is prime for $n = 2, 3, 5$, and 7. Many early workers erroneously believed that $2^n - 1$ was prime whenever $n$ was prime, but in 1536 the factorization

$$2^{11} - 1 = 2047 = 23 \cdot 89$$

was discovered. It seems amazing that this discovery took so long, but cumbersome Roman numerals were still being used by Europeans until about 1400. The reader should note carefully that Proposition 3.1.8 says that the only Mersenne numbers that can possibly be prime are those whose exponent is prime, but it does not say that a prime $n$ guarantees that $2^n - 1$ is prime.

A table of the thirty-two Mersenne primes known in 1992 is given in Table 3.4. Note that $2^{127} - 1$ was the largest known prime for seventy-five years, until the advent of computers. Of course, there are many primes that are not Mersenne primes, but special techniques that exist only for Mersenne numbers make it much easier to check them for primality. Thus, with a few rare exceptions, the largest known prime numbers are, and have always been, those of the form $2^n - 1$. No one knows whether there are infinitely many Mersenne primes. In Table 3.4, the middle column is the number of digits in the Mersenne prime, and the third column is the date of discovery.

The twenty-fifth and twenty-sixth Mersenne primes were discovered by high school students, and they received national television coverage for their efforts. More recently, finding primes has turned into a competition between several companies that manufacture supercomputers. It is known that this list is complete for $n \leq 365,000$; if you want to discover a new Mersenne prime, you will have to look at numbers greater than $2^{365,000}$.

TABLE 3.4. Mersenne Primes.

| $n$ | Digits | Date |
|---|---|---|
| 2 | 1 | B.C. |
| 3 | 1 | B.C. |
| 5 | 2 | B.C. |
| 7 | 3 | B.C. |
| 13 | 4 | 1400's |
| 17 | 6 | 1603 |
| 19 | 6 | 1603 |
| 31 | 10 | 1772 |
| 61 | 19 | 1883 |
| 89 | 27 | 1911 |
| 107 | 33 | 1914 |
| 127 | 39 | 1876 |
| 521 | 157 | 1952 |
| 607 | 183 | 1952 |
| 1279 | 386 | 1952 |
| 2203 | 664 | 1952 |
| 2281 | 687 | 1952 |
| 3217 | 969 | 1957 |
| 4253 | 1332 | 1961 |
| 4423 | 1332 | 1961 |
| 9689 | 2917 | 1963 |
| 9941 | 2993 | 1963 |
| 11213 | 3376 | 1963 |
| 19937 | 6002 | 1971 |
| 21701 | 6533 | 1978 |
| 23209 | 6987 | 1979 |
| 44497 | 13395 | 1979 |
| 86243 | 25962 | 1983 |
| 110503 | 33265 | 1988 |
| 132049 | 39751 | 1983 |
| 216091 | 65050 | 1985 |
| 756839 | 227832 | 1992 |

One result that applies only to Mersenne numbers is the following result, as observed by Fermat.

PROPOSITION 3.1.10. *If p is an odd prime, then any positive divisor of* $2^p - 1$ *must be of the form* $2kp + 1$ *for some integer k.*

In other words, the only possible divisors of $2^p - 1$ are numbers that are 1 more than a multiple of $2p$. Observe how this is satisfied by the factorization of $2^{11} - 1$ as $23 \cdot 89$. If you were trying to factor $2^{11} - 1$, Proposition 3.1.10 tells you that 23 is the first divisor you should try.

We postpone the proof of Proposition 3.1.10 until Section 3.4, at which time we will have learned the requisite tools. We demonstrate the utility of this result by using it to show that $2^{17} - 1$ is prime.

PROPOSITION 3.1.11. $2^{17} - 1$ *is prime.*

*Proof.* $2^{17} - 1 = 131071$. Since $\sqrt{131071} = 362.03$, Proposition 3.1.2 implies that it will suffice to show that 131071 has no prime divisors less than 362. By Proposition 3.1.10, the only possible divisors of $2^{17} - 1$ are of the form $34k + 1$. We list the ten numbers of this form that are less than 362:

$$35, \ 69, \ 103, \ 137, \ 171, \ 205, \ 239, \ 273, \ 307, \ 341.$$

Of these, only 103, 137, 239, and 307 are prime. So all we have to do is to verify that none of these four numbers divides evenly into 131071, and this is easily done. Q.E.D.

The *proper divisors* of a number are the divisors that are less than the number. For example, the proper divisors of 6 are 1, 2, and 3, while the proper divisors of 8 are 1, 2, and 4. A number is called *perfect* if it equals the sum of its proper divisors. Thus 6 is perfect, while 8 is not, since $1 + 2 + 4$ is not equal to 8. The Greeks were interested in perfect numbers for mystical reasons,[3] and the following important result was proved in Euclid's *Elements*.

THEOREM 3.1.12. *If* $2^n - 1$ *is prime, then* $2^{n-1}(2^n - 1)$ *is perfect.*

Thus for every Mersenne prime, one obtains a perfect number by multiplying the Mersenne prime by an appropriate power of 2. For example, the first two Mersenne primes, $2^2 - 1 = 3$ and $2^3 - 1 = 7$, give rise to the perfect numbers $6 = 2 \cdot 3$ and $28 = 2^2 \cdot 7$.

*Proof of Theorem 3.1.12.* Suppose $2^n - 1$ is prime, and call it $p$ for notational simplicity. Then, by Theorem 3.1.1, a complete list of the proper divisors of $2^{n-1}p$ is given by

3. Somewhat later, (c. A.D. 400) Saint Augustine wrote that God created the heavens and earth in six days to signify the perfection of the work.

$$1, \ 2, \ 2^2, \ldots, 2^{n-1}, \ p, \ 2p, \ 2^2 p, \ldots, 2^{n-2} p.$$

Note that we do not list $2^{n-1} p$, since it is not a proper divisor. Thus the sum of the proper divisors is

(3.1.13)          $1 + 2 + \cdots + 2^{n-1} + p(1 + 2 + \cdots + 2^{n-2}).$

Letting $x = 2$ and $b - 1 = k$ in (3.1.9) yields the useful formula,

(3.1.14)          $1 + 2 + \cdots + 2^{k-1} + 2^k = 2^{k+1} - 1.$

Applying this twice to (3.1.13) shows that the sum of the proper divisors of $2^{n-1} p$ is $2^n - 1 + p(2^{n-1} - 1)$. Now we write $p$ as $2^n - 1$ again. Thus under our assumption that $2^n - 1$ is prime, we have shown that the sum of the proper divisors of $2^{n-1}(2^n - 1)$ is

$$(2^n - 1) + (2^n - 1)(2^{n-1} - 1) = (2^n - 1) + 2^{n-1}(2^n - 1) - (2^n - 1)$$
$$= 2^{n-1}(2^n - 1).$$

Thus $2^{n-1}(2^n - 1)$ is perfect. Q.E.D.

The following result, proved by Euler in 1738, shows that the family of perfect numbers given by the Greeks gives all even perfect numbers.

THEOREM 3.1.15. *(Euler). The only even perfect numbers are those given by 3.1.12, that is, numbers $2^{n-1}(2^n - 1)$ when $2^n - 1$ is prime.*

It is widely believed, but still not proved, that there are no odd perfect numbers. It is known that there are no odd perfect numbers less than $10^{100}$, and that any odd perfect number must have at least eleven distinct prime divisors. But this does not constitute a proof. Thus the only known perfect numbers are the thirty-two numbers obtained by Theorem 3.1.12 from the known Mersenne primes.

In order to prove Theorem 3.1.15, we introduce the function $\sigma(n)$, defined to be the sum of all the divisors of $n$, including $n$ itself. For example,

$$\sigma(12) = 1 + 2 + 3 + 4 + 6 + 12 = 28.$$

Note that $n$ is perfect if and only if $\sigma(n) = 2n$, since $n$ is included along with the proper divisors.

The proof of the following lemma is relegated to Exercise 31. These nice formulas are the reason for using the sum of all divisors rather than the sum of all proper divisors.

LEMMA 3.1.16.

   *i. If $p$ is prime, then*

$$\sigma(p^a) = \frac{p^{a+1} - 1}{p - 1};$$

  *ii. If $p$ and $q$ are distinct primes, then*

$$\sigma(p^a q^b) = \frac{p^{a+1} - 1}{p - 1} \cdot \frac{q^{b+1} - 1}{q - 1}.$$

For example,

$$\sigma(12) = \sigma(2^2 3^1) = \frac{2^3 - 1}{2 - 1} \cdot \frac{3^2 - 1}{3 - 1} = 7 \cdot 4 = 28,$$

agreeing with the direct computation made prior to the lemma. An analogous result holds for products of powers of more than two primes. This also implies that if $\alpha$ and $\beta$ are numbers that contain no common prime factors, then $\sigma(\alpha\beta) = \sigma(\alpha)\sigma(\beta)$; in other words, $\sigma$ applied to a product is the product of the $\sigma$'s, provided the factors are relatively prime.

*Proof of Theorem 3.1.15.* Suppose $n$ is an even perfect number. Thus $n = 2^a m$ with $a > 0$ and $m$ odd, and so $\sigma(n) = (2^{a+1} - 1)\sigma(m)$. On the other hand, since $n$ is perfect, $\sigma(n) = 2n = 2^{a+1}m$. Equating these two expressions for $\sigma(n)$ yields

(3.1.17)                 $(2^{a+1} - 1)\sigma(m) = 2^{a+1}m.$

This implies that $2^{a+1} - 1$ must be a divisor of $m$, since it has no common factors with $2^{a+1}$. Thus $m = (2^{a+1} - 1)\ell$, for some integer $\ell$. Substituting this into (3.1.17) and canceling $2^{a+1} - 1$ yields $\sigma(m) = 2^{a+1}\ell$. Hence

$$\sigma(m) = 2^{a+1}\ell = \ell + (2^{a+1} - 1)\ell = \ell + m.$$

Since $\ell$ and $m$ are both divisors of $m$, and their sum equals $\sigma(m)$, they must be the only divisors of $m$. Since 1 is also a divisor of $m$, we must have $\ell = 1$, and therefore $m = 2^{a+1} - 1$ and is prime, as desired. Q.E.D.

### ASIDE: FERMAT'S LAST THEOREM

Pierre de Fermat (1601–1665) was one of the best and most famous mathematicians of his century. His profession was law; he was king's councillor in the local parliament of Toulouse in France. Mathemat-

Pierre Fermat

*Note added in proof*: Fermat's Last Theorem has apparently been proved by Andrew Wiles in June 1993. The proof (200 pages long) is too long to fit in this margin.

ics was an avocation to him. He did not publish papers, and only rarely did he furnish proofs of his statements. His work was known through private correspondence. He made important contributions to the development of calculus, probability, and analytic geometry. But his best work was in number theory. Section 3.4 will be devoted to a theorem he stated without proof. The proof of this "Little Fermat Theorem" was given by Leibniz in 1683.

Fermat's greatest fame resulted from one of his statements that no one has been able to prove or disprove. It was discovered by his son Samuel shortly after Fermat's death. Samuel was going through his father's papers, correspondence, and books in order to publish some sort of Collected Works. There were many notes that Fermat had written in his copy of the Latin translation of the book *Arithmetic* by the ancient Greek, Diophantus. One of the problems in the book was: "Given a number which is a square, write it as a sum of two other squares." In our Theorem 1.2.4, we showed how to generate all solutions of this problem. It is equivalent to the question of what right triangles there are with sides of integer length. In the margin next to this problem, Fermat had written, in Latin: "On the other hand, it is impossible for a cube to be written as a sum of two cubes, or a fourth power to be written as a sum of two fourth powers, or, in general, for any number which is a power greater than the second to be written as a sum of two like powers. I have a truly marvelous demonstration of this proposition which this margin is too narrow to contain."

This mathematical statement has become known as Fermat's Last Theorem, even though it should be called a conjecture, since it has never been proved.

CONJECTURE 3.1.18. *(Fermat's Conjecture or "Last Theorem"). If $n > 2$, then the equation,*

$$(3.1.19) \qquad\qquad a^n + b^n = c^n,$$

*has no solutions in positive integers.*

It is a topic of some debate among historians of mathematics as to whether Fermat actually had a proof of this conjecture. All other statements that he claimed to be able to prove have been proved, so that gives him a good track record. On the other hand, many of the greatest mathematicians during the subsequent 325 years have tried to prove it, without success. Some have thought they could prove the conjecture, and we shall discuss some of these attempts below. Much important mathematics has been developed because of those attempts. But the problem remains unsolved. Virtually every mathematician believes that Fermat's conjecture is true. If it is false, the numbers $a^n$, $b^n$, and $c^n$ will be unimaginably large. The exponent $n$ will have to be at least 150,000. But nevertheless, a tremendous amount of fame awaits the person who is able to prove it, in large part because so many people have tried and failed.

The proof of Fermat's conjecture when the exponent $n$ is equal to 3 is generally attributed to Leonhard Euler, the greatest mathematician of the eigteenth century, whom we discussed in Section 1.3. We say "generally attributed" because Euler's proof was not complete by contemporary standards, and there was an assumption of unique factorization of a certain type of number which, although true in the case required by Euler, was not given the attention by him that it warranted.

Much later, in 1847, Gabriel Lamé announced to the Paris Academy that he could prove Fermat's Last Theorem for all exponents $n$. Immediately after Lamé sketched his argument, Joseph Liouville pointed out the fatal flaw in Lamé's reasoning. Lamé was assuming that certain sets of complex numbers admitted unique factorization into prime factors, similar to the unique factorization of the integers, but this was not a valid assumption in the case he was using. Some historians think that Fermat might have been making the same mistake. In Exercises 4 to 8, we let the interested reader learn about such examples.

The best progress toward a proof of Fermat's Conjecture was made by Ernst Kummer (1810–1893), a German. Although he did at one point submit a false proof of the entire conjecture, he made great strides by introducing the theory of ideals, which compensated somewhat for the lack of unique factorization in certain sets of complex numbers. Using this, he was able to prove Fermat's conjecture whenever the exponent $n$ is what he called a *regular* prime. This implied that the conjecture was true for all exponents $\leq 100$ except 37, 59, and 67, and by other methods he proved the conjecture for these exponents as well.

In 1983, Gerd Faltings, a German, made major progress toward a proof. His work implied that for any exponent $n$ there could be at most a finite number of essential counterexamples to Fermat's conjecture. By "essential counterexamples" we mean that $a$, $b$, and $c$ in (3.1.19) cannot share any common factors. Of course, if you find one set of numbers $a$, $b$, $c$, and $n$ that satisfy (3.1.19), then if you multiply $a$, $b$, and $c$ by the same number, you get another counterexample to Fermat's conjecture, and in this way you could obtain infinitely many counterexamples if you had one. But if this trivial operation is disallowed, Faltings' work shows that there can only be a finite number of solutions for a fixed value of $n$, and that is a major advance. For his work, Faltings was rewarded with a professorship at Princeton and a Fields Medal, which, as mentioned earlier, is the mathematical equivalent of the Nobel Prize.

In 1988 a well-known Japanese mathematician, Yoichi Miyaoka, gave a lecture in Bonn, Germany, in which he claimed to have proved Fermat's conjecture by expanding upon some of Faltings' ideas. Thanks to electronic mail, word traveled very fast to the mathematical community, and it was reported in the popular press. However, within two weeks a fatal flaw in Miyaoka's argument was discovered. In its April 18, 1988, issue, *Time* magazine's Charles Krauthammer wrote a beautiful editorial about mathematics, centered around Miyaoka's attempt. It began: "For one brief shining moment, it appeared as if the 20th century had justified itself. The era of world wars, atom bombs, toxic waste, AIDS, Muzak, and now, just to rub it in, a pending Bush-Dukakis race, had redeemed itself, it seemed. It had brought forth a miracle. Fermat's last theorem had been solved."

The editorial went on to describe the problem and Miyaoka's failed attempt. Then it went on to discuss why such a question, devoid of practical application, should be considered so important. Krauthammer wrote, "That it should have to justify itself by its applications, as a tool for making the mundane or improving the ephemeral, is an affront not just to mathematics but to the creature that invented it. What higher calling can there be than searching for useless and beautiful truths? Number theory is as beautiful and no more useless than mastery of the balance beam or the well-thrown forward pass. And our culture expends enormous sums on those exercises without asking what higher ends they serve."

A prize of 10,000 German marks awaits the person who proves Fermat's Conjecture, provided it is done before September 13, 2007. This prize is called the Wolfskehl Prize and is administered by the University of Göttingen, which receives an average of one proposed solution per week. This prize was left in Wolfskehl's will in 1908 in part because Fermat's Last Theorem saved Wolfskehl's life.[4] A despondent mathematician, he was about to commit suicide when he noticed Kummer's work on Fermat's Conjecture. He became so interested in reading it that he decided not to commit suicide, after all.

---

4. Actually he left 100,000 marks, but after inflation wiped out the value of that award, it was set at its current value.

## Exercises

1. For each number from 300 to 329, write it as a product of prime factors, or else say that it is prime.
2. Find the greatest common divisor of each pair of numbers by factoring:
   a. 235 and 325
   b. 176 and 248
   c. 3000 and 585
   d. 432 and 847
3. Write all the positive divisors of 180, following the procedure used for 168 in the text.
4. For convenience, we will use the word *nimbers* in Exercises 4 to 8 to refer to numbers of the form $a + b\sqrt{-5}$, where $a$ and $b$ are integers. These include the cases where $a$ or $b$ are 0. Nimbers are added by adding coefficients, and multiplied by multiplying term-by-term, using $\sqrt{-5} \cdot \sqrt{-5} = -5$. For example,

$$(3 + 2\sqrt{-5}) + (4 - 5\sqrt{-5}) = (3 + 4) + (2 - 5)\sqrt{-5} = 7 - 3\sqrt{-5}$$
$$(3 + 2\sqrt{-5}) \cdot (4 - 5\sqrt{-5}) = 12 - 15\sqrt{-5} + 8\sqrt{-5} - 10(-5)$$
$$= (12 + 50) + (-15 + 8)\sqrt{-5} = 62 - 7\sqrt{-5}$$

   Calculate the sum $\alpha + \beta$ and product $\alpha\beta$ if $\alpha = 4 - \sqrt{-5}$ and $\beta = 2 + 3\sqrt{-5}$.
5. We define a norm $N(-)$ on nimbers by

$$N(a + b\sqrt{-5}) = a^2 + 5b^2.$$

   For example, $N(3 + 2\sqrt{-5}) = 3^2 + 5 \cdot 2^2 = 29$. The norm is a measure of the size of a nimber. List (with their norm) all nimbers whose norm is less than 10. Don't forget to include nimbers with negative numbers or 0 as coefficients.
6. It is true that if nimbers satisfy $y = \alpha \cdot \beta$, then $N(y) = N(\alpha)N(\beta)$. In other words, the norm of the product is the product of the norms.
   a. Verify this for the multiplication that you performed in #4.
   b.* Prove the multiplicative property of the norm stated in this exercise. [Suggestion: Multiply out $(a+b\sqrt{-5})(c+d\sqrt{-5})$, compute the norm of this product, and compare this answer with $N(a + b\sqrt{-5}) \cdot N(c + d\sqrt{-5})$.]
7. A *prime nimber* is one that cannot be written as a product of other nimbers $\neq \pm1$. The norm can be used to help show that certain

nimbers are prime. For example, to see that $1 + 2\sqrt{-5}$ is a prime nimber, we note that $N(1+2\sqrt{-5}) = 21$, and so if $1+2\sqrt{-5}$ could be factored, the factors would have to have norms 3 and 7, but there are no nimbers with norm 3 or 7.

  a. Show that each of the nimbers you listed in #5 is prime (except for 0 and $\pm 1$).

  b. Show that 29 is not a prime nimber. (Hint: Since its norm is $29 \cdot 29$, you should try to write it as a product of two nimbers of norm 29.)

8. Show that the set of nimbers does not have unique factorization into primes by considering the following examples:

  a. Show that the nimber 6 can be factored in two different ways as a product of prime nimbers. [Hint: $N(6)$ can be written as $4 \cdot 9$ or $6 \cdot 6$. Find nimbers of norm 4 and 9 whose product is 6, then find nimbers of norm 6 and 6 whose product is 6. Note that factoring as $(-2)(-3)$ is not considered to be essentially different than factoring as $2 \cdot 3$.]

  b. Find at least two of the three distinct factorizations of the nimber 21 as a product of prime nimbers.

9.* Show that if one knew that Fermat's Last Theorem were true whenever the exponent $n$ was an odd prime number or 4, then it is true for all exponents $n$. For example, the truth of Fermat's Last Theorem for $n = 3$ implies its truth for $n = 6$ since if $a^6 + b^6 = c^6$, then $A^3 + B^3 = C^3$, where $A = a^2$, $B = b^2$, and $C = c^2$. What you are being asked to do is to generalize this argument to any composite exponent $n$.

10. Perform the Sieve of Eratosthenes on the numbers 2 to 200. What is the largest number whose multiples you have to count? How many primes are there which are $\leq 200$?

11. One can prove that a number is divisible by 11 if and only if the number obtained by starting from the right-most digit and alternately subtracting and adding digits is divisible by 11. For example, 22453 is not divisible by 11 since $3 - 5 + 4 - 2 + 2 = 2$ is not divisible by 11.

  a.* Prove that this works for a 4-digit number $1000a + 100b + 10c + d$.

  b. Use this method to test the following numbers for divisibility by 11: 71544, 23456, and 38291.

12. The algorithm for generating primes that was discussed after the proof of the infinitude of the primes had generated 2, 3, 7, and 43 after three steps. What primes does it generate in the fourth step? (Hint: I said *primes*.)

13.* Mimic the proof of Proposition 3.1.5 to show that there are infinitely many primes of the form $6k + 5$.

14. Explain why Euclid's proof of the infinitude of primes would not have worked if he had used $N = (p_1 p_2 \cdots p_k) + 2$ instead of $(p_1 p_2 \cdots p_k) + 1$. Then show that one can prove that there are infinitely many odd primes by letting $N$ equal 2 more than the product of a hypothetical complete list of all odd primes.

15.* A short formula that yields the primes in order was given by James Jones in "Formula for the $n$th prime number," *Canadian Mathematics Bulletin* 18 (1975): 433–34. It requires use of a modified minus sign, written $\overset{\circ}{-}$, which is defined by $a \overset{\circ}{-} b = a - b$ unless $a - b$ is negative, in which case $a \overset{\circ}{-} b = 0$. It also requires the use of the function $r(x, y)$ which is defined to be the remainder when $x$ is divided by $y$, and to be $x$ if $y = 0$. Then the $n$th prime is given by

$$\sum_{i=0}^{n^2} \left( 1 \overset{\circ}{-} \left( \left( \sum_{j=0}^{i} r((j \overset{\circ}{-} 1)!^2, j) \right) \overset{\circ}{-} n \right) \right).$$

The central part of the formula says to compute the factorial of $(j \overset{\circ}{-} 1)$, square it, and then take the remainder when that is divided by $j$. A convention here is that $0! = 1$. In this complicated expression, you work from the inside out. Thus, for each value of $i$ from 0 to $n^2$, you calculate the sum of the $r$-numbers just mentioned for all values of $j$ from 0 to $i$, and perform two modified subtractions. Then you sum the resulting numbers for all values of $i$.

a. Substitute $n = 2$ into this formula to find the second prime.

b. The fact that makes this formula work is Wilson's Theorem, which says that a number $k \geq 2$ is prime if and only if $(k - 1)! + 1$ is divisible by $k$. Verify this for $k = 2, 3, 4, 5, 6$, and 7.

16. By just counting primes, calculate $\pi(n)$ for each value of $n$ between 25 and 45.

17. Legendre gave a refinement of the Prime Number Theorem, estimating $\pi(n)$ as

$$\frac{n}{\log n - 1.08366.}$$

Calculate Legendre's approximation to $\pi(n)$ for the values of $n$ listed in Table 3.3. Use a calculator, and $\log(10^k) \approx 2.3026k$.

Observe how much closer this approximation is to the actual $\pi(n)$ than is Gauss's formula.

18. Find the largest pair of twin primes that you can.

19. Prove that $\{3, 5, 7\}$ is the only prime triplet. By *prime triplet* we mean three consecutive odd numbers, all of which are prime.

20. Goldbach's Conjecture is another famous unproved conjecture about prime numbers. It states that any even number $> 2$ can be written as the sum of two prime numbers. Verify Goldbach's conjecture for all even numbers less than 100.

21. For each value of $n$ from 2 through 7, decide whether $n! + 1$ is prime or composite.

22. Show that starting with $23! + 2$ is a string of at least twenty-seven consecutive composite numbers.

23. Use the proof of Proposition 3.1.8 to obtain two different explicit factorizations of $2^{21} - 1$ as a product of two integers greater than 1, but not necessarily prime.

24. Prove that if $a > 2$ and $n$ is any positive integer, then $a^n - 1$ is composite.

25. Prove that $2^{13} - 1$ is prime, by mimicking the proof of Proposition 3.1.11.

26. Show that $2^{23} - 1$ is not prime by giving an explicit divisor. Use Proposition 3.1.10 and a calculator.

27. Calculate the perfect numbers corresponding to the first five Mersenne primes.

28. Write out all the proper divisors of the fourth perfect number in an orderly fashion, using the proof of 3.1.12, and verify without significant calculation that this number is perfect.

29. Classify each of the numbers from 2 to 30 as deficient (sum of proper divisors is less than the number), perfect, or abundant (sum of proper divisors exceeds the number). For each, list all proper divisors and their sum.

30.* Prove that if $2^n - 1$ is not prime, then $2^{n-1}(2^n - 1)$ is abundant. Use the proof of 3.1.12 as an aid.

31. Prove Lemma 3.1.16. (Hint: Write the divisors is an orderly fashion, and use (3.1.9) to evaluate sums such as $1 + p + p^2 + \cdots + p^a$.)

32. Use Lemma 3.1.16 to evaluate $\sigma(n)$ for all values of $n$ from 25 to 45. Use your calculations to classify these numbers as deficient, perfect, or abundant.

## 3.2   The Euclidean Algorithm

The Euclidean algorithm is a method of finding the greatest common divisor (GCD) of two numbers without factoring them. For large numbers, it is certainly a more efficient way of finding the GCD than is the factoring method. (See, for example, Exercise 1e, where the GCD would be very difficult to find by factoring, but is easy using the Euclidean algorithm.) The Euclidean algorithm also provides a method of finding integer solutions of certain kinds of equations, which will be useful to us in the following sections. Most of the methods of this section were known to the Greeks and appeared in Euclid's *Elements*.

### *FINDING THE* GCD

We denote by $(a, b)$ the GCD of integers $a$ and $b$. For example, we write $(12, 10) = 2$, since 2 is the GCD of $12(= 2^2 \cdot 3)$ and $10(= 2 \cdot 5)$. The key to the Euclidean algorithm is the fact that if $r$ is the remainder obtained when $a$ is divided by $b$, then $(a, b) = (b, r)$. This says that the GCD of the pair of the smaller numbers $b$ and $r$ is the same as the GCD of the larger numbers $a$ and $b$. We formalize this as a lemma.

LEMMA 3.2.1. *If $a = qb + r$, then $(a, b) = (b, r)$.*

Ordinarily, $q$ will be the quotient and $r$ the remainder when $a$ is divided by $b$. Writing a division fact as an equation as is done in Lemma 3.2.1 will be very useful in this section.

*Proof of 3.2.1.* Since $r = a - qb$, every common divisor of $a$ and $b$ will also be a divisor of $r$. Thus every common divisor of $a$ and $b$ is a common divisor of $b$ and $r$. On the other hand, since $a = qb + r$, every common divisor of $b$ and $r$ is a divisor of $a$, and hence is a common divisor of $a$ and $b$. Thus the set of common divisors of $a$ and $b$ equals the set of common divisors of $b$ and $r$. Therefore, the greatest element in each of these two sets must be equal. Q.E.D.

For example, let $a = 72$, $b = 30$, and $r = 12$, since $72 = 2 \cdot 30 + 12$. The set of common divisors of 72 and 30 is $\{1, 2, 3, 6\}$, as is the set of common divisors of 30 and 12. Thus 6 is the GCD of both the pair 72 and 30, and the pair 30 and 12.

The Euclidean algorithm says you should iterate this procedure until you can't do it anymore. In other words, to find the GCD of $a$

and $b$ with $a > b$, divide $b$ into $a$, and denote the remainder by $r_1$. Then divide $r_1$ into $b$ and denote the remainder by $r_2$. Note that

$$(a, b) = (b, r_1) = (r_1, r_2).$$

Continue this process until a remainder of zero is obtained. The last nonzero remainder will be the GCD of $a$ and $b$.

We illustrate by finding the GCD of 252 and 198. We perform the following sequence of divisions, denoting the remainder as a number written after the symbol rem:

$$252/198 = 1 \quad \text{rem} \quad 54$$
$$198/54 = 3 \quad \text{rem} \quad 36$$
$$54/36 = 1 \quad \text{rem} \quad 18$$
$$36/18 = 2 \quad \text{rem} \quad 0.$$

Since the last nonzero remainder is 18, the GCD of 252 and 198 is 18. We prefer to write these division facts as equations:

$$252 = 1 \cdot 198 + 54$$
(3.2.2)
$$198 = 3 \cdot 54 + 36$$
$$54 = 1 \cdot 36 + 18$$
$$36 = 2 \cdot 18 + 0.$$

Now Lemma 3.2.1 implies the desired result:

$$(252, 198) = (198, 54) = (54, 36) = (36, 18) = (18, 0) = 18.$$

Note how the numbers are getting smaller at each step, until at the end you use the fact that the GCD of a number and 0 is the number itself. Also note that the values of the quotients do not play a significant role in the algorithm.

We formalize this procedure in the following statement. The proof that it works is implicit in the discussion above, with Lemma 3.2.1 being the essence of the proof.

THEOREM 3.2.3. *(Euclidean algorithm). If*

$$a = q_1 b + r_1$$
$$b = q_2 r_1 + r_2$$
$$r_1 = q_3 r_2 + r_3$$
$$\vdots$$
$$r_{k-2} = q_k r_{k-1} + r_k$$
$$r_{k-1} = q_{k+1} r_k + 0,$$

*then* $(a, b) = r_k$.

The usual convention about remainders is that the remainder is nonnegative and is less than the divisor. Such a number is unique, and the Euclidean algorithm will always work if the remainder is chosen according to this scheme. However, if the remainder is quite large relative to the divisor, you can sometimes save a step by choosing the quotient to be 1 larger than you would ordinarily have done, making the remainder a small negative number. Since the GCD of $-r$ and $b$ is the same as the GCD of $r$ and $b$, you can make this negative remainder positive when it is used as the divisor in the next step. We illustrate by finding $(55, 34)$ first by the method of positive remainders, and then by the method of sometimes using negative remainders. In this case, we can save three steps by choosing three negative remainders:

$$55 = 1 \cdot 34 + 21$$
$$34 = 1 \cdot 21 + 13$$
$$21 = 1 \cdot 13 + 8$$
$$13 = 1 \cdot 8 + 5$$
$$8 = 1 \cdot 5 + 3$$
$$5 = 1 \cdot 3 + 2$$
$$3 = 1 \cdot 2 + 1$$
$$2 = 1 \cdot 1 + 0.$$

Thus $(55, 34) = 1$, a fact that could probably be seen more easily by factoring. This is an extreme example of the straight Euclidean algorithm being slow, because all of the quotients are 1. If we allow negative remainders, we obtain

$$55 = 2 \cdot 34 - 13$$
$$34 = 3 \cdot 13 - 5$$
$$13 = 3 \cdot 5 - 2$$
$$5 = 2 \cdot 2 + 1$$
$$2 = 2 \cdot 1 + 0.$$

*WRITING THE GCD AS A COMBINATION OF THE NUMBERS*

The Euclidean algorithm also allows you to write the GCD of two numbers as an explicit combination of them, with integers as coefficients. For example, based upon the first example worked above, (3.2.2), we can find integers $m$ and $n$ so that

(3.2.4)                         $18 = m \cdot 252 + n \cdot 198.$

Of course, one of $m$ and $n$ will have to be negative. Since both 252 and 198 are multiples of 18, any number of the form $m \cdot 252 + n \cdot 198$ will have to be a multiple of 18, and so, once you have seen that 18 can be written in this way, you will know that it is the smallest positive number that can be so written. Finding these numbers $m$ and $n$ will be important to us in solving congruence equations in Sections 3.3 and 4.2.

This procedure requires some practice and patience. Write the division steps in the Euclidean algorithm as equations expressing the remainder as the larger number minus the quotient times the divisor. It is important that you write the quotient *in front of* the divisor. Write these equations for all the division steps except the last one— the one with the zero remainder. Thus the last equation you write will have the GCD on the left-hand side. The example below is obtained from (3.2.2) in this way:

$$54 = 252 - 1 \cdot 198$$
(3.2.5)                         $36 = 198 - 3 \cdot 54$
$$18 = 54 - 1 \cdot 36.$$

One starts with the last equation, and writes the GCD as a combination of successively larger numbers, ending with the two given numbers. In this case, one starts by using the last equation to write 18 as a combination of 54 and 36, then incorporates the next-to-last equation to write it as a combination of 198 and 54, and finally uses the first equation to write it as a combination of 252 and 198, as desired. Explicitly,

$$18 = 54 - 36$$
$$= 54 - (198 - 3 \cdot 54)$$
$$= 4 \cdot 54 - 198$$
$$= 4(252 - 198) - 198$$
$$= 4 \cdot 252 - 5 \cdot 198.$$

Thus in (3.2.4), $m = 4$ and $n = -5$.

Note how coefficients are collected rather than multiplied out. For example, the 4 in the third equation comes about as $1 - (-3)$, the coefficients of 54 in the second equation. Remember not to multiply out any number on the right side of a dot; those numbers are a sort of place-keeper. Frequently you will have to take into account a co-efficient 1 which does not appear explicitly. You are warned against trying to combine steps. For example, it is tempting to use both the first and second equations of (3.2.5) to replace both the 54 and the 36 in the first step, but this won't save any steps (try it!), and will probably cause confusion.

There is more than one possible answer for $m$ and $n$. If your arithmetic was correct, but you didn't follow the algorithm exactly as we have spelled it out, you could conceivably come up with differ-ent answers for $m$ and $n$, and still be correct. This could happen, for example, if you had used negative remainders, or if you tried to combine steps. It is easy to check your answer on these problems; just multiply and add, and see if it checks.

Equations requiring integer solutions are called *Diophantine equa-tions*, after the Greek Diophantus (c. A.D. 250), whose book *Arithmetic* was very influential for many centuries. We have already seen that Fermat wrote his famous marginalia in his copy of this book. The example worked above shows that the Diophantine equation

(3.2.6)                        $18 = 252x + 198y$

has the solution $x = 4$, $y = -5$. The $x$ and $y$ here are the same as $m$ and $n$ above, just renamed with the traditional letters for unknowns. If asked to solve the Diophantine equation,

$$72 = 252x + 198y,$$

it is usually best to note that since $72 = 4 \cdot 18$, a solution to this equation is obtained by multiplying a solution of (3.2.6) by 4, obtaining $x = 16$, $y = -20$. On the other hand, the Diophantine equation

$$6 = 252x + 198y$$

has no solutions because 6 is not a multiple of the GCD $(252, 198) = 18$. Of course, this equation has many solutions that are not integers, but Diophantine means "integer solutions." The following proposition formalizes what we have been saying.

PROPOSITION 3.2.7. *The Diophantine equation*

$$c = ax + by$$

*can be solved if and only if c is a multiple of the GCD $(a, b)$. If $c = k \cdot (a, b)$, a solution can be obtained as $x = kx'$, $y = ky'$, where $x'$ and $y'$ are solutions of the equation*

$$(a, b) = ax' + by',$$

*obtained by starting with the last equation in the Euclidean algorithm used to find $(a, b)$, and one-at-a-time, working backward, replacing the remainder term in an equation of the Euclidean algorithm by the combination of the larger numbers it equals.*

We close this subsection with a final example. Find the GCD of 105 and 88 by the Euclidean algorithm, and find a solution of the Diophantine equation

(3.2.8) $$4 = 105x + 88y.$$

We begin by finding the GCD $(105, 88)$. We save a step of writing by immediately writing the division facts in the form that will be useful for solving the Diophantine equation. For example, the first equation below is obtained by noting that $^{105}/_{88} = 1$ rem 17:

$$17 = 105 - 1 \cdot 88$$
$$3 = 88 - 5 \cdot 17$$
$$2 = 17 - 5 \cdot 3$$
$$1 = 3 - 1 \cdot 2.$$

Since the next step, 2 divided by 1, will have remainder 0, we need not write it. This shows that $(105, 88) = 1$, or, in other words, 105 and 88 are relatively prime. We will solve (3.2.8) by solving the equation $1 = 105x' + 88y'$, and then multiplying a solution by 4. We could actually save a little bit of work by noting that if we started with the third of the four equations in the Euclidean algorithm, we could write 2 as a combination of 105 and 88, and then multiply that solution by 2, thus completely avoiding use of the fourth equation. Since this circumstance is somewhat unusual, we will not utilize it here, although you are welcome to do so. We save a step of writing by immediately combining the last two equations:

$$
\begin{aligned}
1 &= 3 - (17 - 5 \cdot 3) \\
&= 6 \cdot 3 - 17 \\
&= 6(88 - 5 \cdot 17) - 17 \\
&= 6 \cdot 88 - 31 \cdot 17 \\
&= 6 \cdot 88 - 31(105 - 88) \\
&= 37 \cdot 88 - 31 \cdot 105.
\end{aligned}
$$

Thus, multiplying the coefficients by 4 yields $x = 4 \cdot (-31) = -124$ and $y = 4 \cdot 37 = 148$ as a solution of (3.2.8).

LINEAR DIOPHANTINE EQUATIONS

In this optional subsection, we expand a bit upon the method of solving linear Diophantine equations presented in the previous section. A *linear Diophantine equation* is an equation of the form $ax + by = c$, where $a$, $b$, and $c$ are integers, and we require that $x$ and $y$ be integers.

In Proposition 3.2.7, we gave the necessary and sufficient condition for such an equation to have a solution. We explain now how to find all solutions, for it turns out that if there is one solution, then there will be infinitely many. We illustrate with the equation

(3.2.9)                                        $70x + 54y = 4.$

We first find the GCD:

$$
\begin{aligned}
16 &= 70 - 1 \cdot 54 \\
6 &= 54 - 3 \cdot 16 \\
-2 &= 16 - 3 \cdot 6.
\end{aligned}
$$

Since $^6/_2$ is an integer, we are done, and $(70, 54) = 2$. Note that at the last step, we used a negative remainder in order to save a step.

Next, we work our way backwards through these equations in order to write $-2$ as a combination of 70 and 54:

$$\begin{aligned} -2 &= 16 - 3(54 - 3 \cdot 16) \\ &= 10 \cdot 16 - 3 \cdot 54 \\ &= 10(70 - 54) - 3 \cdot 54 \\ &= 10 \cdot 70 - 13 \cdot 54. \end{aligned}$$

Multiplying by $-2 (= {}^4/_{-2})$, we obtain

$$(3.2.10) \qquad\qquad 4 = -20 \cdot 70 + 26 \cdot 54,$$

so that $x = -20$, $y = 26$ is one solution of (3.2.9).

Suppose that $x'$ and $y'$ form a solution of the equation

$$(3.2.11) \qquad\qquad 70x' + 54y' = 0,$$

which is formed from (3.2.9) by replacing the 4 by a 0. Then $-20 + x'$ and $26 + y'$ will form a solution of (3.2.9) since

$$\begin{aligned} 70(-20 + x') + 54(26 + y') &= 70(-20) + 54(26) + 70x' + 54y' \\ &= 4 + 0 = 4. \end{aligned}$$

It is easy to see that any solution of (3.2.9) can be obtained in this way, as the sum of whatever particular solution of (3.2.9) you found originally plus a solution of (3.2.11).

The solutions of (3.2.11) are easily obtained. You should first simplify by dividing through by the GCD of the two coefficients, which in this case is 2, the GCD of 70 and 54. The equation becomes

$$35x' + 27y' = 0,$$

and it is easy to see that all solutions of this equation are given by $x' = -27n$, $y' = 35n$, where $n$ is any integer, positive or negative. Thus all solutions of (3.2.9) are obtained as

$$x = -20 - 27n, \qquad y = 26 + 35n,$$

where $n$ ranges over all integers, positive and negative. For example, if $n = -1$, we obtain the solution $x = 7$, $y = -9$, which involves the smallest values of $x$ and $y$.

This example explains our earlier remark about nonuniqueness of solutions of the types of equations being considered here. The statement that generalizes the above example is

PROPOSITION 3.2.12. *If $x = x_0$, $y = y_0$ is one solution of the Diophantine equation*

$$ax + by = c,$$

*found, for example, by the method of Proposition 3.2.7, then all solutions are given by*

$$x = x_0 - n\frac{b}{(a,b)}$$
$$y = y_0 + n\frac{a}{(a,b)},$$

*where $n$ can be any integer.*

For example, in the above analysis of (3.2.9), we have $x_0 = -20$, $y_0 = 26$, and $(a,b) = 2$.

If $c$ is large enough, we can find solutions of $ax + by = c$ in which $x$ and $y$ are both nonnegative, which would be required in many word problems. The method will be that just discussed, but we must choose $n$ so as to make both $x$ and $y \geq 0$. We offer the following
**Example.** Regular compact disks sell for \$12, while double disks sell for \$25. You spent \$331 altogether. How many disks of each type did you buy?

Let $x$ and $y$ denote the number of regular and double disks, respectively. Then we want integers $x \geq 0$ and $y \geq 0$ satisfying

(3.2.13)                    $12x + 25y = 331.$

It is easy to see that $1 = (25, 12)$ and that $1 = 25 - 2 \cdot 12$ is one way of writing 1 as a combination of 12 and 25. In general, you would obtain this by the Euclidean algorithm, which you could have used here, but in this case the numbers were easy enough to do it by inspection. Multiplying this by 331 yields $x_0 = -662$, $y_0 = 331$ as one solution of (3.2.13). Thus every solution of (3.2.13) is given by $x = -662 + 25n$, $y = 331 - 12n$.[5] We want to choose $n$ so that both of these are nonnegative. This forces $n$ to satisfy both $-662 + 25n \geq 0$ and $331 - 12n \geq 0$. The first says $n \geq {}^{662}/_{25} = 26.5$, while the second

---

5. We have chosen to reverse the $-n$ and $+n$ in applying Proposition 3.2.12 here. This is inconsequential, since $n$ can be chosen to be any integer, positive or negative. We did this because we preferred to have $n$ be positive.

says $n \le {}^{331}/_{12} = 27.6$. The only integer satisfying both of these is $n = 27$, and so $x = -662 + 25 \cdot 27 = 13$, $y = 331 - 12 \cdot 27 = 7$ is the solution.

*ASIDE: PROOF OF THE FUNDAMENTAL THEOREM OF ARITHMETIC*

In this "Aside," we prove the Fundamental Theorem of Arithmetic, the unique factorization of integers into primes, which we stated as Theorem 3.1.1. The proof, which appeared in Euclid's *Elements*, is intimately related with the Euclidean algorithm. A mark of the genius of the Greeks is their realizing that the uniqueness of factorization is a result that requires proof.

The key to the proof is the following innocuous-sounding lemma.

LEMMA 3.2.14. *If a prime $p$ is a divisor of $ab$, then $p$ is a divisor of either $a$ or $b$.*

Before proving the lemma, we note that the conclusion need not be true if $p$ is composite. For example, if $p = 6$, $a = 10$, and $b = 21$, then $p$ is a divisor of $ab = 210$, but $p$ is a divisor of neither $a$ nor $b$. Of course, this does not contradict the lemma, because 6 is not prime. We also note that this lemma is false in the set of nimbers ($= a + b\sqrt{-5}$) discussed in Exercises 4 to 8 of Section 3.1. For example, $p = 3$ is a prime nimber, and is a divisor of $ab$ if $a = 1 + 2\sqrt{-5}$ and $b = 1 - 2\sqrt{-5}$, but 3 is not a divisor of $a$ or $b$. A main reason for the failure of the analogue of Lemma 3.2.14 to be true for nimbers is that there is no analogue of the Euclidean algorithm for nimbers.

*Proof of 3.2.14.* If $p$ is a divisor of $a$, then we are done. So, we assume $p$ is not a divisor of $a$, and will deduce that it must then be a divisor of $b$, which will complete the argument. Since the only divisors of $p$ are 1 and $p$, and $p$ is not a divisor of $a$, the GCD of $p$ and $a$ must be 1. Thus by the Euclidean algorithm, we can find integers $m$ and $n$ so that

$$mp + na = 1.$$

Multiplying this by $b$ yields

$$(3.2.15) \qquad mpb + nab = b.$$

By the hypothesis of the lemma, $ab$ is a multiple of $p$, and hence (3.2.15) expresses $b$ as a sum of two multiples of $p$. Thus $b$ is a multiple of $p$, as desired. Q.E.D.

Now we can easily deduce the following corollary, which says for a product of an arbitrary number of factors what was said about a product of two factors in 3.2.14.

COROLLARY 3.2.16. *If a prime $p$ is a divisor of $a_1 a_2 \cdots a_r$, then $p$ is a divisor of at least one of the $a_i$'s*

*Proof.* We first apply 3.2.14 with $a = a_1$ and $b = a_2 \cdots a_r$ to deduce that either $p$ is a divisor of $a_1$ or else $p$ is a divisor of $a_2 \cdots a_r$. In the first case, we are done, while in the second case we apply 3.2.14 again, this time with $a = a_2$ and $b = a_3 \cdots a_r$. This time we deduce that $p$ is a divisor either of $a_2$ or of $a_3 \cdots a_r$. Continuing this process, we will eventually get to some $a_i$ of which $p$ is a divisor. Q.E.D.

A more elegant way of formulating this proof is by mathematical induction. Corollary 3.2.16 is certainly true when $r = 1$. Assume that 3.2.16 is true when $r$ has a certain value $k$. Then deduce that it is true when $r = k + 1$ by applying 3.2.14 with $a = a_1$ and $b = a_2 \cdots a_{k+1}$ to deduce $p$ is a divisor of either $a_1$ or $a_2 \cdots a_{k+1}$. In the latter case, there are just $k$ factors, and so $p$ is a divisor of one of them by the induction hypothesis. By the Principle of Mathematical Induction, we are done.

We elaborate on why this constitutes a proof. We have noted that the corollary is true when $r = 1$. Thus we may assume that the result is true when $r = 1$, and by the induction step, we may deduce that it is true when $r = 2$. Now we may assume that the result is true when $r = 2$, and by the induction step we may deduce that it is true when $r = 3$. This continues forever, showing that the statement is true for all integers $r$. (See Exercise 10 for more about the Principle of Mathematical Induction.)

Now we can deduce the Fundamental Theorem of Arithmetic. Suppose a number is factored into primes in two different ways. Thus we have

(3.2.17) $$p_1 p_2 \cdots p_k = q_1 q_2 \cdots q_r,$$

where all $p_i$'s and $q_j$'s are primes. Some of the primes may be repeated on either side of the equation. Cancel out all common factors, so that we may assume that we never have $p_i = q_j$. Now $p_1$ is certainly a divisor of the left-hand side of (3.2.17), and so by 3.2.16 it must be a divisor of some $q_j$. Since $q_j$ is prime, we must have $p_1 = q_j$, contradicting the assumption that we had canceled all common factors. Q.E.D.

We close by giving one application of the Fundamental Theorem of Arithmetic. It is a generalization of the argument given in Theorem 1.1.5 that $\sqrt{2}$ is irrational.

PROPOSITION 3.2.18. *If $a$ is an integer, then $\sqrt[n]{a}$ is either irrational or an integer.*

We remind the reader that $\sqrt[n]{a}$, also written $a^{1/n}$ and called the $n$th root of $a$, is the number whose $n$th power equals $a$. This says, for example, that unless $a$ is a perfect cube, then $\sqrt[3]{a}$ is irrational.

*Proof of 3.2.18.* If $\sqrt[n]{a} = u/v$, with $u$ and $v$ relatively prime, then

$$(3.2.19) \qquad u^n = v^n a.$$

By Corollary 3.2.16, every prime factor of $a$ must be a factor of $u$, and hence must appear $n$ times on the left hand side of (3.2.19). Since $u$ and $v$ are relatively prime, this prime is not a factor of $v$, hence it appears $n$ times as a factor of $a$. Doing this for every prime factor of $a$ implies that $a$ is itself an $n$th power, and so $\sqrt[n]{a}$ is an integer. Q.E.D.

### Exercises

1. Find the GCD of each of the following pairs of numbers, using the Euclidean algorithm. In parts a, b, and c, also write the GCD as a linear combination of the two numbers.
   a. 252 and 147
   b. 176 and 105
   c. 600 and 398
   d. 6497 and 5767
   e. 1,456,813 and 1,468,823
2. Find a solution of each of the following equations or explain why no solution exists:
   a. $192x + 147y = 3$
   b. $250x + 162y = 6$
   c. $825x + 588y = 2$
3. A stockbroker sells stocks worth \$79 and \$41. The total amount of his sales is \$6358. How many stocks of each type did he sell? Use the methods of this section to solve this problem.
4. About all that is known about Diophantus is the following, from the *Greek Anthology*. "Diophantus's boyhood lasted one-sixth of his life; his beard grew after one twelfth more; he married after one seventh more, and his son was born five years later; the son

lived half as long as his father, and the father died four years after his son." How long did Diophantus live?

5. Find all solutions of the Diophantine equation

$$70x + 162y = 6.$$

Find the one for which $y$ is closest to zero.

6.* Find the GCD of 70, 98, and 105, using the Euclidean algorithm twice. Then write this GCD as a linear combination of the three numbers, using your two applications of the Euclidean algorithm.

7. Show that $\sqrt{2} + \sqrt{3}$ is irrational. (Hint: If it is rational, then so is its square.)

8. Show that $\log_{10} 2$ is irrational. (Hint: Show that if it equals $u/v$, then $10^u = 2^v$.)

9. Solve the nonlinear Diophantine equation $x^2 + 2y^2 = 102$. (Use trial and error. Including negatives, there are eight solutions.) This illustrates that, unlike linear Diophantine equations, a nonlinear Diophantine equation with one solution need not have infinitely many.

10.* The Principle of Mathematical Induction says that if a statement that involves an integer $r$ is true when $r = 1$, and if the validity of the statement for $r = k$ implies its validity for $r = k + 1$, then the statement is true for all values of $r$. (Truth for $r = 1$ implies truth for $r = 2$ implies truth for $r = 3$, etc.) This principle was used in the alternate proof of the Fundamental Theorem of Arithmetic at the end of this section. Use the Principle of Mathematical Induction to prove

$$1^2 + 2^2 + \cdots + (r - 1)^2 + r^2 = r(r + 1)(2r + 1)/6.$$

## 3.3   Congruence Arithmetic

Congruence is a relationship between integers that generalizes the notion of parity. We say that two numbers have the same parity if both are even or both are odd. For example, 14 and 26 have the same parity, while 7 and 14 do not. One way of determining whether two numbers have the same parity is to subtract one from the other; if the answer is even, they do, and if it is odd, they don't. Here we are using the fact that the difference of two odd numbers is even. We can formalize this a bit more by saying that $a$ and $b$ have the same parity if and only if $a - b$ is a multiple of 2.

Another example that is generalized by the idea of congruence is the notion of weekday. Let us say that two numbers $a$ and $b$ have the same weekday if the $a$th and $b$th days of the year fall on the same day of the week. Thus 18 and 25 and 32 all have the same weekday, since there are 7 days in a week. You can tell whether $a$ and $b$ have the same weekday by checking whether $a - b$ is a multiple of 7.

A third example that is encompassed by congruence is what we might call "clock time." We start measuring time at a certain moment and say that the time of an event is the number of hours elapsed until that event, rounded off to the nearest hour. We say that two events have the same *clock time* if their times differ by a multiple of 12. For example, if one event occurred after 14 hours and the other after 26 hours, a clock that began running when we started measuring time would read 2:00 at both of these events. We really don't have to round off time to the nearest hour. We could deal in hours, minutes, and seconds, but won't, since the notion of congruence, which generalizes this, deals with integers.

We incorporate these examples into the same formalism by saying in the first case,

$$a \equiv b \bmod 2 \text{ if } a - b \text{ is a multiple of 2;}$$

in the second case,

$$a \equiv b \bmod 7 \text{ if } a - b \text{ is a multiple of 7;}$$

in the third case,

$$a \equiv b \bmod 12 \text{ if } a - b \text{ is a multiple of 12;}$$

and, in general,

$$a \equiv b \bmod m \text{ if } a - b \text{ is a multiple of } m.$$

This is read "$a$ is congruent to $b$ mod (or modulo) $m$." Note that this kind of congruence has very little to do with the congruence of geometric figures—both are a kind of equality, but the geometric one is a much stronger form of equality. The number $m$ above is called the *modulus.*

For example, $7 \equiv 13 \bmod 6$, since $13 - 7$ is a multiple of 6, and $25 \equiv 7 \bmod 6$, but $7 \not\equiv 20 \bmod 6$, since $20 - 7$ is not a multiple of 6. It doesn't matter in which order you do the subtraction, since $-d$ is a multiple of $m$ if and only if $d$ is. A number is congruent mod $m$ to all numbers whose distance from it on the number line is a multiple of $m$. This includes negative distances and negative numbers. For example, the following numbers are all $\equiv$ 19 mod 11: 19, 30, 41, 52, . . . ; also, 8, $-3$, $-14$,. . . . (See Fig. 3.3.)

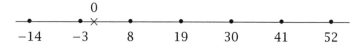

Figure 3.3. Congruence mod 11 on the number line.

These numbers are all congruent (mod 11) to one another, too. The "mod 11" was written in parentheses here because it could have been omitted, since it was clear from the context that the modulus was 11. We will sometimes omit writing "mod $m$" when the value of $m$ is clear from the context. The number line makes it clear that two numbers that are congruent to the same thing are congruent to each other. This can also be proved easily from the definition.

When a number $a$ is divided by $m$, the remainder $r$ is congruent to $a$ mod $m$. This is true since the equation $a = qm + r$ makes it clear that $a - r$ is a multiple of $m$. Since the remainder $r$ always satisfies $0 \le r \le m - 1$, it can be characterized as the only number congruent to $a$ mod $m$ which lies in the range 0 to $m - 1$. This number is also called the *least nonnegative residue* of $a$ mod $m$.

**Example.** Find the least nonnegative residue of 1000 mod 7. Find the least nonnegative residue of $-1000$ mod 7.

To do the first part, we just note that when 1000 is divided by 7, the quotient is 142, and the remainder is 6. Thus $1000 \equiv 6$ mod 7, and 6 is the answer. From this, we easily deduce that $-1000 \equiv -6$ mod 7. One way to see this is to note that the distance on the number line between $-1000$ and $-6$ is the same as that between 1000 and 6, which was just shown to be a multiple of 7. But $-6$ isn't the answer to the second question, because $-6$ isn't nonnegative. Instead, we want to find a number in the range from 0 to 6 that is congruent to $-6$ mod 7. The easiest way to do this is to notice that adding 7 does not change the congruence class, and will bring the $-6$ into the desired range. Thus

$$-1000 \equiv -6 \equiv 1 \text{ mod } 7,$$

and 1 is the answer to the second question.

## ADDITION AND MULTIPLICATION

Although the idea of congruence is a very natural and simple one, it required the genius of Gauss to formalize it. One reason for its usefulness is that addition, subtraction, and multiplication can all be performed in congruence arithmetic. This is formalized in the following proposition.

PROPOSITION 3.3.1. *If* $a \equiv b$ mod $m$ *and* $c \equiv d$ mod $m$, *then* $a + c \equiv b + d$ mod $m$, $a - c \equiv b - d$ mod $m$, *and* $ac \equiv bd$ mod $m$.

*Proof.* We prove that it works for multiplication, and leave the easier proofs for addition and subtraction as an exercise. The hypothesis says that $a - b = km$ and $c - d = \ell m$, for some integers $k$ and $\ell$. Using these equations to rewrite $a$ and $c$, we obtain

$$ac - bd = (b + km)(d + \ell m) - bd = (kd + b\ell + k\ell m)m,$$

which exhibits $ac - bd$ as a multiple of $m$. Q.E.D.

One implication of this proposition is that we can make addition and multiplication tables for mod $m$ arithmetic just using the numbers from 0 to $m - 1$ (See Table 3.5). Since every number is congruent to exactly one of these, these tables will contain all possible facts about addition and multiplication in mod $m$ arithmetic. To form the table, just perform ordinary addition and multiplication, and then simplify mod $m$. We illustrate with $m = 6$.

TABLE 3.5. Tables of Arithmetic Mod 6.

| + | 0 | 1 | 2 | 3 | 4 | 5 |
|---|---|---|---|---|---|---|
| 0 | 0 | 1 | 2 | 3 | 4 | 5 |
| 1 | 1 | 2 | 3 | 4 | 5 | 0 |
| 2 | 2 | 3 | 4 | 5 | 0 | 1 |
| 3 | 3 | 4 | 5 | 0 | 1 | 2 |
| 4 | 4 | 5 | 0 | 1 | 2 | 3 |
| 5 | 5 | 0 | 1 | 2 | 3 | 4 |

| × | 0 | 1 | 2 | 3 | 4 | 5 |
|---|---|---|---|---|---|---|
| 0 | 0 | 0 | 0 | 0 | 0 | 0 |
| 1 | 0 | 1 | 2 | 3 | 4 | 5 |
| 2 | 0 | 2 | 4 | 0 | 2 | 4 |
| 3 | 0 | 3 | 0 | 3 | 0 | 3 |
| 4 | 0 | 4 | 2 | 0 | 4 | 2 |
| 5 | 0 | 5 | 4 | 3 | 2 | 1 |

For example, the table says that in mod 6 arithmetic $4 \times 5 \equiv 2$. To verify this, one multiplies 4 times 5 in ordinary arithmetic, obtaining 20, and then takes the remainder when 20 is divided by 6.

A very important application of Proposition 3.3.1 is that if you are working in mod $m$ arithmetic, a number can be replaced by any simpler number to which it is congruent. This is especially convenient in multiplication. We give several examples.

**Example.** What is the remainder when $123 \cdot 257 \cdot 425$ is divided by 7?

*Solution.* Instead of multiplying them out, replace each factor by the remainder when it is divided by 7. Thus $123 \equiv 4$, $257 \equiv 5$, and $425 \equiv 5$, all mod 7. Thus the product is congruent to $4 \cdot 5 \cdot 5 \equiv 100$. The remainder when 100 is divided by 7 is 2, and so 2 is the answer to the original question. You might write it as

$$123 \cdot 257 \cdot 425 \equiv 4 \cdot 5 \cdot 5 = 100 \equiv 2 \bmod 7.$$

The "$=$" in the middle could have been written as "$\equiv$". Equality is a strong form of congruence; numbers which are equal are congruent in any modulus.

**Example.** What is the remainder when

$$1! + 2! + 3! + \cdots + 99! + 100!$$

is divided by 12? (Recall that $n!$ is the product of all positive integers equal to or less than $n$.)

*Solution.* Each term from $4!$ on is divisible by 12, since it contains as a factor $4 \cdot 3$. Thus each of those terms is congruent to 0 mod 12. Thus the big sum is congruent to $1 + 2 + 6 + 0 + 0 + \cdots$, and so the answer is 9.

Sometimes it is convenient to replace numbers by small negative numbers to which they are congruent. For example, to find the remainder when $37 \cdot 77$ is divided by 39, it is easiest to note that $37 \equiv -2 \bmod 39$ and $77 \equiv -1 \bmod 39$, because they are respectively 2 and 1 less than multiples of 39. Thus,

$$37 \cdot 77 \equiv (-2)(-1) = 2 \bmod 39,$$

and so 2 is the answer to the question. You just have to develop some experience in seeing when it is useful to use negative numbers. It will happen when a given number is slightly less than a multiple of the modulus. Of course, the answer to this problem could have been obtained without using this trick, but our observation allowed us to obtain the answer with a minimum of calculation.

Proposition 3.3.1 can be used to prove general statements about all integers. For example, we can prove that if $a$ is any odd integer, then $a^2 \equiv 1 \bmod 8$. There are several ways to prove this result. The one which most directly uses 3.3.1 is to say that this is a statement about mod 8 arithmetic, and so $a$ may be replaced by either 1, 3, 5, or 7. Since it is easily verified that the squares of each of these four numbers is 1 greater than a multiple of 8, we are done. Another way to prove it is to say that $a$ can be written as $2b + 1$ for some integer $b$, and so

$$a^2 = (2b + 1)^2 = 4b^2 + 4b + 1.$$

It is not immediately obvious that $4b^2 + 4b$ is a multiple of 8 until you notice that $b^2 + b$ is always even, because it is the sum of either two odd or two even integers. Thus $a^2$ is 1 greater than 4 times an even number, and hence is congruent to 1 mod 8.

EXPONENTIATION

The following proposition follows easily from Proposition 3.3.1, by just iterating the multiplication. In Exercise 12, you are asked to write out the proof.

PROPOSITION 3.3.2. *If $e$ is a positive integer, and $a \equiv b$ mod $m$, then $a^e \equiv b^e$ mod $m$.*

There are many applications of this simple proposition. Many of these involve finding the remainder when $a^e$ is divided by $m$ for some specific values of $a$, $e$, and $m$. One thing to look for is to see if some small power of $a$ is congruent to 1 or $-1$ mod $m$. This is nice because multiplying by 1 and $-1$ is so simple. For example, what is the remainder when $2^{13}$ is divided by 33? The easiest way to get the answer is to notice that $2^5 = 32 \equiv -1$ mod 33. Then

$$2^{13} = 2^5 \cdot 2^5 \cdot 2^3 \equiv (-1)(-1)8 = 8 \text{ mod } 33,$$

and so the answer is 8. If asked about the remainder when $2^{99}$ is divided by 33, note that

$$2^{99} = (2^5)^{19}2^4 \equiv (-1)^{19}16 \equiv -16 \equiv 17,$$

and so the answer is 17. You will have to obtain some facility with the basic laws of exponents,

$$a^b a^c = a^{b+c}, \qquad (a^b)^c = a^{bc},$$

to work these problems. The key in evaluating $2^{99}$ was to note that since $2^5$ had such a nice value, you want to break the ninety-nine 2's that are multiplied together into groups of five, and the way to do this is with nineteen groups of five, with four 2's left over. Exercises 13, 14, and 15 can be worked by this method.

Some nice applications of congruence arithmetic are obtained by noting that the last digit of a number is congruent to the number mod 10, and the number obtained by looking at the last two digits of a number is congruent to the number mod 100. One very nice application of this is the following result, which was suspected by the Greeks but not proved until at least the time of Euler. It could be phrased as "every even perfect number ends in 6 or 28."

THEOREM 3.3.3. *If $P$ is an even perfect number, then either $P \equiv 6$ mod 10 or $P \equiv 28$ mod 100.*

*Proof.* We first recall Euler's theorem, 3.1.15, that all even perfect numbers are of the form $2^{n-1}(2^n - 1)$ with $n$ and $2^n - 1$ both prime. The perfect number obtained when $n = 2$ is 6, and since all other primes are odd, it suffices to prove:

    i. if $n \equiv 1$ mod 4, then $2^{n-1}(2^n - 1) \equiv 6$ mod 10.
    ii. If $n \equiv 3$ mod 4, then $2^{n-1}(2^n - 1) \equiv 28$ mod 100.

We will prove (ii), and leave the easier (i) as an exercise.

Write $n = 4e + 3$. Then the number whose mod 100 value we want to determine is

$$(3.3.4) \quad 2^{4e+2}(2^{4e+3} - 1) = 2^2(2^4)^e(2^3(2^4)^e - 1) = 4 \cdot 16^e(8 \cdot 16^e - 1).$$

The slickest way to the proof is to note that

$$(3.3.5) \qquad\qquad\qquad 16^e \equiv 16 \text{ mod } 20.$$

This can be proved by first noting that $16^2 = 256 \equiv 16$ mod 20, and the result follows from this by repeated multiplication. For example,

$$16^3 = 16^2 16 \equiv 16 \cdot 16 \equiv 16 \text{ mod } 20,$$

where we have used the result for $16^2$ in the last two $\equiv$'s. A more formal proof of (3.3.5) is by the Principle of Mathematical Induction.

Thus in (3.3.4), $16^e$ can be replaced by $20q + 16$ for some number $q$ whose value we will see is irrelevant. The expression (3.3.4) is now evaluated, mod 100, as follows:

$$
\begin{aligned}
4(20q + 16)(8(20q + 16) - 1) &\equiv (20q + 16)4(60q + 27) \\
&\equiv (20q + 16)(40q + 8) \\
&\equiv 60q + 40q + 28 \\
&\equiv 28.
\end{aligned}
$$

Here we have always dropped terms that are multiples of 100, and dropped the 100's digit off numbers. For example, the $40q + 8$ in the middle line is obtained from $4 \cdot 60q + 4 \cdot 27 = 240q + 108$ by dropping the 200 and 100. Q.E.D.

In practice, when doing exponentiation problems, it is rare that you will readily get to a number that is congruent to 1 or $-1$ mod $m$. In Section 4.2, we will need to raise large numbers to large powers in congruence arithmetic with a large modulus. The most efficient way to do this in general is to do repeated squaring to obtain $a^2$, $a^4$, $a^8$, etc., always using the modulus to keep the numbers as close to zero as possible. The desired power $a^n$ can be written as a product of these by writing the exponent $n$ as a sum of distinct 2-powers, which can always be done, and then multiplying together in mod $m$ arithmetic the appropriate 2-power powers already determined.

We illustrate with an example that will be required in an example in Section 4.2: What is the least nonnegative residue mod 143 of $2^{103}$? Even a computer has difficulty working with a number as large as $2^{103}$, and so computer algorithms to do this calculation will probably proceed in the same way as our hand calculation.

We begin by doing repeated squaring in mod 143 arithmetic to obtain $2^2$, $2^4$, $2^8$, $2^{16}$, $2^{32}$, and $2^{64}$. This is as far as we will need to go because $2^{128}$, the next square, is larger than our desired number. The reason for proceeding in this way is that repeated squaring is the quickest way to get up to large powers. The occurrence of the 2 in $2^{103}$ is completely irrelevant to the decision to use repeated squaring.

Always simplifying mod 143, we obtain $2^2 = 4$, $2^4 = (2^2)^2 = 4^2 = 16$ (no simplification due to modulus yet), and

$$2^8 = (2^4)^2 = 16^2 = 256 \equiv 113$$
$$2^{16} = (2^8)^2 \equiv 113^2 = 12769 \equiv 42$$
$$2^{32} = (2^{16})^2 \equiv 42^2 = 1764 \equiv 48$$
$$2^{64} = (2^{32})^2 \equiv 48^2 = 2304 \equiv 16.$$

Note how each line is obtained by squaring the result of the previous line and reducing mod 143.

Most simple calculators do not have a straightforward way of determining the remainder when a number such as 12769 is divided by a number such as 143. The calculator will tell you that $12769 \div 143 = 89.293706$. To find the remainder, you should first multiply the integer part of the quotient, 89, by 143, yielding 12727.

From here, there are several ways to see that the remainder is 42. One is to note by hand or head subtraction that $12769 - 12727 = 42$. Another is to have the calculator perform this subtraction for you. A third, which involves less button pushing than the second, is to subtract 12769 from the 12727 that is already displayed on your calculator. This gives $-42$ as the answer, but you know that you want to make it positive since you were doing the subtraction in the wrong order. Another way to find the remainder is to multiply the decimal part .293706 by 143.

Returning to the exponentiation problem, we next note that

(3.3.6)                     $$2^{103} = 2^{64} \cdot 2^{32} \cdot 2^4 \cdot 2^2 \cdot 2^1.$$

By the law of exponents, this is equivalent to expressing 103 as a sum of 2-powers as $103 = 64 + 32 + 4 + 2 + 1$. We will see this in another context in the "Aside" at the end of this section when we talk about the binary expansion of a number. To obtain this expression of 103, note that 128 is too large, and so 64 is the largest 2-power less than or equal to 103. Subtracting, we obtain $103 = 64 + 39$. Now we do the same thing with 39, writing it as $32 + 7$, and so $103 = 64 + 32 + 7$. Finally, we write 7 as a sum of 2-powers as $4 + 2 + 1$.

Next we substitute into (3.3.6) the mod 143 values of the 2-power powers determined above, obtaining mod 143

$$2^{103} \equiv 16 \cdot 48 \cdot 16 \cdot 4 \cdot 2.$$

Note that $2^8$ and $2^{16}$ are not used directly in finding this number, but they were necessary en route to $2^{32}$ and $2^{64}$. Finally we have to multiply out these five numbers mod 143. One should always be on the lookout for shortcuts, and always reduce mod 143. The multiplication can be done in any order, and here we note that $16 \cdot 16 \equiv 113$ was calculated earlier in determining $2^8$. Using this, and multiplying out $4 \cdot 2$ gets us down to $113 \cdot 48 \cdot 8$. With a calculator, you might just multiply these three numbers together and find the remainder when the product is divided by 143. Alternatively, multiply $48 \cdot 8 = 384 \equiv 98$, and then $113 \cdot 98 = 11074 \equiv 63$, which is the final answer. If you are working without a calculator, you might want to replace the 113 by $-30$ and replace the 98 by $-45$. This makes the numbers a little bit smaller. The results of the multiplication can be summarized as

$$2^{103} \equiv 16 \cdot 48 \cdot 16 \cdot 4 \cdot 2 \equiv 113 \cdot 48 \cdot 8 \equiv 113 \cdot 98 \equiv 63.$$

So the answer to the question, "What is the least nonnegative residue mod 143 of $2^{103}$?" is 63. If you think this was a lot of work, you should bear in mind that $2^{103}$ is a number with 31 digits.

*CONGRUENCE EQUATIONS*

In ordinary arithmetic, if you know that $ca = cb$ for some $c \neq 0$, then you can deduce that $a = b$. Formally, the way that you make this deduction is to multiply the first equation by $1/c$, to cancel the $c$. The important thing about $1/c$ is that it is a number which when multiplied by $c$ gives 1. Indeed, the argument could be expressed by the string of equations

$$a = 1 \cdot a = \tfrac{1}{c} \cdot ca = \tfrac{1}{c} \cdot cb = 1 \cdot b = b.$$

The ability to cancel $c$ is important in solving the linear equation $cx = d$, whose solution is of course $x = 1/c \cdot d$.

You can't always do this kind of cancelling in congruence arithmetic, and you cannot always solve linear equations in congruence arithmetic. For example, $2 \cdot 2 \equiv 2 \cdot 5$ mod 6, but $2 \not\equiv 5$ mod 6, so you can't always cancel a 2 in mod 6 arithmetic. Related to this is the fact that the equation

$$2x \equiv 1 \text{ mod } 6$$

has no solution. However, you can do the cancelling if the number being cancelled and the modulus are relatively prime, that is, if their GCD is 1, and you can solve a linear equation if the coefficient of $x$ is relatively prime to the modulus. Moreover, you can always cancel in congruence arithmetic if you allow yourself to change the modulus appropriately. This is the content of the following theorem.

THEOREM 3.3.7.

i. *The congruence equation*

(3.3.8) $$cx \equiv d \text{ mod } m$$

*can be solved if and only if $d$ is a multiple of the GCD $(c, m)$. In particular, (3.3.8) can be solved for any $d$ if $c$ and $m$ are relatively prime.*

ii. *If $ca \equiv cb$ mod $m$, then $a \equiv b$ mod $m/(c,m)$. In particular, if $c$ and $m$ are relatively prime, then $c$ can be cancelled from equations in mod $m$ arithmetic.*

*Proof.* Solving the congruence equation (3.3.8) is equivalent to solving the Diophantine equation

$$cx + my = d$$

for $x$ and $y$. This is true because $my$ just measures the multiple of $m$ by which $cx$ and $d$ differ. Proposition 3.2.7 stated that this Diophantine equation can be solved if and only if $d$ is a multiple of $(c, m)$, and it showed how to find the solution using the Euclidean algorithm. This proves (i).

We just prove (ii) in the important special case when $(c, m) = 1$, that is, $c$ and $m$ are relatively prime. Then, by part (i), we can find a number $s$ satisfying $sc \equiv 1 \bmod m$. This $s$ is just a value that worked for the unknown $x$ in (3.3.8) when $d = 1$. Now we can do the cancellation just as we did in ordinary arithmetic at the beginning of this section. Working mod $m$, multiply the given fact, $ca \equiv cb$, by $s$, obtaining

$$a \equiv 1 \cdot a \equiv sca \equiv scb \equiv 1 \cdot b \equiv b.$$

The proof in the general case is similar. Q.E.D.

We illustrate this theorem with two examples.

**Example.** Solve the congruence equation $23x \equiv 1 \bmod 56$.

*Solution.* This equation can be solved provided the GCD $(23, 56) = 1$, which is quite apparently true since 23 is prime and not a divisor of 56. The solution is found using the Euclidean algorithm. As usual, we write the division facts as equations for the remainders:

$$10 = 56 - 2 \cdot 23$$
$$3 = 23 - 2 \cdot 10$$
$$1 = 10 - 3 \cdot 3.$$

From this we obtain

$$1 = 10 - 3(23 - 2 \cdot 10)$$
$$= 7 \cdot 10 - 3 \cdot 23$$
$$= 7(56 - 2 \cdot 23) - 3 \cdot 23$$
$$= 7 \cdot 56 - 17 \cdot 23.$$

Thus $x \equiv -17$ is a solution of the equation posed in the example, since $23(-17)$ differs from 1 by a multiple of 56. We usually prefer that the solutions be written in the form of least nonnegative residue. For negative numbers, this can be found by adding multiples of the modulus until a positive number is obtained. In this case, we add 56 to $-17$, obtaining $x = 39$ as the desired answer.

Examples of the above type will be very important in Section 4.2. The number $x = 39$ found as a solution of the congruence equation $23x \equiv 1 \bmod 56$ is called an *inverse* of 23 in mod 56 arithmetic.

**Example.** Solve the congruence equation

$$(3.3.9) \qquad\qquad 32x \equiv 6 \bmod 70.$$

*Solution.* It is easy to factor the numbers 32 and 70, and from this see that their GCD is 2. Since 6 is a multiple of 2, Theorem 3.3.7 guarantees that this equation will have a solution. The best way to solve the equation is to begin by dividing the entire equation, including the modulus, by 2, the GCD of 32 and 70. This is a valid thing to do, since saying that $32x - 6 = 70k$ is the same as saying that $16x - 3 = 35k$. Here $k$ is telling what multiple of the modulus equals the difference of the two numbers. Now we run the Euclidean algorithm on 16 and 35:

$$3 = 35 - 2 \cdot 16$$
$$1 = 16 - 5 \cdot 3$$

and hence

$$1 = 16 - 5(35 - 2 \cdot 16)$$
$$= 11 \cdot 16 - 5 \cdot 35.$$

Thus $11 \cdot 16 \equiv 1 \bmod 35$, and so a solution of the equation

$$(3.3.10) \qquad\qquad 16x \equiv 3 \bmod 35$$

is given by multiplying the 11 by 3, obtaining $x = 33$ as a solution of (3.3.10). As already observed, this solution will also be a solution of the original equation, (3.3.9). However, (3.3.9) has a second solution. If we add the modulus, 35, to our solution of (3.3.10), we do not obtain a new solution of (3.3.10), since 33 and 68 are the same number in mod 35 arithmetic. However, the new number, 68, being a solution of (3.3.10), is also a solution of (3.3.9), and since (3.3.9) is an equation in mod 70 arithmetic, the solutions $x = 33$ and $x = 68$ are distinct solutions. Thus the answer to (3.3.9) is $x = 33$ or 68.

This is the recommended way of solving $cx \equiv d \bmod m$ when $(c, m)$ is greater than 1, and we summarize it in a proposition.

PROPOSITION 3.3.11. *If $d$ is a multiple of $(c, m)$, then the equation $cx \equiv d \bmod m$ has exactly $(c, m)$ distinct solutions. To find them, solve the equation*

$$\frac{c}{(c, m)} x \equiv \frac{d}{(c, m)} \bmod \frac{m}{(c, m)}$$

*using the Euclidean algorithm. If $x_0$ is the (unique) solution of this, then the solutions of the given equation are*

$$x_0, \quad x_0 + \frac{m}{(c,m)}, \quad x_0 + 2\frac{m}{(c,m)}, \quad \ldots, \quad x_0 + ((c, m) - 1)\frac{m}{(c,m)}.$$

If you solve $cx \equiv d \bmod m$ with $(c, m) > 1$ by the Euclidean algorithm without first dividing through by $(c, m)$, you will get one of the solutions. If you then think to vary that solution by multiples of $m/(c,m)$, you will obtain all solutions. Or, if you are just asked to find **a** solution, rather than all solutions, you can get by without dividing by $(c, m)$. But dividing by this number gives you smaller numbers to work with, and so it is a good thing to do for that reason, too. If you don't know the GCD $(c, m)$ in advance, you will find it when you run through the first pass of the Euclidean algorithm. In the above example, if you did not realize that $(32, 70) = 2$, the first pass through the Euclidean algorithm would have told you that this GCD is 2, while the second pass would have yielded $2 = 11 \cdot 32 - 5 \cdot 70$. At this point, you could divide the 2, 32, and 70 by 2, and proceed as above.

*ASIDE: BASES OF NUMBER SYSTEMS*

In the familiar decimal notation, the digits of a number tell how it is composed of powers of 10. For example, the number we write as 4035 equals

$$4 \cdot 10^3 + 0 \cdot 10^2 + 3 \cdot 10^1 + 5 \cdot 10^0.$$

Remember that the 0th power of any nonzero number equals 1. Historically, we use base 10 because we have 10 fingers. The Babylonians used base 60, although presumably they did not have 60 fingers. The Mayans used base 20, and computers use base 2. The number system to the base 2, called the binary system, is particularly important. Numbers in it are composed only of 0's and 1's, and so it is conveniently implemented by on-off switches. Because the base 2 representations of numbers are very long, computers often use base

8 or 16 to convey their binary numbers to their human users. In this "Aside," we will learn about arithmetic in different bases.

Let $b$ be any integer satisfying $b \geq 2$. Numbers written in base $b$ use as digits symbols representing all integers from 0 to $b - 1$. If $b > 10$, we need to make up new symbols for the digits greater than 9. The largest $b$ we will use in this book is 12, and then we use $t$ to represent 10 and $e$ to represent 11. The letters $t$ and $e$ are used because they are the first letters of the words "ten" and "eleven." A number whose base $b$ representation has last digit $a_0$, next-to-last digit $a_1$, etc., represents the number

$$\cdots + a_2 b^2 + a_1 b^1 + a_0 b^0.$$

This number is written $(\cdots a_2 a_1 a_0)_b$. For example,

$$
\begin{aligned}
(212011)_3 &= 2 \cdot 3^5 + 1 \cdot 3^4 + 2 \cdot 3^3 + 0 \cdot 3^2 + 1 \cdot 3 + 1 \\
&= 2 \cdot 243 + 81 + 2 \cdot 27 + 3 + 1 \\
&= 625(= (625)_{10}).
\end{aligned}
$$

As another example, we count the first thirteen positive integers in bases 2 and 12:

| 1 | 10 | 11 | 100 | 101 | 110 | 111 | 1000 | 1001 | 1010 | 1011 | 1100 | 1101 |
|---|----|----|-----|-----|-----|-----|------|------|------|------|------|------|
| 1 | 2  | 3  | 4   | 5   | 6   | 7   | 8    | 9    | t    | e    | 10   | 11.  |

The numbers in the first row above should really all have subscript 2, and those in the second row should have subscript 12. We will follow the convention that a number written without a subscript is a base 10 number unless the number is specifically designated as being in a different base, as in some worked-out long division and multiplication problems below. The reader should not confuse arithmetic in base $b$ with arithmetic mod $b$. The latter, with which we dealt throughout the main part of this section, throws out a lot of information about the numbers involved, while the former, which is the topic of this "Aside," contains all the information of the numbers, but viewed from a different perspective.

Another example of converting into base 10 is

$$
\begin{aligned}
(2te)_{12} &= 2 \cdot 12^2 + 10 \cdot 12 + 11 \\
&= 288 + 120 + 11 = 419 = (419)_{10}.
\end{aligned}
$$

Next we show how to go the other way, converting a base 10 number $n$ into base $b$. Thus we want to find numbers $a_0, a_1, \ldots$, each in the range 0 to $b - 1$, so that $(n)_{10} = (\cdots a_2 a_1 a_0)_b$, or equivalently,

(3.3.12)                    $n = \cdots + a_2 b^2 + a_1 b^1 + a_0.$

There are two methods.

**Method 1.** First note that $a_0$ must be the remainder when $n$ is divided by $b$. This is clear from (3.3.12) since all other terms on the right hand side are multiples of $b$. Let $n_1$ be the quotient in this division. In (3.3.12), we have $n_1 = \cdots + a_2 b^1 + a_1$. Thus $a_1$ is the remainder when $n_1$ is divided by $b$. Let $n_2$ be the quotient in this division, $a_2$ the remainder when $n_2$ is divided by $b$, etc.

We illustrate by converting $(69)_{10}$ to base 5:

$$69/5 = 13 \text{ rem } 4. \text{ Therefore } a_0 = 4.$$
$$13/5 = 2 \text{ rem } 3. \text{ Therefore } a_1 = 3.$$
$$2/5 = 0 \text{ rem } 2. \text{ Therefore } a_2 = 2.$$

The procedure stops because the next step would be $0/5$. Therefore, $(69)_{10} = (234)_5$.

The second method generates the number from left-to-right rather than right-to-left. It is slightly more cumbersome computationally, because it involves working with larger numbers.

**Method 2.** Suppose $b^k$ is the largest power of $b$ which satisfies $b^k \leq n$. Then let $a_k$ be the quotient in $n/b^k$, and let $n'$ be the remainder. The desired number will have $a_k$ in the $b^k$-position. The next nonzero digit will be obtained by finding the largest power of $b$ that satisfies $b^\ell \leq n'$. It will necessarily be true that $\ell < k$. The second nonzero digit will be $a_\ell$ in the $b^\ell$-position, where $a_\ell$ is the quotient in $n'/b^\ell$. We apply the same procedure to the remainder in this division, and continue in this manner until we get a remainder less than $b$, which goes in the right-most position.

We illustrate with the same example as above, converting $(69)_{10}$ to base 5. We note that $5^2 = 25$ and $5^3 = 125$. Thus $5^2$ is the largest power less than or equal to 69, and since $^{69}/_{25} = 2 \text{ rem } 19$, our desired number has 2 in the $5^2$-position. Now we compute $^{19}/_5 = 3 \text{ rem } 4$, and so the 3 goes in the $5^1$-position, and 4 goes in the rightmost position. Thus, as before, we obtain $(234)_5$ as the answer.

One can do arithmetic directly in base $b$. You have to remember that there can be no digits equal to or greater than $b$. Whenever you get such a digit, you must subtract $b$ from it and carry 1 onto the next digit to the left. This is just what we do when carrying in ordinary base 10 arithmetic.

**Example.** Perform the following base-5 addition problem working directly in base 5, and then check your answer by converting all numbers to base 10:
$$(1234)_5 + (2012)_5 = \quad .$$

Write the second number under the first, and work from right to left as in ordinary addition: $2 + 4 = 6 = 1 + 1 \cdot 5$, so put down 1 and carry 1. Or you can use the base-5 addition fact, $2 + 4 = 11$. Now $1 + 3 + 1 = 5 = 0 + 1 \cdot 5$, so put down 0 and carry 1. In the last two columns, simple addition works, and the final answer is $(3301)_5$. To perform the check (which is much more tedious than the calculation itself), we calculate

$$(1234)_5 = 125 + 2 \cdot 25 + 3 \cdot 5 + 4 = (194)_{10}$$
$$(2012)_5 = 2 \cdot 125 + 1 \cdot 5 + 2 = (257)_{10}$$
$$(3301)_5 = 3 \cdot 125 + 3 \cdot 25 + 1 = (451)_{10},$$

and indeed, in base 10, $194 + 257 = 451$.

If you work a lot in base $b$, you will just learn the addition and multiplication facts. In base 2, multiplication is completely trivial, and in base 3 there is only one nontrivial multiplication fact to remember, namely $2 \times 2 = 11$, the base-3 representation of the number 4.

**Example.** Work the following base-3 multiplication problem:

$$(212)_3 \times (201)_3 = \quad .$$

The format is the same as for ordinary base-10 multiplication—you multiply the top number by each digit of the bottom, arrange the products appropriately, and add them. The hardest part here is $212 \times 2$, where you twice use $2 \times 2 = 11$, and twice carry 1. The worked-out answer is

$$
\begin{array}{r}
212 \\
201 \\
\hline
212 \\
1201\phantom{0} \\
\hline
121012\,.
\end{array}
$$

Division in base $b$ also follows a format similar to that of ordinary base-10 division. We let the following base-5 division example serve as a guide:

$$\begin{array}{r}
311 \\
14)\overline{10412} \\
\underline{102} \\
21 \\
\underline{14} \\
22 \\
\underline{14} \\
3
\end{array}$$

The familiar rules for divisibility by 2, 3, 5, and 10 given in Proposition 3.1.3 generalize to base $b$ in the following way.

PROPOSITION 3.3.13.
> i. If $d$ is a divisor of $b$, then a base-$b$ number is divisible by $d$ if and only if its last digit is divisible by $d$.
> ii. If $d$ is a divisor of $b - 1$, then a base-$b$ number is divisible by $d$ if and only if the sum of its digits is divisible by $d$.

In the familiar example $b = 10$, part (i) holds for $d = 2$ or 5, and part (ii) holds for $d = 3$ or 9. For a less familiar example, consider base-8. Proposition 3.3.13 says that a base 8 number is divisible by 2 if and only if its last digit is 0, 2, 4, or 6, that it is divisible by 4 if and only if its last digit is 0 or 4, and that it is divisible by 7 if and only if the sum of its digits is a multiple of 7. For example, $(52)_8$ is divisible by 2 and 7, but not by 4. We can check by converting to base 10, where we obtain $5 \cdot 8 + 2 = 42$, which is indeed divisible by 2 and 7, but not by 4.

### Exercises

1. Which of the numbers 9, 23, 35, and 42 have the same parity? Which have the same weekday? Which have the same clocktime?
2. Suppose we wanted to measure time in minutes. What would then be the criterion for two events to have the same clocktime?
3. Prove that if $a \equiv b \bmod m$ and $b \equiv c \bmod m$, then $a \equiv c \bmod m$.
4. For which positive integers $m$ is it true that $17 \equiv 5 \bmod m$?
5. What is the least nonnegative residue mod 17 of each of the following numbers: 100, 1000, $-100$, $-1000$?
6. Prove the addition part of Proposition 3.3.1.
7. Make addition and multiplication tables for mod 4 arithmetic, and for mod 5 arithmetic similar to Table 3.5. Point out a qualitative difference in the two multiplication tables.
8. What is the remainder when $6002 \cdot 7142 \cdot 4751$ is divided by 6?
9. What is the remainder when

$$1^5 + 2^5 + 3^5 + \cdots + 98^5 + 99^5$$

is divided by 4? (Hint: The resemblance to a worked Example involving a sum of factorials is only rather superficial. Here you should look for a pattern in the mod 4 values of the numbers being added.)

10. What is the remainder when $3699 \cdot 7397$ is divided by 37? What is the least nonnegative residue mod 23 of $228 \cdot 45 \cdot 2297$? You should be able to work out this problem in your head.

11. Prove that for all integers $a$, $a^5 \equiv a$ mod 5. (Hint: You need to consider only five possible values of $a$. Explain.)

12. Prove Proposition 3.3.2. You can do it using 3.3.1 and the Principle of Mathematical Induction, or by using a factorization of $a^e - b^e$ similar to (3.1.12).

13. What is the remainder when $5^{11} \cdot 3^{14}$ is divided by 26? (Hint: $5^2 = 25$, and $3^3 = 27$.)

14. What is the least nonnegative residue mod 65 of $4^{80}$? You should be able to do this with virtually no calculation.

15. What are the remainders when $2^{50}$ and $34^{65}$ are divided by 7? You should be able to do this with virtually no calculation.

16.* Give a formula for the last two digits of $16^e$ for any positive integer $e$. Prove that your formula always works. (Hint: It will depend on the mod 5 value of $e$. Your "formula" can just be a list of these five cases.)

17.* Prove part (i) in the proof of Theorem 3.3.3.

18. Use the Principle of Mathematical Induction to prove (3.3.5).

19. What is the remainder when $2^{44}$ is divided by 89? Use repeated squaring. If you are not using a calculator, you might profit at one point by noting that $78 \equiv -11$.

20. What is the least nonnegative residue mod 79 of $3^{100}$?

21. Explain using just elementary ideas of divisibility, i.e., without referring to any theorems, why the equation $6x \equiv 1$ mod 8 has no solutions.

22. Solve the equation $32x \equiv 1$ mod 45. Your answer should be a number between 0 and 44.

23. Solve $7x \equiv 3$ mod 24.

24. Find all solutions of $8x \equiv 6$ mod 70.

25. Find all solutions of $14x \equiv 10$ mod 54.

26. Convert the following numbers to base 10: $(100011011)_2$, $(456)_8$, $(t0t)_{12}$.

27. Convert the base-10 number 4567 to bases 3, 8, and 12.

28. Perform the base-7 addition problem $(4506)_7 + (3443)_7$, working completely in base 7.

29. Since multiplication by 0 and 1 is simple, the only nontrivial multiplication facts in base 4 are $2 \times 2$, $2 \times 3$, and $3 \times 3$. What do they equal? Use these to perform the multiplication $(3201)_4 \times (213)_4$.

30. Perform the division $(1100101101)_2 \div (1001)_2$, working entirely in base 2, and then check your answer by converting all numbers to base 10.

31.* Prove Proposition 3.3.13.

32. Primes $p$ such that $2p + 1$ is also prime are called *Sophie Germaine primes*, after a well-known female mathematician of the nineteenth century. They arise in studies of Mersenne primes and Fermat's Last Theorem. It is not known whether there are infinitely many of them.

    a. List eight Sophie Germaine primes.

    b.* Prove that the only twin primes that are both Sophie Germaine primes are 3 and 5. (Hint: If $p$ and $p + 2$ are twin primes with $p > 3$, what must be the mod 3 value of $p$?)

33. Prove that any number is congruent mod 9 to the sum of its digits. You may restrict your proof to 4-digit numbers, if you wish, to keep the notation simple. This result generalizes Proposition 3.1.3iv, and forms the basis for the method of "casting out 9s" in arithmetic. Show with virtually no calculation that the equation

$$8563 \times 5743 = 49176309$$

must be incorrect.

## 3.4   The Little Fermat Theorem

The Little Fermat Theorem is a theorem that was stated by Fermat in the 1600s and has many important applications today. It forms the theoretical underpinning of the method of cryptography that will be discussed in Section 4.2. Cryptography is the study of methods of sending secret codes. The method to be discussed in Section 4.2 was developed in the late 1970s and is widely used in top-secret applications today, yet it is based upon mathematical facts that have been known for three hundred years. It utilizes the topics of all four sections of chapter 3. As we shall stress later, this is a principal embodiment of the main theme of this book: unexpected applications of theoretical mathematics.

As mentioned in Section 3.1, Fermat did not usually present proofs of his stated theorems, and this one was no exception. Unlike the famous Fermat's "Last Theorem," discussed in the "Aside" to Section 3.1, this statement is easily proved, and we will give two proofs of it. It is called "Little" to distinguish it from other results of Fermat, especially the unproved Last Theorem.

We now state the Little Fermat Theorem.

THEOREM 3.4.1. *Let $p$ be a prime number.*
  *i. For any integer $a$,*

$$a^p \equiv a \bmod p.$$

  *ii. If $a$ is not a multiple of $p$, then*

$$a^{p-1} \equiv 1 \bmod p.$$

The two parts of this theorem are equivalent. It is enough to prove either part, for each easily implies the other. To see this, we first note that part (i) implies part (ii) by cancelling a factor of $a$, which, by 3.3.7, can be done as long as $a$ and $p$ are relatively prime, which means that $a$ is not a multiple of $p$, since $p$ is prime. On the other hand, part (ii) implies that part (i) is true when $a$ is not a multiple of $p$ by simply multiplying the conclusion of part (i) by $a$, while part (i) is certainly true if $a$ is a multiple of $p$, for then both $a^p$ and $a$ are congruent to 0 mod $p$. We emphasize that what we have said here is not a proof of Theorem 3.4.1. It just says that if you prove one of the two statements, then you will have also proved the other.

Note that, for a fixed prime $p$, Theorem 3.4.1 appears to be a statement about infinitely many values of $a$, but can be verified by just verifying it for $a$ equal to all numbers from 1 to $p-1$. This is because it is a statement about mod $p$ arithmetic, and every integer which is not a multiple of $p$ is congruent to a number between 1 and $p-1$. For example, when $p = 5$, it says that $a^4 \equiv 1 \bmod 5$ whenever $a$ is not a multiple of 5, but it is enough to check that it is true for $a = 1, 2, 3$, and 4. This is easily done as follows, with all congruences mod 5:

$$1^4 = 1 \equiv 1, \quad 2^4 = 16 \equiv 1, \quad 3^4 = 81 \equiv 1, \quad 4^4 = 256 \equiv 1.$$

Then, for example, the statement is true for $a = 17$ since $17^4 \equiv 2^4$, and we have already checked that $2^4 \equiv 1$. Of course, verifications are not really required once the theorem has been proved, for the proof is a guarantee that it will always work. Nevertheless, in Exercise 1 you are asked to verify it when $p = 7$, as an aid to your understanding and appreciation of the theorem.

*TWO PROOFS*

We will give two proofs of the Little Fermat Theorem, even though only one is required to establish its validity. However, both of these arguments are very elegant, and they are very different from one another. They provide a fine example of beauty in mathematics—how a nonobvious and far-reaching result can be derived from simple principles.

*First Proof of Theorem 3.4.1.* We will assume $a$ is not a multiple of $p$, and prove part (ii). The first step is to note that the $p - 1$ numbers $a, 2a, 3a, \ldots, (p - 1)a$, when considered mod $p$, are just a rearrangement of the numbers $1, 2, \ldots, p - 1$. For example, if $p = 5$ and $a = 4$, the numbers 4, 8, 12, and 16 are congruent to 4, 3, 2, and 1, respectively. To see why this is always true, we first observe that, for $i$ taking values from 1 through $p - 1$, the numbers $i \cdot a$ cannot be congruent to 0 (why not?) or to one another mod $p$. The latter is true because if $i \cdot a \equiv j \cdot a$ mod $p$, then $(i - j)a$ must be a multiple of $p$. Since $a$ is not a multiple of $p$, Proposition 3.2.14 implies that $i - j$ must be a multiple of $p$, but that cannot be true if $i \neq j$ and both are in the range 0 to $p - 1$. Since there are $p - 1$ numbers $i \cdot a$ mod $p$ and $p - 1$ numbers (1 to $p - 1$) to which they may be congruent, and they are all congruent to different ones of these, each number from 1 to $p - 1$ must be obtained once.

Therefore, if the numbers $a, 2a, 3a, \ldots, (p - 1)a$ are multiplied together mod $p$, the answer is the same as if $1, 2, \ldots, p - 1$ are multiplied together mod $p$. Restated,

$$a \cdot (2a) \cdot (3a) \cdots ((p - 1)a) \equiv 1 \cdot 2 \cdots (p - 1) \bmod p,$$

since the two sets of numbers represent the same $p - 1$ numbers mod $p$. The left-hand side can be rewritten as $1 \cdot 2 \cdots (p - 1) \cdot a^{p-1}$, the exponent $p - 1$ being the number of factors of $a$. Thus, ignoring a 1 on the left-hand side,

$$2 \cdot 3 \cdots (p - 1) \cdot a^{p-1} \equiv 1 \cdot 2 \cdots (p - 1) \bmod p.$$

Each of the factors $2, \ldots, p - 1$, being relatively prime to the modulus $p$, can be cancelled by Proposition 3.3.7, yielding the desired result

$$a^{p-1} \equiv 1 \bmod p.$$

This completes the first proof. We remark that the only ingredients in it were (1) the ability to cancel a number that is relatively prime to the modulus, which is a consequence of the Euclidean algorithm, and (2) what is sometimes called the Pigeon Hole Principle, that if $n$ things go into $n$ boxes with no two things going into the same box, then every box will be filled.

Our second proof also requires only two basic tools, and given them, it is even slicker. One of these is the Principle of Mathematical Induction, which we introduced in Exercise 10 of Section 3.2. The other is the Binomial Theorem, which is part of a good high school algebra course. It states:

THEOREM 3.4.2. *(Binomial Theorem). For any positive integer $n$,*

$$(x + 1)^n = x^n + \binom{n}{n-1}x^{n-1} + \cdots + \binom{n}{2}x^2 + \binom{n}{1}x + 1,$$

*where*

$$\binom{n}{i} = \frac{n!}{i!(n-i)!} = \frac{n(n-1)\cdots(n-i+1)}{i(i-1)\cdots 1}.$$

We omit the proof of this standard result.

*Second Proof of Theorem 3.4.1.* We will prove part (i), recalling that it suffices to prove either part. Note that the conclusion, $a^p \equiv a \bmod p$, is clearly true when $a = 1$. We assume the result is true for $a$ and will deduce that it must then be true for $a + 1$. Once this deduction is completed, the theorem follows by the Principle of Mathematical Induction. We remind the reader why this works. We know the theorem is true for $a = 1$. The induction step shows that if it is true for $a = 1$, then it must also be true for $a = 2$. Thus it is true for $a = 2$. Now the induction step shows that if it is true for $a = 2$, then it must also be true for $a = 3$. Hence it is true for $a = 3$. This procedure continues to prove the validity for every value of $a$.

So now we are assuming $a^p \equiv a \bmod p$, and want to prove $(a + 1)^p \equiv a + 1 \bmod p$. To do this, we expand $(a + 1)^p$ by the binomial theorem. The last term is 1, and the first term is $a^p$, which by our induction hypothesis is congruent to $a \bmod p$. The intermediate terms are of the form $(p!/i!(p-i)!)a^i$, where $i$ is some number between 1 and $p - 1$. For each of these values of $i$, the coefficient is a multiple of $p$, because a factor of $p$ appears in the numerator $(p!)$ but not in the denominator $(i!(p - i)!)$. Thus all the intermediate terms are congruent to 0 mod $p$. Hence, using in the second step the assumption that $a^p \equiv a$, we obtain $(a + 1)^p \equiv a^p + 1 \equiv a + 1 \bmod p$, as desired. Q.E.D.

*APPLICATIONS*

The most important applications of the Little Fermat Theorem are to primality testing, which will be discussed later in the subsections following this one, and public-key cryptography, the subject of Section 4.2. In this subsection, we give two other applications.

The Little Fermat Theorem is frequently useful in congruence exponentiation problems when the modulus is prime. The following example is typical.

**Example.** What is the remainder when $3^{1000}$ is divided by 17?

*Solution.* The Little Fermat Theorem tells you how to find a power of 3 that is congruent to 1 mod 17. Then you proceed as in Section 3.3. By Theorem 3.4.1, $3^{16} \equiv 1$ mod 17. Thus we should divide the one thousand 3's being multiplied together into groups of sixteen. This is accomplished by dividing 1000 by 16, getting 62 with remainder 8. Thus there are sixty-two groups of sixteen 3's, with eight 3's left over. This reduces you to calculating $3^8$ mod 17, which is done by any method, such as repeated squaring. Since $3^4 = 81 \equiv -4$, we obtain mod 17

$$3^{1000} = (3^{16})^{62}3^8 \equiv 1^{62}3^43^4 \equiv (-4)(-4) = 16.$$

Since 16 is in the range 0 to 16, it is the answer.

The second application in this subsection is to proving another of Fermat's theorems, which we stated as Proposition 3.1.10. Recall that Proposition 3.1.10 was used as an aid in determining whether or not various Mersenne numbers are prime.

*Proof of Proposition 3.1.10.* We want to prove that if $p$ is prime, then any divisor of $2^p - 1$ is of the form $2kp + 1$ for some integer $k$. It suffices to prove this for prime divisors of $2^p - 1$. (Why? See Exercise 9.) So suppose $q$ is a prime number that is a divisor of $2^p - 1$. We will show in the next paragraph that $p$ is a divisor of $q - 1$. Since $q$ is a divisor of $2^p - 1$, it must be odd. Hence $q - 1$ is even, so that 2 is also a divisor of $q - 1$. Thus $2p$ is a divisor of $q - 1$, as desired.

In this paragraph we finish the proof by proving that $p$ is a divisor of $q - 1$. The proof is by *reductio ad absurdum*: we assume that $p$ is not a divisor of $q - 1$ and will derive a contradiction. If $p$ is not a divisor of $q - 1$, then they are relatively prime. (This uses the fact that $p$ is prime.) Thus, by 3.2.7, we can find integers $m$ and $n$ such that

$$mp + n(q - 1) = 1.$$

Either $m$ or $n$ must be negative. Suppose $m$ is negative. (A similar argument works if, instead, $n$ is negative.) Thus $-m$ is positive. Since $q$ is given to be a divisor of $2^p - 1$, we have $2^p \equiv 1 \bmod q$, and hence $2^{-mp} \equiv 1^{-m} \equiv 1 \bmod q$. By the Little Fermat Theorem, $2^{q-1} \equiv 1 \bmod q$, and hence $2^{n(q-1)} = (2^{q-1})^n \equiv 1 \bmod q$. Thus

$$2 = 2^1 = 2^{mp+n(q-1)} \equiv 2^{n(q-1)} \equiv 1 \bmod q,$$

a contradiction. Here we have multiplied by $2^{-mp} \equiv 1$ in the middle step. This completes the proof of Proposition 3.1.10.

PSEUDOPRIMES

The Little Fermat Theorem provides a method of showing that a number is not prime. If a number $n$ has the property that, for some integer $a$ which is not a multiple of $n$, $a^{n-1} \not\equiv 1 \bmod n$, then $n$ is not prime. This is just the contrapositive of Theorem 3.4.1. The simplest number to use for $a$ is 2. So if you have a number $n$ that you suspect might be composite, one way to test it is to calculate $2^{n-1} \bmod n$. If it is different from 1, then you can assert that $n$ is composite, while if it equals 1, you can draw no conclusion.

The ancient Chinese thought that this was a perfect test for primality. They thought that an odd number $n$ is prime if and only if $2^{n-1} \equiv 1 \bmod n$. Note that the Little Fermat Theorem only asserts the implication from left to right. It is noncommittal about the implication in the other direction. The Chinese were wrong; they just didn't look at enough numbers. For all numbers through 340, the Chinese were right, but 341 is a composite number that satisfies $2^{340} \equiv 1 \bmod 341$. You will verify that, with hints, in Exercise 11.

An odd composite number $n$ that satisfies $2^{n-1} \equiv 1 \bmod n$ is called a *pseudoprime to the base 2*. This is a number for which the Chinese test for primality fails. There is no need to restrict to 2 the base used in the exponentiation. A composite number $n$ which satisfies the primality test

$$(3.4.3) \qquad\qquad b^{n-1} \equiv 1 \bmod n$$

for some number $b$ that is relatively prime to $n$ is called a *pseudoprime to the base $b$*.

A number $n$ that is prime will pass the primality test (3.4.3) for all bases $b$ which are relatively prime to $n$. The number 341 passed for $b = 2$ but fails for $b = 7$. (See Exercise 13.) Thus the contrapositive of the Little Fermat Theorem can be used to show that 341 is composite.

It would be nice if every composite number failed the primality test (3.4.3) for some base $b$. Unfortunately, that is not the case.

DEFINITION 3.4.4. *A* **Carmichael number** *is a composite number that passes the primality test (3.4.3) for every base b which is relatively prime to b.*[6]

The smallest Carmichael number is 561, as you will verify in Exercise 15. Carmichael numbers are quite rare; there are only 1547 such numbers which are less than 10,000,000,000. However, it was proved in the summer of 1992 by three mathematicians at the University of Georgia that there are infinitely many Carmichael numbers (see A. Granville, "Primality testing and Carmichael numbers," *Notices of American Mathematical Society*, September 1992). This gives an excellent illustration of progress in mathematics. It is the second example of a problem which was discussed as being unsolved in an early draft of this book, but was solved before the book went to press. (The other is the determination of the Hausdorff dimension of the boundary of the Mandelbrot set discussed in Section 5.3.)

The existence of Carmichael numbers limits the usefulness of the Little Fermat Theorem as a test for primality. If you have a number that you suspect might be composite, check whether it passes the primality test (3.4.3) for several bases $b$. If it fails even one, then it is composite, but if it passes them all, you can draw no conclusion. In our "Aside," we will discuss modifications of this test that are more effective.

*ASIDE: PRIMALITY TESTING*

We shall see in Section 4.2 that finding large prime numbers is important in applications to cryptography. Finding effective tests for primality is a very active area of research, both in universities and in places such as the National Security Agency, a branch of the United States government involved in research related to cryptography, which employs more mathematicians than any other employer in the United States. In this "Aside," we discuss briefly a test that can quickly show that a number has a very high probability of being prime, and how this test will give a fairly quick guarantee that a number is prime, provided a certain theorem in complex analysis is true.

The following test for primality was proposed by G. L. Miller in 1975.

---

6. R. D. Carmichael, "On composite numbers $P$ which satisfy the Fermat congruence $a^{P-1} \equiv 1 \bmod P$," *American Mathematical Monthly* 19 (1912): 22–27.

MILLER'S TEST 3.4.5. *If $n$ is an odd number with $n - 1 = 2^s t$ with $t$ odd, we say that $n$ passes Miller's test to the base $b$ if either*
    *i. $b^t \equiv 1 \bmod n$, or*
    *ii. $b^{2^j t} \equiv -1 \bmod n$ for some $j$ with $0 \le j < s$.*

The following result is not difficult to prove. (See Exercises 19 and 20.)

THEOREM 3.4.6. *If $n$ is prime and $b$ is not a multiple of $n$, then $n$ passes Miller's test for the base $b$.*

For example, we show that $n = 13$ passes Miller's test to the base 2. We have $n - 1 = 2^2 \cdot 3$, so that $s = 2$ and $t = 3$ in Miller's test. Criterion (i) and criterion (ii) with $j = 0$ both fail because $2^t = 8 \not\equiv \pm 1 \bmod 13$. However, criterion (ii) with $j = 1$ passes, since $2^6 = 64 \equiv -1 \bmod 13$.

What makes this better than the primality test based on the Little Fermat Theorem, which was discussed in the preceding subsection, is the following result, which is more difficult to prove.

THEOREM 3.4.7. *If $n$ is an odd composite number, then $n$ passes Miller's test for at most $(n - 1)/4$ bases $b$ with $1 \le b \le n - 1$.*

See Exercise 16 for an example. This means that if we use Miller's test instead of (3.4.3) we are guaranteed to eventually detect every composite number. A prime number will pass Miller's test for every base $b$, while a composite number will fail for most bases $b$.

If $n$ is a large number with no obvious divisors such as 2, 3, or 5, and you want to test it for primality, try Miller's test for a few values of $b$. If it fails for any of them, then $n$ is composite. But what if it passes for all those bases $b$ that you tried? Unless you tried more than $(n - 1)/4$ values of $b$, you cannot say for sure that $n$ is prime. And trying that many values of $b$ would take a lot of time. It would be much quicker just to check it by the naïve method of testing divisibility by numbers through $\sqrt{n}$ learned in Section 3.1.

If the number $n$ is in fact composite, then the probability that $n$ passes Miller's test with a randomly selected base $b$ is less than $1/4$, by Theorem 3.4.7. The probability that $n$ passes Miller's test for $k$ randomly selected bases $b$, without ever failing, is less than $(1/4)^k$. This follows from a basic law of probability about repeated independent events. For example, with $k = 20$, the probability that a composite number $n$ passes Miller's test for twenty straight values of $b$ is $(1/4)^{20}$, which is approximately .000000000001. This is an extremely small probability. So in this case you are not guaranteed that $n$ is prime, but it is the safest bet you will ever be able to make. Carrying out Miller's test for twenty values of $b$ might seem like a lot of work, but if $n$ is very large (say 100 digits, which is the order of magnitude used in high-security cryptography applications), this will be much faster than trying to divide $n$ by all numbers through $\sqrt{n}$.

The above paragraph showed how Miller's test can, in a reasonable amount of time, show that a number has a very high probability of being prime[7], but the paragraph that preceded it said that it is not feasible to use Miller's test to guarantee that the number is prime. However, there is a famous conjecture in complex analysis, the Generalized Riemann Hypothesis, closely related to the classical Riemann Hypothesis discussed in Section 3.1, which, if true, would make Miller's test a feasible test for guaranteeing primality. To see this, we state the following deep result, also proved by Miller in 1975.[8]

THEOREM 3.4.8. *If the Generalized Riemann Hypothesis is true, then for every composite positive integer $n$, there is a prime base $b$ with $b < 2(\log n)^2$ such that $n$ fails Miller's test for base $b$.*

This greatly limits the number of values of $b$ that you have to check in order to say that $n$ is prime if it keeps passing Miller's test. Recall that $\log 10^k \approx 2.3k$. Thus if $n$ has 60 digits, so that $n \approx 10^{60}$, and if also $n$ passes Miller's test for each of the approximately 3600 prime numbers $b$ which are less than $2(2.3 \cdot 60)^2 \approx 38,000$, then $n$ is guaranteed to be prime, provided the Generalized Riemann Hypothesis is true. If that seems like a lot of work, compare it with the effort involved in seeing if $n$ is divisible by any of the numbers less than $\sqrt{n} \approx 10^{30}$. The Miller's test calculation is feasible for a modern computer, but the standard divisibility test is not.

## Exercises

1. Verify by direct calculation that Theorem 3.4.1ii is true when $p = 7$, using $a = 1, 2, 3, 4, 5$, and 6. Although you may do it by brute force, it is nicer to do some simplifications mod 7 at intermediate steps to minimize your work. For example, the verification when $a = 6$ can be done very simply.

---

7. Some care is required to make this precise. One can show that if a randomly selected large integer $n$ passes Miller's test for $k$ randomly selected bases $b$, then the probability that $n$ is prime is at least $1 - {(\log n)}/{4^k}$, which can be made as close to 1 as desired by choosing $k$ large enough.

8. G. L. Miller, "Riemann's hypothesis and tests for primality," *Journal of Computer Systems Science* 13 (1976): 300–17.

2. Verify directly that if $p = 7$ and $a = 4$, then the numbers $a$, $2a$, $3a, \ldots, (p-1)a$ are a rearrangement of the numbers $1, 2, \ldots, p-1$ mod $p$.

3. Determine what goes wrong with the first proof of the Little Fermat Theorem when $p = 8$ and $a = 3$. In particular, which of the following three steps is the first one to be incorrect, and what specific statement in the proof fails because 8 isn't prime?
   a. The numbers $a$, $2a, \ldots, (p - 1)a$ are a rearrangement of $1, 2, \ldots, p - 1$.
   b. $2 \cdots (p - 1) \cdot a^{p-1} \equiv 1 \cdot 2 \cdots (p - 1)$ mod $p$.
   c. $a^{p-1} \equiv 1$ mod $p$.

4. Same as number 3, using $p = 8$ and $a = 2$. Give a general explanation of why this case fails earlier than the previous case. Give a general criterion in terms of $p$ and $a$ for when the kind of failure embodied in this problem occurs.

5. What goes wrong with the second proof of the Little Fermat Theorem when $p = 8$ and $a = 3$? What specific statement in the proof fails because 8 isn't prime?

6. Expand $(x + 1)^n$ for $n = 1, 2, 3, 4, 5, 6, 7$, and 8. Circle the coefficients that are multiples of $n$.

7. Write the coefficients to your answers in Exercise 6 in successive rows, each centered at the same place. Each number should not be directly below a number in the previous row, but rather below the space between two adjacent numbers in the preceding row. These numbers form a triangle (with a top 1 missing), called *Pascal's triangle*. Each number should equal the sum of the two numbers closest to it in the row above it. Use this fact to write the next two rows of Pascal's triangle, yielding the coefficients in $(x + 1)^9$ and $(x + 1)^{10}$.

8. Use the Little Fermat Theorem to find the remainder when $3^{1002}$ is divided by 11.

9. Prove the second sentence in the proof of Theorem 3.1.10, given in the "Applications" subsection of this section. You must show that if, for some fixed prime number $p$, all prime divisors of a number are of the form $2kp + 1$, then all divisors of that number are of that form.

10. The argument in the last paragraph of the proof of 3.1.10 is a special case of the fact that the GCD of $2^a - 1$ and $2^b - 1$ is $2^{(a,b)} - 1$, where $(a, b)$ denotes the GCD. Verify this in the following cases:
   a. $a = 6, b = 8$.
   b. $a = 6, b = 9$.
   c. $a = 8, b = 12$.

11. Show that $2^{340} \equiv 1$ mod 341. Since $341 = 11 \cdot 31$, this makes 341

a pseudoprime to the base 2. To show the desired congruence, it is enough to show that $2^{340} \equiv 1 \bmod 11$ and $2^{340} \equiv 1 \bmod 31$. Explain why. The Little Fermat Theorem is helpful in showing each of these.

12. For all odd numbers $n$ from 3 to 19, show that $2^{n-1} \equiv 1 \bmod n$ if and only if $n$ is prime. Do not use the Little Fermat Theorem in your calculations.

13. Show that $7^{340} \not\equiv 1 \bmod 341$, and hence 341 fails the primality test (3.4.3) when $b = 7$. (Hint: First calculate $7^3 \bmod 341$. What is $2^{10} \bmod 341$?)

14. Show that 91 is a pseudoprime to the base 3. (Hint: Argue similarly to Exercise 11.)

15. Show that $561 (= 3 \cdot 11 \cdot 17)$ is a Carmichael number. (Hint: It is enough to show that if $b$ and 561 are relatively prime, then $b^{560} \equiv 1 \bmod p$ for $p = 3, 11$, and 17. Explain why. Use the Little Fermat Theorem to help you show these are true.)

16. Show that the composite number $n = 15$ passes Miller's test for $b = 1$ and 14, and fails for all $b$ from 2 to 13. [This amounts to calculating $b^7 \bmod 15$, and you can save a little work by noting that $14^7 \equiv (-1)^7$, etc.]

17.* Show that 561 fails Miller's test with base 2. Use repeated squaring. (The significance is that 561 is the smallest composite number which (3.4.3) fails to establish as composite. This exercise shows that Miller's test does prove 561 composite.)

18. Show that 25 passes Miller's test with base 7. Find a number $b$ for which 25 fails Miller's test with base $b$.

19. Prove that if $p$ is prime and $a^2 \equiv 1 \bmod p$, then $a \equiv \pm 1 \bmod p$. (Hint: $a^2 - 1 = (a - 1)(a + 1)$.)

20.* Prove Theorem 3.4.6. (Hint: Repeated squarings get you from $b^t$ to $b^{2^s t}$. The last number is 1 mod $n$. Use Exercise 19 to deduce that the last of these numbers not equal to 1 must be $-1$.)

# 4. Cryptography

## 4.1 Some Basic Methods of Cryptography

*Cryptography* is the study of methods of encrypting (also called "enciphering") secret messages so that they can be decrypted (or deciphered) by the designated receiver, and not by others who might intercept the message. *Cryptanalysis* is the study of how to decipher intercepted messages without knowing the key—or, in other words, of how to break the code. The *key* is the specific information about how the message was encrypted, or how to decrypt it. *Cryptology* encompasses both cryptography and cryptanalysis.

These sciences have always been very important to the military. In today's computer age, they are important also for many reasons related to financial affairs and personal data. For example, the Personal Identity Number (PIN) that one enters into an Automatic Teller Machine to obtain cash from one's bank account is encrypted in the computer so that someone who has access to the bank's files would be unable to learn others' PINS. Data bases of tax and credit information are encrypted in order to protect the individual's rights.

Another place where cryptography is used is in lottery tickets. The tickets contain not only the identifying number that must match the winning number in order to win the lottery, but also an encrypted form of the identifying number that is used to prevent forgery of the winning tickets.

In this section we will learn several basic methods of cryptography that have been used in the past. In the next section, we will study a method called "Public Key Cryptography" used in many applications today; it is based on the principles of number theory we learned in Chapter 3.

### SUBSTITUTION CIPHERS

Many ciphers are performed by permuting the alphabet, which means that letters are substituted for other letters according to some rule. One such method was used by Julius Caesar. For him, a message was encrypted by replacing each letter by the letter that was three farther along in the alphabet, with the last three letters of the alphabet being encrypted into the first three. A table presenting this correspondence is given below.

Thus, for example, the message

THIS IS CAESARS WAY

would be encrypted as

WKLVLVFDHVDUVZDB.

Note that we have eliminated spacing between words in order to make the task of the would-be cryptanalyst more difficult. The designated recipient would know that to decipher the message each letter must be pushed back by three. Equivalently, the above table should be read from bottom to top. The presumption is that the recipient, having decrypted the message to "THISISCAESARSWAY," would be able to insert spaces appropriately, assuming there was no dance called the Caesar Sway.

Of course, there is nothing magical about the number 3. The alphabet could be shifted by any number. We could write this as a mathematical formula once we associate numbers with letters. The standard way to do this is by associating the number $i$ to the $i$th letter of the alphabet, where the numbering starts with 0. Then Caesar's cipher could be written

$$y \equiv x + 3 \bmod 26,$$

where $y$ is the encryption of the letter $x$, and, more generally, for any number $d$, there is a cipher given by

(4.1.1)                           $y \equiv x + d \bmod 26.$

By switching the number $d$, the people involved in the message-sending could have a bit more security than Caesar's scheme, which always used $d = 3$. Here the number $d$ becomes the key, and transmission of the key becomes a worry. One method could be to agree in advance that $d$ will be the product of the digits of the day of the month.

For example, if this were your agreement, and on the twenty-fifth day of the month you received the message "WOBBIMRBSCDWKC," then you would know to go back ten letters in the alphabet from each letter in the message. Table 4.1 shows the numbers associated to the letters of the alphabet.

TABLE 4.1. Correspondence between Letters and Numbers.

| A | B | C | D | E | F | G | H | I | J | K | L | M | N | O | P | Q | R | S | T | U | V | W | X | Y | Z |
|---|---|---|---|---|---|---|---|---|---|---|---|---|---|---|---|---|---|---|---|---|---|---|---|---|---|
| 0 | 1 | 2 | 3 | 4 | 5 | 6 | 7 | 8 | 9 | 10 | 11 | 12 | 13 | 14 | 15 | 16 | 17 | 18 | 19 | 20 | 21 | 22 | 23 | 24 | 25 |

You want to subtract 10 mod 26 from the numbers corresponding to each letter received. Thus the first letter of the message is the letter corresponding to $22 - 10 = 12$, and so the first letter is M, and the entire message says "MERRY CHRISTMAS."

If the interceptor of a message knows that the message has been enciphered by a simple shift (4.1.1), it is not difficult to determine the message. One way would be to just try each of the twenty-six possible values of $d$, and see which gives a plausible message. Somewhat more efficient is to try to guess the value of $d$ by using frequency analysis. The first guess would be that the most frequently occurring letter in the intercepted message should be the encryption of the letter E, because E is the most frequently occurring letter in most English text. Both of the messages we have seen so far have been atypical, in that the most common letter in the first was S, and in the second it was R.

Most books on cryptanalysis list a frequency distribution for the occurrence of letters in English text. Different books list slightly different percentages, which can be accounted for by normal statistical variation or by their using different types of texts as the sources. In Table 4.2, we list the distribution from [Konheim]. The letters are listed in order of frequency of occurrence, followed by their percentage of occurrence.

TABLE 4.2. Percentage of Occurrence of Letters.

| Letter | Percent | Letter | Percent |
|--------|---------|--------|---------|
| E | 13.04 | U | 2.49 |
| T | 10.45 | M | 2.49 |
| A | 8.56 | Y | 1.99 |
| O | 7.97 | G | 1.99 |
| N | 7.07 | P | 1.99 |
| R | 6.77 | W | 1.49 |
| I | 6.27 | B | 1.39 |
| S | 6.07 | V | 0.92 |
| H | 5.28 | K | 0.42 |
| D | 3.78 | X | 0.17 |
| L | 3.39 | J | 0.13 |
| F | 2.89 | Q | 0.12 |
| C | 2.79 | Z | 0.08 |

**Example.** You intercept the message "KYVTRKZEKYVYRK," which you expect to have been enciphered by a simple shift. What is the message?

*Solution.* The commonly occurring letters are K with 4, Y with 3, and R and V with 2 each. Your first guess should be that K is the encipherment of E, since E is often the most commonly occurring letter. This would say that $d = 6$ in (4.1.1), although what is perhaps easier is just to write beneath the alphabet a shifted copy of the alphabet with K beneath E. This gives your best guess for the key.

A B C D E F G H I J K L M N O P Q R S T U V W X Y Z
G H I J K L M N O P Q R S T U V W X Y Z A B C D E F

Reading this from bottom to top, we find that the decrypted message would be

ESPNLETYESPSLE,

which doesn't make sense as a message. Your next guess should be either that E encrypts to Y (the second most commonly occurring letter in the message), or that T (the second most commonly occurring letter in the English language) encrypts to K. If you try the first of these, you will again obtain gibberish, while the latter ($d = 17$) causes the alphabets to line up as below, a proposed encryption table.

A B C D E F G H I J K L M N O P Q R S T U V W X Y Z
R S T U V W X Y Z A B C D E F G H I J K L M N O P Q

Applying the decryption form of this table to the given message yields

THECATINTHEHAT,

the name of a children's book by Dr. Seuss, which must be the message.

The simple shift method we have been considering is clearly not very secure. There are various ways to modify it so that it becomes more secure. The simplest modification is to apply an affine transformation, rather than a simple shift, to the alphabet. This means that, under the usual association of numbers to letters, the enciphering formula would be

(4.1.2) $$y \equiv ax + b \bmod 26$$

for some numbers $a$ and $b$. In order for this to have a unique deciphering, it is necessary that $a$ and 26 be relatively prime. The word "affine transformation" which occurs here will occur again in the "Aside" to Section 5.1. It is a rule that transforms each number $x$ into $ax + b$ for some fixed numbers $a$ and $b$.

The partners in the message could have some agreement about how the numbers $a$ and $b$ would be chosen, or they could be transmitted by a messenger. For example, they might agree that $a$ will be the sum of all digits of the day of the month and the number of the month (or if this sum is not relatively prime to 26, then take as $a$ the smallest number greater than this sum that is relatively prime to 26), while $b$ is the product of all these digits. Thus on January 15, the enciphering would be given by

(4.1.3)                    $$y \equiv 7x + 5 \bmod 26,$$

since $1 + 1 + 5 = 7$, while $1 \cdot 1 \cdot 5 = 5$.

It is a bit harder to write out the encrypting and decrypting tables for this affine transformation method than it was for the simple shift considered above. One way to make an encipherment table for (4.1.3) would be to start by saying that A($= 0$) enciphers to F($= 5$), and then count cyclically around the alphabet by sevens, starting with F. Thus B would encrypt to the seventh letter after F, which is M, etc. This yields the following encryption table:

```
A B C D E F G H I J K L M N O P Q R S T U V W X Y Z
F M T A H O V C J Q X E L S Z G N U B I P W D K R Y
```

The reader should understand why this method of forming the encryption table yields the same table as would be obtained by substituting into (4.1.3). It is because increasing $x$ by 1 in (4.1.3) increases $y$ by 7. As a check, we note that if $x = 13$ (corresponding to N), then

$$y = 7 \cdot 13 + 5 \equiv 96 \equiv 18 \bmod 26,$$

which does indeed correspond to S, as the table says.

If you are the partner on the receiving side of this cipher, you can just read this table backwards to decrypt the messages you receive. For example, if you receive

BIFUIMZLMJSVIZLZUUZD,

the first letter, B, decrypts to S, because S is above B in the above table, and the entire message decrypts to "STARTBOMBINGTOMORROW," which the Allied forces in Saudi Arabia would have had no trouble reading. This message would have been a bit tricky for an eavesdropper using frequency analysis to cryptanalyze, since it has no E's and its most frequent letter, O, is only the fourth most common letter in general.

Cryptanalysts who intercept messages they suspect may have been encrypted by the affine transformation method will have to guess two of the letters in the correspondence in order to determine the entire key. For example, when they finally get around to the guess that maybe the most common letter in the enciphered message ($Z = 25$) might correspond to $O(= 14)$, and a second most common encrypted letter ($I = 8$) might correspond to $T(= 19)$, they would have to solve two equations in mod 26 arithmetic for $a$ and $b$:

$$14a + b \equiv 25 \bmod 26$$
(4.1.4)                           $$19a + b \equiv 8 \bmod 26.$$

The first of these equations is obtained by substituting $x = 25$ (for Z) and $y = 14$ (for O) into (4.1.2), and the second is obtained similarly.

Although we did not learn how to do this in Chapter 3, there are standard methods for solving pairs of equations in congruence arithmetic, totally analogous to solving ordinary systems of equations. (See Exercise 6.) An alternative for the cryptanalyst would be to try the 312 different possible pairs of values of $a$ and $b$ with $a$ relatively prime to 26 to see which gives nongibberish as the decryption of the intercepted message.

The next step in complication of an enciphering scheme is to use an arbitrary permutation of the alphabet. In this case, the key is some shuffling of the letters of the alphabet in such a way that each letter corresponds to exactly one letter. The difficulty with this method is the transmission of the key. It is not as easily describable as the one or two numbers that determined the transformations in the two methods considered previously. Although such an enciphering scheme is more secure than the previous two methods, it can easily be cryptanalyzed by frequency analysis, and so you cannot use the same permutation forever. Much work has been done on the cryptanalysis of this method. The cryptanalyst looks not only for frequently occurring letters, but also for things such as frequently occurring three-letter combinations, which have a good chance of being THE. Cryptanalyzing messages that have been encoded by this method is similar to solving cryptogram puzzles in the newspaper.

Another complication that can make cryptanalysis more difficult for an eavesdropper is to have the number $d$ by which a letter is shifted vary from letter to letter. For example, the partners might agree that messages will be encoded by adding the numerical values of the letters in the word "CHILDREN," written over and over again, to the letters of the message being encrypted. Since the numbers associated to the successive letters of CHILDREN are, respectively, 2, 7, 8, 11, 3, 17, 4, and 13, this means that the formula

$$y \equiv x + 2 \bmod 26$$

holds for the 1st, 9th, 17th, 25th, etc., letters of the message, while

$$y \equiv x + 7 \bmod 26$$

holds for the 2nd, 10th, 18th, etc. For example, if we want to encipher the message

TRANSFER THE MONEY

we can write out the required shifts as follows:

message T R A N S   F E R T H   E M O N E Y
letter added C H   I L D   R E N C H   I L D R E N
encryption V Y   I Y V W   I E V O M   X R E I L

Thus VYIYVWIEVOMXREIL is the encrypted message. What this amounts to is addition in mod 26 arithmetic, except that we use letters as replacements for the numbers. For example, the first letter, T, of the actual message is shifted by 2, which corresponds to the C, yielding V as its encryption. This corresponds to the mod 26 addition $19 + 2 = 21$. It is useful to make a $26 \times 26$ table depicting the addition formulas for the letters of the alphabet. The reader is encouraged to do this in Exercise 11. There will be rows as well as columns labeled by the letters A through Z. In the position in the T row and C column will appear the letter V, corresponding to the addition fact used above.

The communication of what word, or eight letters, to use for shifting in the way that CHILDREN was used above is easily carried out without use of a messenger. For example, you could have an agreement that you use the first eight letters on the third page of the previous day's *New York Times*.

There are many other ways of complicating the substitution method so as to make a cryptanalyst's job more difficult. One of these, our final example, involves encrypting pairs of letters rather than single letters. One way of doing this is to agree on four numbers $a$, $b$, $c$, and $d$, and use the formula

$$y_1 \equiv ax_1 + bx_2 \bmod 26$$
$$y_2 \equiv cx_1 + dx_2 \bmod 26$$

(4.1.5)

to encipher the pair of letters $x_1 x_2$ to the letters $y_1 y_2$. For example, suppose you had agreed on the numbers $a = 1$, $b = 3$, $c = 7$, and $d = 12$, and your message began with the letters FU, corresponding to the numbers $x_1 = 5$ and $x_2 = 20$, using Table 4.1. Then the first letter you would transmit would be N, corresponding to the number $y_1 = 1 \cdot 5 + 3 \cdot 20 = 65 \equiv 13$, and the second letter would be P, corresponding to the number $y_2 = 7 \cdot 5 + 12 \cdot 20 = 275 \equiv 15$. The third and fourth letters you transmit would be calculated similarly, based on the third and fourth letters in your message.

To decipher messages, you will be receiving the letters corresponding to the numbers $y_1$ and $y_2$ and want to solve (4.1.5) for $x_1$ and $x_2$. This can be done as long as $ad - bc$ is relatively prime to 26. The solving of these equations can be easily mechanized.

The adversaries will not initially know the numbers $a$, $b$, $c$, and $d$. If the text is long enough, they may be able to determine these numbers by using frequency analysis of letter pairs. Tables similar to our 4.2 exist for letter pairs. The most common are TH, IN, and ER.

### THE ENIGMA CIPHERS

The Enigma was a type of electrical machine used for enciphering and deciphering by the Germans in World War II. A group of British cryptanalysts, led by the mathematician Alan Turing, developed methods of deciphering intercepted messages, and this played a major role in the Allied victory. The messages, many of which dealt with the positions and intended positions of German ships and submarines, were sent by ordinary radio telegraph transmission, and so were easily intercepted. The Germans thought that the method of encrypting used by these Enigma machines was so complex that no one could decipher it without knowing the key, and so this source of Allied information remained a secret throughout the war. Turing's efforts were featured in a play by Hugh Whitmore, *Breaking the Code*, which was very popular in London and New York in 1989.

Peter Hilton, who later did important work in algebraic topology, was still a teenager when he joined this group of cryptanalysts. In his article, "Reminiscences of Bletchley Park, 1942-1945," written in 1988, he vividly describes the excitement of providing to British officials ultrasensitive German military plans within hours of their actual transmission.

The Enigma machine, pictured schematically in Figure 4.1, consisted of three (later five) rotors connected to an input device similar to a typewriter keyboard and an output device. The output device was a system of lights that indicated the letter that was the encryp-

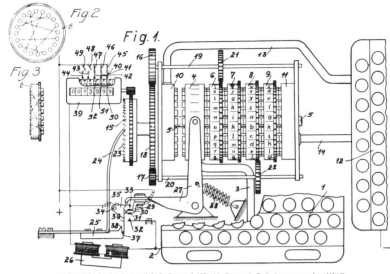

Arthur Scherbius' "Engima," U.S. Patent 1,657,411. Rotors 6, 7, 8, 9 are set to key NIAG.
Figures 2 and 3 schematically show rotor connections

Figure 4.1. An Enigma machine.

tion of the typed letter, or, later, it was a sort of printer. Each of these rotors was a disk, similar in size and shape to a hockey puck. On each side of one of these disks were twenty-six electrical contacts, while inside each disk were wires connecting each position on one side of the disk to one of the positions on the other side. The three rotors were connected in series, and each could turn. When a letter was typed, the rotors would be lined up in such a way that each of the twenty-six positions of the right side of the first rotor would be connected electrically to the adjacent position on the left side of the second rotor, with a similar connection from the right side of the second rotor to the left side of the third rotor. On the right side of the third rotor were wires connecting pairs of positions to one another.

In Figure 4.2, we have made a schematic drawing of the three rotors for a six-letter alphabet. The actual rotors had twenty-six letters instead of our six. The rotors in our drawing are disproportionately fatter than in an actual machine. In this drawing, we assume that the rotors have been turned so that the up-positions are, respectively, 3, 6, and 4. There is also a ring with the letters of the alphabet that is associated to the numbers on the left side of the first rotor. If the letter A is pressed on the typewriter, current goes into the left rotor at position 3 and out at position 4. It then goes into the second rotor at position 1 and out at 6, and into the third rotor at 4 and out at 5.

The wire at the right then sends the current into the right side of the third rotor at 6, out the left side of the third rotor at 1, then right to left in the second rotor from 3 to 2, and through the first rotor from 5 to 2. Thus the output will be the letter corresponding to position 2, which in this case will be F.

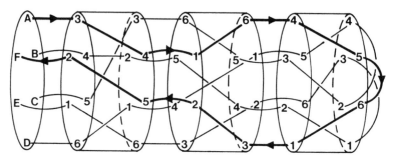

Figure 4.2. A wiring of an Enigma-type machine.

Thus the setting in Figure 4.2 encrypts A to F. Following the same darkened lines in the opposite order shows that this setting also encrypts F to A, or, equivalently, it decrypts A to F. This was an important feature of the Enigma machines—at a given setting, a machine can be used for either encrypting or decrypting. However, this feature also made the work of the cryptanalyst somewhat easier, since the symmetry of these ciphers decreased the number of possibilities that had to be considered.

What made this more than just a simple substitution cipher was that after each letter was encrypted, one of the rotors turned, odometer-style—that is, the left rotor would turn one notch after every letter, the middle rotor would turn one notch after every 26 letters, and the right rotor would turn one notch after every 676 $(= 26^2)$ letters. In Figure 4.3, we have shown how the rotors of Figure 4.2 would be positioned one letter later, after the left rotor had moved one notch. Now A will be enciphered to E.

The Germans had many Enigma systems, including different ones for their navy, army, and air force, and still different ones for the various geographical theaters of the war. All Germans who might be involved in sending or receiving naval messages would have a naval Enigma machine, and each of these would have its three rotors wired in the same way. The three rotors could be put on in any order, and the order was changed daily.[1] Also communicated daily was a basic setting of the three rotors for the day.[2] These communications were

1. Ultimately, different rotors were made, and a different selection of three out of eight was made each day.
2. By "setting," we mean the numbers that appear at the top of the rotors.

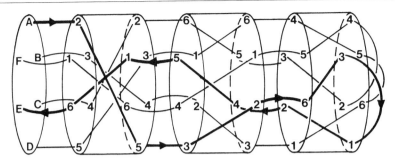

Figure 4.3. The way Figure 4.2 appears after turning one notch.

secure. A man enciphering a message could choose any initial setting, but he would use the basic setting of the day to encrypt in the first part of the message the information that told what he was using as the initial setting. The person receiving the message would know the basic setting, use it to decipher the encrypted initial setting, and then set his machine accordingly to decipher the message.

The British cryptanalysis group obtained a reconstruction of an Enigma machine from a member of the Polish resistance, which gave them the wiring of the rotors. What they didn't know was the basic setting of the day or the order of the rotors. Because they received thousands of intercepted messages each day, they had much raw material to study. A Polish cryptanalysis group had devised machines that could consider, one at a time, all possible basic settings for the messages of a given day. These were applied to some of the earliest messages of the day, and the basic setting was determined by seeing what gave sensible information for the initial setting information and consequent message. Turing's British group improved on these machines—for example, by making them work faster. A machine called the COLOSSUS[3] that was developed by Turing's group was a big step toward the modern computer. Other complications had to be considered, such as figuring out how this setting information was encoded, and responding to changes in the way the Germans utilized the machines (such as adding rotors).

### FOCUS: ALAN TURING AND HIS "MACHINE"

Alan Turing (1912-1954) was the leader of the group of British mathematicians who broke the German code during World War II, a fact that played a major role in the defeat of the Germans. This effort was veiled in secrecy even long after the war. (In fact, even in his 1988 article, Peter Hilton was not at liberty to say as much as he would

3. The COLOSSUS was actually used for the decryption of a different German encryption system called the Fish.

Alan Turing

have liked.) Thus Turing received no public recognition for this work during his lifetime, which, as we shall see later, ended at an early age.

Turing did important theoretical work in 1936 while still a graduate student, long before his cryptanalysis. For this work, which we will describe in some detail, he is frequently given the title, "Father of the Computer." He solved a famous problem in mathematical logic that had been posed by the great German mathematician David Hilbert in 1928. This problem, the *Entscheidungsproblem*, asked whether there was a procedure by which one could tell whether or not any given mathematical statement was provable. It had already been established by Gödel[4] that there were some statements, called "independent," that could be neither proved nor disproved. But Gödel did not establish a procedure by which one could tell whether or not a given statement is independent.

Turing was influenced by his professor, Max Newman, who lectured on this problem and innocently used the phrase "mechanical process" for the desired procedure. This led Turing to formulate the idea of an abstract computer that could carry out any procedure followed when one does mathematics. This abstract machine is now called a "Turing machine," and, as we shall indicate later, it was very influential in the work of Turing and others on the first actual computers of the late 1940s.

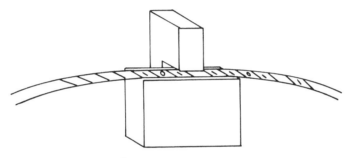

Figure 4.4. A Turing machine.

A Turing machine is a theoretical device that processes an imaginary infinite strip of paper marked off into squares (Fig. 4.4). Each square of the paper may contain a single symbol chosen from a finite alphabet. This "alphabet" need not be our ordinary alphabet; it may consist of numbers or other symbols. The device will at any time be in one of a finite number of states, or configurations. A "state" is a setting of the internal parts of the machine that tells it how to respond to the next input it receives. The strip passes through the device in such a way that the machine can read the content of one square, called the "input." Depending on the state of the device and the contents of the square it reads, the device does three things:

1. It may erase the content of the square and then may write some other letter of the alphabet on that square, called the *output*.

4. See the "Aside" of Section 2.1.

2. It may move the strip one position in either direction; this has the effect at the next step to make the device focus on the square to the left or to the right of the current square. We can think of the machine as moving its focus along the strip.
3. It may change its state.

We illustrate by describing a Turing machine that adds two numbers. We will use as our alphabet just the number 1, and think of a number $n$ as being represented by 1's in $n$ consecutive squares. For example, the machine does not think of the number of fingers on your hand as the symbol 5, but rather as 11111, in successive squares. The numbers to be added will be separated on the strip by a single blank square. The machine will initially focus on a square to the left of all the 1's. It will move to the right through the first block of 1's, put a 1 in the blank square between the two blocks of 1's, move through the second block of 1's, and, when it gets to the end of this, it will erase the last 1, and stop. Figure 4.5 shows a sample beginning-and-ending diagram, corresponding to $3 + 4 = 7$.

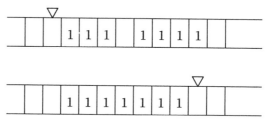

Figure 4.5. Input and output tapes for a Turing machine.

A machine to accomplish this procedure can be made with four states, corresponding roughly to

State 1.  . . . has not yet encountered a 1. This is the initial state of the machine.
State 2.  . . . is moving through the first block of 1's.
State 3.  . . . is moving through the second block of 1's.
State 4.  . . . has moved through both blocks of 1's.

See Table 4.3 to see how the machine should behave for each of the eight possible pairs (Input, State). "Output" is what it writes in the square it reads. Frequently, Output equals Input, which means that it leaves the square unchanged. "Left" or "Right," for the motion of the machine, indicates whether it focuses its attention on the next square to the left or to the right. The reader is asked in Exercise 16 to verify that the machine behaves as we said it will.

Alan Turing showed that Hilbert's *Entscheidungsproblem* was

TABLE 4.3. Behavior of Turing Machine for Addition.

| Input | State | Output | Motion | New State |
|-------|-------|--------|--------|-----------|
| blank | 1 | blank | right | 1 |
| 1 | 1 | 1 | right | 2 |
| blank | 2 | 1 | right | 3 |
| 1 | 2 | 1 | right | 2 |
| blank | 3 | blank | left | 4 |
| 1 | 3 | 1 | right | 3 |
| blank | 4 | blank | no | stop |
| 1 | 4 | blank | no | 4 |

equivalent to the question of whether there was a way of determining if any particular Turing machine would ever stop, which is sometimes called the Halting Problem. En route to obtaining a negative answer to this, he showed that there was a *universal Turing machine*, a single machine that could simulate the behavior of any Turing machine. Using this universal machine, he showed that if the halting question were always decidable, then one would obtain a contradiction somewhat similar to Russell's Paradox in set theory, discussed in the "Aside" to Section 2.1. Thus the answer to Hilbert's *Entscheidungsproblem* was negative.

Turing's idea of a single machine that could perform any computation was fundamental when, at the end of World War II, many different universities, corporations, and government agencies throughout the world rushed to develop the first practical electronic computers. Between 1945 and 1950, Turing was employed by three such groups in England. His expertise, both theoretical and practical (the latter a consequence of his cryptanalytic work), was invaluable. Some claim that during this period he invented the art of computer programming. The development of physical computers from Turing's work on the abstract *Entscheidungsproblem* is another example of the theme of this book—unexpected applications of work in abstract mathematics.

Turing must share the credit for the idea and early development of the computer with many others. Notable among these are the British mathematician Charles Babbage (1791–1871)—who had the idea of a machine that could perform any computation a century before Turing—and his colleague Ada, Countess of Lovelace and daughter of the poet Lord Byron. Ada developed some programming techniques used today, and a modern programming language is named Ada in her honor. Babbage's idea was somewhat less universal than Turing's, and his vision of a physical model was mechanical rather than electrical or electronic. The physical model of his idea was never completed because of inadequate engineering techniques. A working model of Babbage's design was constructed in 1991 by scientists at London's Science Museum to show that Babbage's principles were sound.

John von Neumann (1903–1957) emigrated from Europe to the United States at the time of the Nazi takeover. Von Neumann was an extremely powerful mathematician who made important contributions to pure mathematics, quantum physics, meteorology, computer science, and economics. He was the leader in the development at the Institute for Advanced Study in Princeton of one of the first American electronic computers, which was then used in the development of the hydrogen bomb. Although von Neumann is given much credit for the development of computers, he reportedly said "the fundamental conception is owing to Turing."[5]

Another aspect of computer science in which Turing was ahead of his time was that of Artificial Intelligence (AI), the question of whether a computer can think like a person. For example, Turing pioneered chess-playing by computers, with the emphasis on having computers learn by playing. Although they were just a pipe dream in Turing's time, by 1990 chess programs could beat all but the top few players in the world. The most successful is a custom-built chess-playing computer named Deep Thought built by a group of graduate students at Carnegie-Mellon University.

Turing's most lasting mark on AI is what is still called the Turing Test. He posited that AI will have been achieved when a computer can answer questions so skillfully that the interrogator will not know whether the questions are being answered by a computer or by a human. This test is still far from being passed.

Turing's life ended early, at the age of 41, as a result of cyanide poisoning, apparently self-inflicted.[6] Although the reasons for the apparent suicide are not completely clear, a contributing cause seems to have been his arrest for homosexuality in 1952. Turing's house had been robbed by an acquaintance of a sex partner of his, and in explaining to the police how he knew who the thief was, Turing made no secret of his relationship. He was so open because he thought, incorrectly, that the laws against homosexuality were about to change. His sentence was to undergo a year of hormone treatments. His death

5. Hodges 1983, p. 304

6. His mother, Sara Turing, wrote a biography, *Alan M. Turing* (Cambridge, U.K.: W. Heffer, 1959), in which she strongly denied suicide.

did not occur until a year after the end of this undesirable treatment, when his life seemed to have returned pretty much to normal. He was doing interesting new research in mathematical biology at the time of his death. Partly because of the mystery behind Turing's death, Hodges subtitled his biography "The Enigma."

I. J. Good, one of Turing's coworkers on the Enigma cryptanalysis, wrote: "It was only after the war that we learned he was a homosexual. It was lucky the security people didn't know about it early on, because if they had known, he might not have obtained his clearance, and we might have lost the war."[7]

## Exercises

1. Encipher "CRYPTOGRAPHY IS FUN" using (4.1.1) with $d = 5$.
2. Decipher "TMMTVDTMWTPG" if it has been enciphered by (4.1.1) with $d = 19$.
3. Apply frequency analysis to decipher the intercepted message

   DRKYVDRKZTJTREYVCGJFCMVKYVNFICUJGIFSCVDJ,

   if you know that it has been encrypted by a simple shift (4.1.1).
4. Apply frequency analysis to decipher the intercepted message

   SQUIQHICUJXETMQIJEEIYCFBUJERUKIUVKB,

   if you know that it has been encrypted by a simple shift (4.1.1).
5. Compare the frequency of occurrence of letters in something substantial that you have written with that of Table 4.2.
6.* Solve the system of equations (4.1.4) in mod 26 arithmetic. (Hint: Subtract one from the other to eliminate $b$.)
7. Explain why 312 is the number of pairs $a$ and $b$ with $a$ relatively prime to 26.
8. You receive the message "XLQEGQGQRNOURQY," which has been encrypted by the formula $y \equiv 3x + 4 \bmod 26$. What is the message?
9. Explain why $a$ in (4.1.2) must be relatively prime to 26.
10.* Use frequency analysis to decrypt the following message, which has been encoded by a permutation of the alphabet. To make it easier, we have included breaks between words. (Hint: The frequency of letters in the encrypted message is not unusual, at

---

7. Metropolis, Howlett, and Rota 1980. p.34

least as far as the two most frequently occurring letters are concerned.)

GA GAXA ZMUKN OAUIMRA DJFAW FVA JXAEMAWUN

QAFVDY WAAYR ZDWPAX FAHF

11. Make a $26 \times 26$ table depicting the addition formulas for the letters of the alphabet. This is easily done by just writing the alphabet over and over again, shifting it by one more space each time.

12. Suppose that you and your partner have agreed to use the letters CALIFORN to determine by what amount the letters in a message are to be shifted, similarly to the way CHILDREN was used in the text. You receive the message UUCZJBURT. What is your partner telling you? (Note: The last occurrence of CALIFORN will not use all the letters.)

13. In the example in the text using (4.1.5) to encipher pairs of letters, suppose the next two letters in the message are ND. Using the same values of $a$, $b$, $c$, and $d$ as in the text, what two letters would you transmit?

14.* Using the same method of enciphering as in the previous exercise, including the same values of $a$, $b$, $c$, and $d$, suppose you receive the letters FP. What was the message? (Hint: Multiply one of the equations by a number so that subtracting will eliminate one of the variables.)

15. Suppose the Enigma machine in Figure 4.3 has its first and second rotors turn one more position (clockwise). Draw a diagram of the resulting positions, and determine how A would be encoded.

16. Assume that the Turing machine described in the text begins at the blank square just to the left of the 1's, and is in State 1. Follow the rules for the behavior of the machine, and tell at each step what the position and state of the machine are, and what the strip looks like.

17.* What will the Turing machine described in the text do if there is more than one blank space between the two batches of 1's? Devise a Turing machine that will "add" two batches of 1's separated by an arbitrary batch of spaces. (Hint: One way to do it would be to have it erase the initial 1, then fill in a 1 when it gets to a first blank, and if more blanks intervene, have it go back to the start, and repeat the steps. The number of blanks separating the two batches of 1's will be one less than it was before.)

## 4.2 Public Key Cryptography

For most of the methods of cryptography discussed in the previous section, one of the biggest problems is transmission of the key. This information, which is required of the receiver of the messages in order to decipher received messages, must be changed frequently in order to confuse those who intercept these messages, and it must be transmitted in a secure manner.

In the late 1970s, methods of cryptography were developed in which no transmission of a key would be required. In this method, called Public Key Cryptography, a person who wants to receive messages makes public some numbers (the key) to all those who might be sending him messages. These numbers tell them how to encrypt the messages they wish to send to him. He knows a secret number that enables him to decipher the messages he receives.

For the most commonly used method of public key cryptography, called the RSA method, the numbers actually used will be about one hundred digits long; therefore computers are required to do the enciphering and deciphering. Those intercepting the message could decrypt it if they knew the secret number. In principle this number is deducible from the public numbers, but in practice many months of computer time would be required to find the secret number.

In this section, we will discuss the mathematics underlying both the RSA method and its historical predecessor, the Knapsack Method, both of which are based on elementary principles of number theory that we learned in Chapter 3. We will also discuss issues related to the security of the RSA system.

### *THE* RSA *METHOD*

The RSA method is named after its discoverers, R. L. Rivest, A. Shamir, and L. M. Adleman, who discovered it at MIT in 1978. It depends for its security on the fact that it is easy to find large prime numbers, but if they are multiplied together it is difficult to retrieve the prime factors from the knowledge of their product. In practice, the prime numbers involved have at least fifty digits, but even with much smaller numbers you can get some appreciation that finding factors is much more difficult than finding primes. It is not difficult to check that 191 and 283 are both prime, but if you are given the number 54,053 and asked to factor it, you will probably have a great deal of difficulty realizing it is $191 \cdot 283$ unless you have a computer.

This procedure is implemented in the following way. We use the first-person form of address to designate the person who will be receiving messages. First, we all have a standard way of translating messages into numbers; Table 4.1 will suffice for our purposes, except that we must denote A through J by 00 through 09 rather than 0 through 9. If we didn't do that, it would not be clear whether 12 referred to M (which it does) or BC (which should be represented as 0102).

My secret numbers are two large prime numbers $p$ and $q$, and a number $N$ which is relatively prime to the product $(p-1)(q-1)$; this says that $N$ has no common factors with either $p-1$ or $q-1$. The numbers that I make public are the product $pq$ and a positive number $M$ satisfying $MN \equiv 1 \bmod (p-1)(q-1)$. I make them public by writing in a public directory that anyone who wants to send me a message represented by a number $x$ should transmit to me the least nonnegative residue of $x^M \bmod pq$. Here $M$ and $pq$ will be explicit numbers. The number $x$ must satisfy $0 \le x < pq$, and the number transmitted will lie in the same range. This necessitates that the number corresponding to our message be broken into chunks that are each less than $pq$, and that the calculation be performed on these one at a time. When I receive a number $y$ which I know is the encrypted form of part of someone's message, I decipher it by calculating the least nonnegative residue of $y^N \bmod pq$. This will be the original number $x$ that my correspondent wished to convey to me. Here $N$ is my secret exponent.

**Example.** (Details presented later.) My secret numbers are $p = 11$, $q = 13$, and $N = 7$. Note that 7 has no common factors with $p-1$ or $q-1$. I privately calculate, using the Euclidean algorithm, that the smallest positive number $M$ that satisfies $7M \equiv 1 \bmod 120$ is $M = 103$, and I make public the following announcement: "Anyone who wants to send me a message, represented by a 2-digit number $x$ in the standard way, should transmit the least nonnegative residue of $x^{103} \bmod 143$." If someone wants to send me a message CT, represented by the numbers 02,19, they calculate $2^{103} \equiv 63 \bmod 143$ and $19^{103} \equiv 72 \bmod 143$. (These are quite tedious to calculate by hand, but easy for a computer. You can do them by hand by the method of repeated squaring learned in Section 3.3. In fact, $2^{103} \bmod 143$ was the example worked in that section.) They send the message 63,72. When I receive this message, I calculate $63^7 \equiv 2 \bmod 143$ and $72^7 \equiv 19 \bmod 143$, and so I know the original message was $02, 19 = CT$.

Note that anyone who has access to this public directory can send me messages, but people who intercept messages being sent to me cannot decipher them because they do not know my secret exponent $N = 7$. We will discuss below how they might be able to determine $N$, which is all they would need in order to decipher intercepted messages.

When working problems of this type, you must remember when to use $pq$ as the modulus, and when to use $(p-1)(q-1)$. You use $pq$ as the modulus when doing the exponentiation, and $(p-1)(q-1)$ when doing the Euclidean algorithm. Note that none of the numbers $p$, $q$, or $(p-1)(q-1)$ is made public, only the product $pq$. As we discuss in the next paragraph, an interceptor who can factor the number $pq$ to find $p$ and $q$ will be able to decipher intercepted messages just as well as we can. Thus the security of this method depends on the numbers $p$ and $q$ being too large for potential interceptors to discover them as factors of their product $pq$.

In the above example, someone intercepting the message who realizes that 143 could be factored as $11 \cdot 13$ would know that $p$ and $q$ are 11 and 13, so that $p-1$ and $q-1$ are 10 and 12. This eavesdropper also knows the public exponent $M = 103$ and that my secret exponent $N$ satisfies $MN \equiv 1 \bmod (p-1)(q-1)$. Putting in the numbers, this becomes

$$103N \equiv 1 \bmod 120.$$

This is the kind of congruence equation we learned how to solve in Section 3.3. The Euclidean algorithm is applied to the numbers 103 and 120, working down to the GCD, which is 1, and then backing up the equations to write 1 as a combination of 103 and 120. The coefficient of 103 is my secret exponent 7. Of course, it is easy to factor 143 as $11 \cdot 13$, but in practice much larger numbers are used, and the success of the method depends on the fact that it is extremely difficult for even the most powerful computers to factor very large numbers.

Now we give some details from the above example:

1. *Finding M*. Once I have selected $p = 11$, $q = 13$, and $N = 7$, I must find the number $M$ which satisfies $M \cdot 7 \equiv 1 \bmod 120$. The Euclidean algorithm lasts only one step: $120/7 = 17$ rem 1. Thus $1 = 120 - 17 \cdot 7$ shows that $(-17) \cdot 7 \equiv 1 \bmod 120$, so that we could take $M = -17$, except that $M$ must be positive. Since this is an equation in mod 120 arithmetic, we add 120 to get $M = -17 + 120 = 103$.

2. *Encoding 19*. We must calculate $19^{103} \bmod 143$. This is similar to the calculation of $2^{103} \bmod 143$ performed in Section 3.3.

Always working mod 143, we have

$$19^2 = 361 \equiv 75$$
$$19^4 \equiv 75^2 = 5625 \equiv 48.$$

Since 48 was obtained as one of the answers when calculating $2^{103}$ in Section 3.3, we can use those calculations to read off

$$19^8 \equiv 48^2 \equiv 16$$
$$19^{16} \equiv 16^2 \equiv 113$$
$$19^{32} \equiv 113^2 \equiv 42$$
$$19^{64} \equiv 42^2 \equiv 48.$$

Hence, decomposing 103 as a sum of 2-powers as was done in Section 3.3, we have

$$19^{103} = 19^{64}19^{32}19^419^219^1 \equiv 48 \cdot 42 \cdot 48 \cdot 75 \cdot 19$$
$$\equiv 16 \cdot 42 \cdot (-5) \equiv -80 \cdot 42$$
$$\equiv -71 \equiv 72.$$

The step that went from five factors to three involved using the calculation of $48^2 \equiv 16$ done previously, and $75 \cdot 19 = 1425$, which is 5 less than $10 \cdot 143$.

3. *Deciphering.* Always working mod 143, we have

$$63^2 = 3969 \equiv 108$$
$$63^4 \equiv 108^2 = 11664 \equiv 81$$
$$63^7 = 63^463^263^1 \equiv 81 \cdot 108 \cdot 63 \equiv 25 \cdot 63 \equiv 2.$$

Similarly,

$$72^2 = 5184 \equiv 36$$
$$72^4 \equiv 36^2 = 1296 \equiv 9$$
$$72^7 = 72^472^272^1 \equiv 9 \cdot 36 \cdot 72 \equiv 9 \cdot 18 \equiv 19.$$

We could do $36 \cdot 72$ with essentially no calculation since we had just done $36 \cdot 36 \equiv 9$, and $36 \cdot 72$ will be twice as large.

4. *Intercepting.* As we saw above, if you can factor 143, then all you need to do to find the secret exponent $N$ is to solve

the equation $103N \equiv 1 \bmod 120$. Two steps of the Euclidean algorithm gets you to the GCD of 1, and to the equations

$$17 = 120 - 1 \cdot 103$$
$$1 = 103 - 6 \cdot 17$$

Combining them yields $1 = 103 - 6(120 - 103) = 7 \cdot 103 - 6 \cdot 120$, and so $N = 7$.

You should have been asking yourself by now: Why does this method work? In other words, if a number is enciphered by raising to the $M$th power mod $pq$, why does raising the transmitted number to the $N$th power mod $pq$ retrieve the original number? If $x$ is the number representing the message, then $y \equiv x^M \bmod pq$ is the encrypted message, and $y^N \equiv (x^M)^N \bmod pq$ is the deciphering of the encrypted message. That this equals $x$ is the content of the following result.

THEOREM 4.2.1. *If $p$ and $q$ are distinct prime numbers, and $MN \equiv 1 \bmod (p-1)(q-1)$, then $(x^M)^N \equiv x \bmod pq$.*

*Proof.* Choose the positive integer $k$ so that $MN = k(p-1)(q-1) + 1$, which can be done since $MN \equiv 1 \bmod (p-1)(q-1)$. Thus,

$$(x^M)^N = x^{MN} = x^{k(p-1)(q-1)+1} = x^{k(p-1)(q-1)} \cdot x,$$

and so we want to show that

(4.2.2) $$x^{k(p-1)(q-1)} \cdot x \equiv x \bmod pq.$$

There are three cases:

   Case 1. Suppose $x$ is divisible by both $p$ and $q$. Then both sides of (4.2.2) are congruent to 0 mod $pq$.
   Case 2. Suppose $x$ is divisible by $q$ but not by $p$. (A similar argument works if $x$ is divisible by $p$ but not by $q$.) The Little Fermat Theorem, 3.4.1, with $a = x^{k(q-1)}$, says that

(4.2.3) $$(x^{k(q-1)})^{p-1} \equiv 1 \bmod p.$$

   You can always multiply a congruence equation, including the modulus, by a number. (See Exercise 1.) Thus (4.2.3) implies

(4.2.4) $$x^{k(p-1)(q-1)} \cdot q \equiv q \bmod pq.$$

Since $x$ is assumed to be a multiple of $q$, we can write $x = qx'$, for some integer $x'$. Multiplying (4.2.4) by $x'$ yields the desired result (4.2.2) in this case.

Case 3. Suppose $x$ is not divisible by $p$ or $q$. The Little Fermat Theorem with $a = x^{k(q-1)}$ says that $(x^{k(q-1)})^{p-1} \equiv 1 \bmod p$, while the Little Fermat Theorem using $q$ as the prime and $a = x^{k(p-1)}$ says $(x^{k(p-1)})^{q-1} \equiv 1 \bmod q$. These two equations say that $x^{k(p-1)(q-1)} - 1$ is a multiple of both $p$ and $q$, and, since $p$ and $q$ are distinct primes, this implies that it is a multiple of $pq$. Hence so is $x(x^{k(p-1)(q-1)} - 1)$, which is a restatement of our desired conclusion, (4.2.2). Q.E.D.

As you can see from the example worked above, the tedious part of the RSA algorithm is the exponentiation in congruence arithmetic. Actual users of the RSA algorithm have computer chips made specially to perform the RSA algorithm efficiently, but as we will see in the subsection called "Security," the time required to do these exponentiations limits the usefulness of the RSA algorithm. Access to a computer would be useful to the reader in working the exercises based on the RSA algorithm. All exercises can be worked by hand or with the aid of a calculator. See the discussion prior to (3.3.6) for hints on how to use a calculator to do such calculations. The advantage that computers have over calculators here is that they have a MOD operator in virtually any programming language, while modular arithmetic on a simple calculator is awkward.

In a basic language such as BASIC, you can write things such as

$$36 * 72 \bmod 143,$$

and the computer will immediately respond with 18. BASIC will not, however, calculate

(4.2.5) $$19^{103} \bmod 143,$$

because the number $19^{103}$ is too large for this computer language to work with. The way that BASIC evaluates (4.2.5) is to first try to evaluate $19^{103}$ and then take the remainder upon division by 143. The computer can help you with each step in the repeated squarings that we did in order to compute (4.2.5) earlier. There is a problem with many simple programming languages in that they are quite limited with the size of the integers with which they will work. Many of them have a LongInteger type, but this still requires some care. Many computer algebra packages such as MAPLE and MATHEMATICA have special modular exponentiation statements that automatically perform the repeated squaring algorithm, so that a statement similar to (4.2.5) will work in those languages.

## SIGNATURES

One nice feature of the RSA algorithm is that it lets you "sign" your message in a secure way. Here we assume that there is a group of people on a computer network using the RSA method to send messages to one another. Each has published his or her numbers $M$ and $pq$ in a common directory. Alice wants to send a message to Bob in such a way that he will know it is really coming from her, and not from somebody else on the system pretending to be her.

She encrypts her message in the way described in the previous subsection, by breaking her message up into blocks represented by numbers that are smaller than Bob's published number $(pq)_B$ and raising these to Bob's published exponent $M_B$, always working mod $(pq)_B$. We use the subscript $B$ to indicate Bob's numbers, as opposed to Alice's, which will be indicated with subscript $A$. After somehow indicating that the message is over and the signature is starting, she breaks her name up into numbers and carries out the following procedure on each of the numbers $x$.

She first computes $z = x^{N_A} \mod (pq)_A$, where $N_A$ is her secret exponent and $(pq)_A$ her public modulus. She then encrypts this number $z$ using Bob's public instructions. Thus she transmits the number $y \equiv z^{M_B} \mod (pq)_B$. We have assumed here for simplicity that the number $z$ is smaller than Bob's public modulus $(pq)_B$. If not, then $z$ must be broken up into blocks, and the above procedure applied to each of them.

When Bob receives the message, he deciphers it in the usual way, by raising to the $N_B$ power mod $(pq)_B$. But with each number $y$ of the signature, he carries out the following two steps. He first computes $w \equiv y^{N_B} \mod (pq)_B$. Then he applies Alice's public numbers $M_A$ and $(pq)_A$ to $w$ by calculating $w^{M_A} \mod (pq)_A$. This will equal one of the numbers $x$ corresponding to Alice's name.

We explain why this procedure works. The four steps are, in order:

$$z \equiv x^{N_A} \mod (pq)_A$$
$$y \equiv z^{M_B} \mod (pq)_B$$
$$w \equiv y^{N_B} \mod (pq)_B$$
$$\text{output} \equiv w^{M_A} \mod (pq)_A.$$

By applying Theorem 4.2.1 to the two middle steps, we see that $w = z$ and hence

$$\text{output} = z^{M_A} \mod(pq)_A.$$

Now we apply Theorem 4.2.1 to this new equation and the first of the four steps above, and find that the output is indeed $x$.

The reason this signature is guaranteed to have come from Alice is that she used her secret exponent in the first step of the encryption. Presumably, no one else knows this number. And it is the only number which is inverse to the $M_A$-exponentiating that Bob does at the last stage. There is no need to go through this extra step on the whole message, but without doing this for the signature, other people on the network could send a message to Bob and say that it was from Alice.

### SECURITY

We saw above that an eavesdropper who can factor the public number $pq$ can find $(p-1)(q-1)$ and then apply the Euclidean algorithm to $(p-1)(q-1)$ and the public exponent $M$ to find the private exponent $N$. This person will be able to decipher intercepted messages just as easily as the originator of the cipher. Thus it is imperative that the numbers used be large enough to make the factoring of $pq$ impractical.

In practice, this means that $p$ and $q$ must be primes of at least sixty digits, and with rapid advances in the computer factorization of integers, to be really safe, $p$ and $q$ should have about eighty-five digits. With numbers this large, the encryption and decryption of long messages becomes very slow, even on the fastest computers. Consequently, RSA is often used in practice just for the transmission of the key of a simpler, faster cipher, of a type similar to some of those studied in Section 4.1, which is used for transmission of the entire message. As we saw there, transmission of the key is the biggest problem for such a cipher, so it is very worthwhile to have a secure public way to transmit.

The large prime numbers $p$ and $q$ that must be chosen by the originator of an RSA cipher must be selected so that they are not easily guessed. The methods of primality testing we studied in Section 3.4 are useful for this selection. The person choosing the prime determines approximately what value he or she wants the prime to have, eliminates obvious nonprimes such as multiples of 2, 3, and 5, and then uses Miller's test to check candidates for primality. For cryptography purposes, it is enough to have a very large (e.g.,

.99999999999999) probability of being prime, without having a rigorous guarantee. The Prime Number Theorem, 3.1.7, tells roughly how many numbers you will have to check before finding a prime.

At places such as the National Security Agency, much research is being done on methods of factoring numbers because of the implications for cryptography. This is mostly classified research, and some mathematicians in academia have been upset when the government restricts publication of their research in number theory because the methods of cryptography must remain secret.

The great British mathematician G. H. Hardy took great pride in the pureness of his work, much of which was in number theory. In 1940, in his delightful essay, "A Mathematician's Apology," he wrote: "There is one comforting conclusion which is easy for a real mathematician. Real mathematics has no effects on war. No one has yet discovered any warlike purpose to be served by the theory of numbers or relativity, and it seems unlikely that anyone will do so for many years. . . . So a real mathematician has his conscience clear; there is nothing to be set against any value his work may have; mathematics is . . . a 'harmless and innocent' occupation."[8] Not long after Hardy's statement, relativity was having a great impact on war via applications of Einstein's formula $E = mc^2$, and in less than forty years, number theory had developed applications of sufficient weight to cause them to become secret information.

Great strides have been made in factoring in recent years. This is in part due to improvements in computer hardware and networking capabilities, but also to more sophisticated mathematical techniques. In 1988 major newspapers ran prominent articles reporting that a 100-digit number had been decomposed into its prime factors, which had 41 and 60 digits. This was done by feeding parts of the problem to hundreds of small computers. One of the leading experts, Arjen Lenstra, was quoted as saying: "A few years ago, people suggested that 100 digits would be safe. Now maybe it will take 150."[9]

It didn't take long for the 150-digit barrier to fall. As reported in the *New York Times* on June 20, 1990 (see article), a similar effort succeeded in factoring a 155-digit number into its three prime factors. This suggests that increasingly larger numbers are going to have to be used by RSA cryptographers to insure the security of their messages. Certainly, a massive effort was required to obtain the factorization of this huge number, but progress is being made so quickly that one expert said: "If national security were hidden behind a 150-

---

8. Hardy 1967, p. 140.
9. Allentown (Pa.) *Morning Call*, October 12, 1988.

# Biggest Division a Giant Leap in Math

**By GINA KOLATA**

In a mathematical feat that seemed impossible a year ago, a group of several hundred researchers using about 1,000 computers has broken a 155-digit number down into three smaller numbers that cannot be further divided.

The number is about 50 digits longer than any that mathematicians have reported being able to break down in the same way, an unusually long leap in this area of mathematics.

The latest finding could be the first serious threat to systems used by banks and other organizations to encode secret data before transmission, cryptography experts said yesterday.

These systems are based on huge numbers that cannot be easily factored, or divided into numbers that cannot be divided further. For example, the factors of 10 are 2 and 5.

**First Break in the System**

This is the first time that mathematicians have factored a number of the size used in these coding systems, said Dr. Arjen Lenstra, director the project who is at Bellcore Inc., in Morristown, N.J., the research arm of the Bell operating companies.

To break the huge number into three smaller numbers, which are 7, 49 and 99 digits long, the mathematicians had to find a new method because the one used in recent years was not up to the job. If someone had asked him to break up a 155-digit number a year ago, Dr. Lenstra said, "I would have said it was impossible."

Dr. Andrew Odlyzko, a mathematician at the American Telephone and Telegraph Bell Laboratories in Murray Hill, N.J., said: "This is a great achievement. From the standpoint of computational number theory, it represents a breakthrough."

The number itself was famous among mathematicians as a factoring challenge. In October 1988, mathematicians reported the factoring of a 100-digit number. It is a rule of thumb in mathematics that for every 10-digit increase in the size of a number, the amount of computing needed to factor it increases 10 fold. Until now, factoring advances had come in increments of 10 digits or less.

**Secrets Are at Stake**

But the practical importance of the result, experts said, is what it might mean to cryptography. In 1977, a group of three mathematicians devised a way of making secret codes that involves scrambling messages according to a mathematical formula based on factoring. Now, such codes are used in banking, for secure telephone lines and by the Defense Department.

In this system, a string of letters in the message are replaced a number. That number is multiplied by itself many times, making a bigger number that helps mask the message. Then the big number is divided by a large number whose factors are secret. The remainder of that division — the amount left over — is the coded message. It can only be decoded by a person who knows the secret factors of the large number.

In making these codes, engineers have to strike a delicate balance when they select the numbers used to scramble messages. If they choose a number that is easy to factor, the code can be broken. If they make the number much larger, and much harder to factor, it takes much longer for the calculations used to scramble a message.

For most applications outside the realm of national security, cryptographers have settled on numbers that are about 150 digits long, said Dr. Gus Simmons, a senior fellow at Sandia National Laboratories in Albuquerque, N.M., who advises the Defense Department on how to make coding secure.

**Broader Application Seen**

Dr. Lenstra, who also led in the breaking of the previous 100-digit number, said: "For the first time, we have gotten into the realm of what is being

## Factoring a 155-Digit Number: The Problem Solved

13,407,807,929,942,597,099,574,024,998,205,846,127,
479,365,820,592,393,377,723,561,443,721,764,030,073,
546,976,801,874,298,166,903,427,690,031,858,186,486,
050,853,753,882,811,946,569,946,433,649,006,084,097

**equals**

2,424,833

**times**

7,455,602,825,647,884,208,337,395,736,200,454,918,
783,366,342,657

**times**

741,640,062,627,530,801,524,787,141,901,937,474,059,
940,781,097,519,023,905,821,316,144,415,759,504,705,
008,092,818,711,693,940,737

*Source: Mark Manasse, Ph.D.*

---

*Mixed blessing: an advance that could imperil secrets.*

---

used in cryptography. It means it is impossible to gurantee security."

Although the number the group factored had a special mathematical structure, Dr. Lenstra and his colleagues say the factoring method can be modified so it would have broad application.

Others are more circumspect. Dr. Simmons said that although he agrees that the method is generally applicable, he is waiting to see whether it can break down other numbers quickly enough to be practical. The method, he said, "may become of concern to cryptographers, but that depends on how efficiently it can be implemented."

Nonetheless, Dr. Simmons said, he would not feel comfortable advising the use of a 150-digit number to maintain security. "If national security were hidden behind a 150-digit number,

*Fred R. Conrad/The New York Times*

If asked to break up a 155-digit number a year ago, "I would have said it was impossible," said Dr. Arjen Lenstra of Bellcore Inc. in Morristown, N.J.

we're getting very close to a situation where it would be feasible to factor that," he said. "Do I advise the Government to use bigger numbers? You bet."

The newly factored number was the largest number on a list mathematicians keep of the 10 Most Wanted Numbers, which are large numbers that are set up as a challenge to factoring experts. And it is so large that it is inconceivable to even think of factoring it without special mathematical tricks.

**Answer in a Few Months**

Dr. Mark Manasse of the Digital Equipment Corporation's Systems Research Center in Palo Alto, Calif., calculates that if a computer could perform a billion divisions a second, it would take 10 to the 60th years, or 10 with 59 zeros after it, to factor the number simply by trying out every smaller number that might divide into it easily. But with a newly discovered factoring method and with a world-wide collaborative effort, the number was cracked in a few months.

The new factoring method was discovered last year by John Pollard of Reading, England, and Dr. Hendrik Lenstra Jr. of the University of California at Berkeley, the brother of Dr. Arjen Lenstra,

Then Dr. Manasse and Dr. Arjen Lenstra recruited computer scientists and mathematicians from around the world to help in the factoring effort. Each person who agreed to help got programs sent electronically to their computers and a piece of the problem to work on.

**Like a Jigsaw Puzzle**

It was like solving a giant, and twisted, jigsaw puzzle, Dr. Manasse said. Each computer was set to work doing the mathematical equivalent of sorting through a box with about 50 million pieces "including all sorts of useless stuff that look like jigsaw pieces that are not," Dr. Manasse said, adding: "Each person has to find the real pieces in the box. Some boxes don't have any and some have just one or two."

After about a month, the researchers got back the equivalent of about two and a half million pieces of the puzzle. To speed up the search and the final putting together of the pieces that would allow them to factor the number, the researchers used a powerful computer at the Universtiy of Florida that finished the job for them in three hours.

The current factoring landmark is the latest in a series of what to mathematicians have been breathtaking feats. In 1971, mathematicians scored a coup by factoring a 40-digit number. Ten years ago, a 50-digit number was thought to be all but impossible to factor. Then, with advances in research that led to unexpected shortcuts, 60-, 70- and 80-digit numbers fell. A year and a half ago, the 100-digit number was cracked.

digit number, we're getting very close to a situation where it would be feasible to factor that. Do I advise the Government to use bigger numbers? You bet!"

The number whose factors were found was a special one to mathematicians—it was a Fermat number $2^{2^9} + 1$. Recall from Theorem 2.2.15 that the only constructible regular polygons with a prime number of sides are those whose number of sides is a prime number of the form $2^{2^m} + 1$. These numbers have long been known to be prime for $m \leq 4$, and are now known to be composite for $5 \leq m \leq 21$, but finding the factors is much more difficult than merely establishing that they are composite. This number was on top of the list of "10 Most Wanted Numbers," the biggest challenges to mathematicians and computer scientists who specialize in factoring. Because of its special form, some special techniques could be applied to it, but it is believed by most workers that it is a harbinger of the factoring of other large numbers.

In the "Aside" at the end of this section, we will discuss the possibility of finding fast ways of doing factoring in general. This discovery could render the RSA method useless. In the next subsection, we will consider the first method of public key cryptography, which preceded the RSA method by a year. It was at first felt to be secure, but within five years of its invention, an effective method of cryptanalyzing it was discovered.[10]

### THE KNAPSACK METHOD

In this subsection, we will briefly consider the Knapsack Method of public key cryptography, which was devised by Stanford computer scientists R. Merkle and M. Hellman in 1977. It is important historically as the first method of public key cryptography, but it is no longer considered to be secure because of the advances that were subsequently made in breaking this code.

In this method, a message first has to be broken down into a string of 0's and 1's. There are standard ways of doing this—by associating a binary number to each letter of the alphabet. Thus we will assume that the information someone wants to transmit is a sequence of 0's and 1's, and that it is common knowledge how to translate back and forth between such sequences and messages.

To originate a knapsack cipher, I make public a carefully selected sequence of integers $b_1, \ldots, b_N$. Anyone wishing to send a message to me first divides the sequence of 0's and 1's that corresponds to

10. By A. Shamir, one of the inventors of RSA.

the message into blocks of length $N$. Let $(\epsilon_1, \ldots, \epsilon_N)$ be one such block. Here each $\epsilon_i$ is 0 or 1. For this block, the person transmits the sum

$$(4.2.6) \qquad S = \epsilon_1 b_1 + \epsilon_2 b_2 + \cdots + \epsilon_N b_N.$$

Because of the way in which I selected the numbers $b_i$, I will be able to deduce the numbers $\epsilon_i$ from the number $S$. Thus I will know the message.

Before I explain how I obtained my numbers, let us see what is happening from the point of view of the sender of a message. Suppose my published numbers are

$$b_1 = 13, \ b_2 = 129, \ b_3 = 39, \ b_4 = 136, \ b_5 = 111, \ b_6 = 132,$$

and the sender wants to communicate to me the sequence 100110. He or she communicates the number,

$$1 \cdot 13 + 0 \cdot 129 + 0 \cdot 39 + 1 \cdot 136 + 1 \cdot 111 + 0 \cdot 132 = 13 + 136 + 111 = 260.$$

Note that someone who knows the published numbers and intercepts the transmitted number 260 and wants to figure out the sequence 100110 is faced with the problem of determining which of the six published numbers can be added together to yield exactly 260 as a sum. If that person can see that the sum of the 1st, 4th, and 6th of the published numbers equals 260, then she or he will know that the message had 1's in the 1st, 4th, and 6th positions, and 0's elsewhere. That may not seem so hard to do, but in practice the list is much longer, and the numbers much larger. This is called the Knapsack Method because you can think of the published numbers as being in a knapsack, and the problem is to draw out of that knapsack a combination of those numbers equaling the transmitted number.

Now we explain how the numbers $b_1, \ldots, b_N$ were selected in a way so that the numbers $\epsilon_1, \ldots, \epsilon_N$ corresponding to any number $S$ could be determined. I begin by privately selecting a sequence of numbers $a_1, \ldots, a_N$ which has the property that each number $a_i$ is greater than the sum of its predecessors. For example, I might choose

$$a_1 = 3, \ a_2 = 5, \ a_3 = 9, \ a_4 = 19, \ a_5 = 38, \ a_6 = 80.$$

Then I choose a number $m$ satisfying $m > 2a_N$, and a pair of positive numbers $v$ and $w$ satisfying

(4.2.7) $$wv \equiv 1 \bmod m,$$

with both $v$ and $w$ less than $m$. This should be done by first choosing $m$, then $v$ relatively prime to $m$, and then using the Euclidean algorithm to find $w$ satisfying (4.2.7). For example, in this case I might choose $m = 161$ and $v = 25$. Applying the Euclidean algorithm to 161 and 25 shows me that $w = 58$ satisfies (4.2.7).

My public numbers $b_i$ are chosen to be the least nonnegative residue of $wa_i \bmod m$. For example, if the $a_i$'s are as above, then to find $b_1$, I calculate the remainder when $58 \cdot 3$ is divided by 161. This is $174 - 161 = 13$. The reader should verify that the other numbers $b_i$ above equal the remainder when $58a_i$ is divided by 161. I can also reorder the numbers $b_i$ in order to make things more difficult for a cryptanalyst. Of course, I must then remember the way in which they were reordered. We will not bother with this fine point here.

When I receive a number $S$, I first multiply it by my secret number $v$ and reduce mod $m$. In our example, $S \cdot v = 260 \cdot 25 = 6500$, which when divided by 161 gives a remainder of 60. Note that (4.2.6) becomes
$$S \equiv \epsilon_1 a_1 w + \cdots + \epsilon_N a_N w \bmod m,$$

and so

$$S \cdot v \equiv \epsilon_1 a_1 wv + \cdots + \epsilon_N a_N wv \bmod m$$
$$\equiv \epsilon_1 a_1 + \cdots + \epsilon_N a_N \bmod m,$$

since $wv \equiv 1$. In our example, this says

$$60 \equiv 3\epsilon_1 + 5\epsilon_2 + 9\epsilon_3 + 19\epsilon_4 + 38\epsilon_5 + 80\epsilon_6 \bmod 161,$$

and since all numbers involved are less than 161, congruence may be replaced by equality, that is,

$$60 = 3\epsilon_1 + 5\epsilon_2 + 9\epsilon_3 + 19\epsilon_4 + 38\epsilon_5 + 80\epsilon_6.$$

This is another knapsack problem, but it is a much easier one, because each of the numbers in the knapsack is greater than the sum of its predecessors. To find what sum of 3, 5, 9, 19, 38, and 80 equals 60, start with the biggest number and work down, selecting a number as long as it doesn't make the sum too large. Thus we don't choose 80 because it is too large, but we do choose 38, leaving $60 - 38 = 22$

more to be selected. We choose the 19, and then need only three more, and so we skip the 9 and 5 and choose the 3. Thus we have seen that the 1st, 4th, and 5th coefficients must be 1, and that is indeed the message 100110 that was being conveyed to us.

The point of this method is that an arbitrary knapsack problem can be very difficult to solve, but a knapsack problem in which each number is greater than the sum of its predecessors is very easy. By use of inverses in congruence arithmetic, we can disguise an easy knapsack problem as a hard one, but we are the only ones who know how to remove the disguise.

We recapitulate the steps in the Knapsack Method:

1. Select numbers $a_1, \ldots, a_N$ such that each number is greater than the sum of its predecessors. Also choose $m > 2a_N$ and numbers $v$ and $w$ satisfying (4.2.7).
2. Compute numbers $b_i$ equal to the remainder when $wa_i$ is divided by $m$.
3. Write in a public directory: "To convey to me a message, divide it into sequences $(\epsilon_1, \ldots, \epsilon_N)$ of 0's and 1's of length $N$, and, for each sequence, transmit the number $S = \epsilon_1 b_1 + \cdots + \epsilon_N b_N$."
4. When you receive a number $S$, let $R$ denote the remainder when $Sv$ is divided by $m$. Then starting with $a_N$ and working your way down through the smaller $a_i$'s, you can easily determine which $a_i$'s can be added together to equal $R$. These $a_i$'s will correspond to the $\epsilon_i$'s that equaled 1 in the message that was being conveyed to you.

*ASIDE: COMPUTATIONAL COMPLEXITY*

Essential to the security of the RSA method of cryptography is the presumption that finding factors of integers will continue to be a difficult problem. Certainly advances such as those described above will continue to be made, but these can be defended against by making the numbers longer, as long as the advances in factoring are not so great as to change the problem from what is considered "hard" to what is considered "easy."

The "easy" problems are those that can be solved in *polynomial time*, which are said to be in class $P$. These are the problems for which there are numbers $C$ and $e$ such that if the input to the problem has length $n$, then the problem can be solved in $Cn^e$ steps. This is like the leading term of a polynomial. By the *length* of a problem involving numbers, we mean the number of digits in the numbers, and by the number of *steps* we mean bit operations, that is, manipulations with

0's and 1's. The Euclidean algorithm for the GCD is an example of a problem in class $P$, for it can be performed in $Cn^2$ steps for some number $C$, where $n$ is the total number of digits in the two numbers. Another problem that was surprisingly shown to be in class $P$ is the knapsack problem for knapsacks that have been formed in the way that $(b_1, \ldots, b_N)$ was formed from $(a_1, \ldots, a_N)$ in the preceding subsection.

Primality testing is probably in class $P$. It is known that if the Generalized Riemann Hypothesis is true, then Miller's Test, which we discussed in the "Aside" of Section 3.4, will determine whether an $n$-digit number is prime in $Cn^5$ steps for some number $C$. Bounds can be given for the numbers $C$ involved in each of these estimates of the number of steps required, but their values are not considered to be very important. The important thing is that as the length of the input numbers grows, the number of steps required grows only with some power of that length, which is considered to be a manageable form of growth.

The big question for the RSA algorithm is whether factoring is in class $P$. It is empirically known that factoring seems harder than primality testing, but it is still conceivable that an algorithm for factoring in polynomial time might be discovered. The consequence of such a discovery would be that the RSA method would probably be deemed impractical as a method of cryptography, because in order to make it secure, the numbers used would have to be so large that enciphering and deciphering would be so slow as to render the system useless.

There is a class of problems, called $NP$ for "Nondeterministic Polynomial," which includes not only all problems in class $P$, but also many other problems such as the General Knapsack Problem and the problem of factoring. Roughly speaking, a problem is in class $NP$ if a purported solution can be checked for validity in polynomial time. It is clear that each of the two problems just mentioned has this property. The difficulty is in discovering the collection from the knapsack with the desired sum, or the numbers whose product is the given number. Once you have found a candidate, it is a simple matter to check whether it works. The name "Nondeterministic" comes from the fact that they are problems which can be solved in polynomial time by a machine that makes lucky guesses.

The most important question in the field of Computational Complexity is whether $P = NP$, in other words, whether all problems whose answers can be *checked* in polynomial time can actually be *solved* in polynomial time. Virtually all workers in the field think this is false, but a proof that it is false has not been found. Such a proof would require showing that there is a problem in $NP$ that cannot be solved in polynomial time.

It is known that certain problems, such as the General Knapsack Problem, have the property that if they can be solved in polynomial time, then so can every $NP$ problem. This was proved in 1971 by S. A. Cook, by showing that every $NP$ problem can be transformed in polynomial time to the General Knapsack Problem. Thus a polynomial-time solution to the General Knapsack Problem would give a polynomial-time solution of every $NP$ problem, but no one expects this to happen. The General Knapsack Problem is not the only problem with this universal property. The problems that have this property are called $NP$-complete. They are in a certain sense the hardest $NP$ problems. Cook's theorem could be stated in the following way.

THEOREM 4.2.8. $P = NP$ if and only if the General Knapsack Problem can be solved in polynomial time.

Factoring is not thought to be $NP$-complete. Hence it is conceivable that a polynomial-time algorithm could be found for factoring but not for the General Knapsack Problem.

### Exercises

1. Prove that if $a \equiv b$ mod $m$, then $ka \equiv kb$ mod $km$. (This simple fact was used in our proof of Case 2 of Theorem 4.2.1.)

2. Use $p = 7$, $q = 11$, and $N = 7$ in the RSA algorithm. Show that the Euclidean algorithm would yield $M = -17$, but then it can be taken as 43, since we are working mod 60. What instructions would you make public? Show that 8 would be encrypted as 50, that is, $8^{43} \equiv 50$ mod 77. Show that 50 is deciphered correctly, that is, $50^7 \equiv 8$ mod 77.

3. Use $p = 5$, $q = 17$, and $N = 3$ in the RSA algorithm. Find $M$. Determine how the number $x = 8$ would be encrypted, and show that this encrypted number would be deciphered correctly.

4. Suppose that you intercept the number 9 being transmitted to someone who uses the RSA algorithm and has published, "Anyone who wants to convey to me the number $x$ should transmit the least nonnegative residue of $x^{11}$ mod 65." What number was being conveyed? [Hint: You must factor 65, determine $(p-1)(q-1)$, and then $N$, and finally calculate $9^N$ mod 65.]

5. If you worked through all the steps in "3. Deciphering" in the first RSA example in the text, perhaps you noticed that on three occasions the remainder happened to equal the quotient. This

was in $5184/143$, $1296/143$, and $11664/143$. Why does the quotient equal the remainder in these cases? Your answer should have to do with a common property shared by 5184, 1296, and 11664.

6. Explain why the secret exponent in the RSA algorithm must have no common factors with $p - 1$ or $q - 1$.

7. You initiate an RSA system with secret primes 11 and 23. Choose an appropriate value for your secret exponent $N$. (There is more than one possible answer here.) What do you write in the public directory? How do you decipher the numbers you receive? (For this last question, just tell what calculation you would perform. You don't have to tell how you would do it.)

8. You intercept the number 2 being transmitted to an RSA-person who has written, "To convey to me the number $x$, transmit $x^7 \mod 77$." Determine what number the transmitter wished to convey.

9. Suppose Alice's numbers are $(pq)_A = 77, M_A = 17,$ and $N_A = 53,$ while Bob's are $(pq)_B = 119, M_B = 5,$ and $N_B = 77.$
   a. Verify that these are legitimate choices, that is, that $MN \equiv 1 \mod (p - 1)(q - 1)$.
   b. The $L$ in Alice's name corresponds to the number $x = 11.$ If she wants Bob to know that it is she who is transmitting this number, and follows the procedure of the "Signatures" subsection, what number will she transmit corresponding to the letter $L$?
   c. Follow the steps Bob would carry out in deciphering the number he receives in (b).

10. Find the numbers $\epsilon_i$ satisfying

$$R = 2\epsilon_1 + 5\epsilon_2 + 9\epsilon_3 + 20\epsilon_4 + 60\epsilon_5 + 100\epsilon_6$$

(a) if $R = 87$; (b) if $R = 74$; (c) if $R = 131$.

11. You are setting up a knapsack cipher system with

$$a_1 = 2, \ a_2 = 5, \ a_3 = 9, \ a_4 = 20, \ a_5 = 60, \ a_6 = 100,$$

and $m = 211$.
   a. Verify that $w = 10$ and $v = 190$ satisfies (4.2.7).
   b. Determine the numbers $b_1, \ldots, b_6$ that you would make public.
   c. You receive the number 384. What was the message? Use step (4) of our summary of the Knapsack Method to obtain your answer.
   d. Check that your answer to (c) is compatible with your answer to (b).

# 5. Fractals

## 5.1 Fractal Dimension

Fractals are objects, either in mathematics or in the real world, that have a fractional dimension. In this section we will explain what this means by giving a variety of examples and focusing our discussion on their fractal dimension. In the "Aside" of this section and in Section 5.3, we will discuss in more detail how some very beautiful fractals can be generated on a computer.

Many of the mathematical ideas now used in the study of fractals were developed in the early years of the twentieth century, as were some of the most important examples of fractals. These ideas lay outside the mainstream of mathematics for many decades because mathematicians preferred to study smooth phenomena, of which fractals are the antithesis. During the 1950s, 1960s, and 1970s, Benoit Mandelbrot, a mathematician at IBM in Yorktown Heights, New York, wrote a series of papers in which he showed how many phenomena in nature have these fractal properties, and he resurrected and supplemented the work of earlier mathematicians.

### THE KOCH CURVE

Most fractals have the property of self-similarity. This means that small parts of them, when magnified, resemble the whole thing. Sometimes this resemblance is only approximate, and sometimes it is precise. An example in which the self-similarity is perfect is the Koch curve, which was first described by the Swedish mathematician Helge von Koch in 1904.

The Koch curve is defined to be the limit of a sequence of curves. The sequence starts with a straight line segment. The result of each step will be a curve that is formed of many straight segments connected together. To obtain the next step, every straight segment is modified by having its middle third replaced by the top two sides of an equilateral triangle sitting above it. We can picture this by saying that at each step every segment —— is replaced by ⌐𝝠⌐. Figure 5.1 shows the first six steps in the formation of a Koch curve.

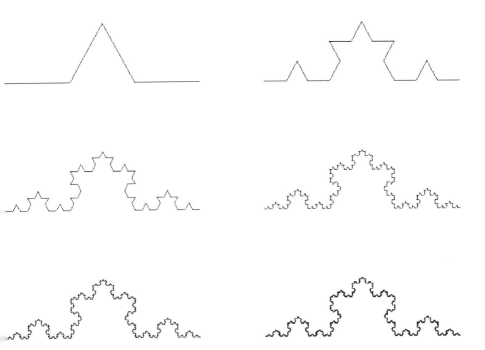

Figure 5.1. First six steps of the Koch curve.

You can have a computer draw the various steps in the construction of the Koch curve, or you can personally draw them, with much care, with pencil and ruler. In either case, after about six steps, you will not be able to notice the difference between the successive steps in the formation of the curve. This is because the lengths of the segments at each step are one-third times the length of the segments at the previous step. Hence, at the sixth step, the segments are $(1/3)^6 = 1/729$ times the length of the original segment. A computer screen, like a television screen, is composed of a grid of little rectangles called *pixels*. This grid is often either 640 by 200 or 640 by 480. When any line on the screen has its length multiplied by $1/729$, the result will be at most about one pixel long. When drawing the curve yourself, suppose you started with a line 10 inches long. After six steps, you will be drawing lines $1/73$ of an inch long, which will test anyone's artistic ability.

Mathematically, however, the curve is still far removed from its final form at the sixth step. Note that at each step the length of the curve is multiplied by $4/3$. Thus, after six steps the length of the curve will be $(4/3)^6$ times as long as the original segment. This is about 5.6 times as long. The ultimate Koch curve, however, has had this length multiplied by $4/3$ infinitely often, and so it is infinitely long. Thus an infinite amount of length has been added, but the additions have been so short that they cannot be discerned on a computer screen or in the work of a most diligent artist.

Although we cannot physically see them, these tiny bendings in the Koch curve are every bit as real as the first few, that are so apparent in the pictures. Mathematical formulas can be given for all of the points in the curve. The bendings after the sixth step can't be seen because they involve changes in some decimal place after the third decimal place, and these changes are too small to be noticed. The infinite length that is being added is just a series of little wrinkles in the curve that are extremely close to the best approximation we can see. It is almost as if these wrinkles are adding thickness to the curve, making it almost 2-dimensional. Fractal dimension gives a precise measure of the extent to which this thickening is happening.

We emphasize that the Koch curve is nevertheless still a curve. There are continuous functions $x(t)$ and $y(t)$ giving the $x$ and $y$ coordinates of the point that is a fraction $t$ of the way along the curve. Here $t$ is going from 0 to 1. Another property of curves which the Koch curve shares is that if you remove one point from it, it becomes disconnected. Note in Figure 5.2 that removal of a single point from the curve divides it into separated parts, whereas this is not the case for a 2-dimensional region. Defining "separated" precisely requires some care, and we shall not do so here. Readers who have studied calculus will appreciate that the Koch curve is an example of a continuous curve that is nowhere differentiable. This means that it is never smooth enough to have a tangent at a point.

Figure 5.2. Removing a point from a curve disconnects it.

Before we discuss some of the ways of defining fractal dimension, we close this subsection by pointing out another pleasing way of envisioning the Koch curve. This is as the boundary of the Koch Island or Koch Snowflake, which are two words for the same thing. It is obtained by starting with a solid equilateral triangle instead of a line segment, and then following a procedure similar to the one above.

Figure 5.3. Six steps in forming the Koch Snowflake.

Since we are now considering solid figures, we just adjoin an equilateral triangle to the middle third of each segment on the boundary at each step of the construction. The first six steps are pictured in Figure 5.3.

The boundary of the Snowflake is three copies of the Koch curve. This Snowflake has finite area (see Exercise 1), but its boundary has infinite length.

### SIMILARITY DIMENSION

The Koch curve is exactly self-similar. As Figure 5.4 suggests, it can be subdivided into four parts, each of which is exactly like the whole thing with all of its lengths divided by 3, or it can be subdivided into sixteen parts, each of which is exactly like the whole thing with all of its lengths divided by 9.

Figure 5.4. Self-similarity of the Koch curve.

We can use some rather uninteresting examples of mathematical objects that are exactly self-similar to help us get a handle on what should be the relationship between the self-similarity dimension of the object and the two quantities relevant to the self-similarity:

$d$ = self-similarity dimension, yet to be defined

$N$ = number of parts, each exactly proportional to the whole object, into which a self-similar object can be divided

$s$ = the scaling factor, which is the ratio of lengths in the whole object to corresponding lengths in the parts in the subdivision.

For example, in the Koch curve, we can take $N = 4$ and $s = 3$, or $N = 16$ and $s = 9$. We will figure out what the appropriate value of $d$ should be.

The uninteresting examples are a line, square, or cube. As indicated in Figure 5.5, these objects are clearly self-similar.

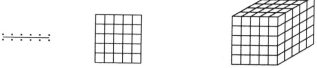

Figure 5.5. Self-similarity of the line, square, and cube.

We see that a line can be divided into 5 parts, each exactly like the line except that their lengths are only $1/5$ that of the line. A square can be divided into 25 parts, each exactly like the whole thing except that lengths are divided by 5. And a cube can be divided into $5^3 = 125$ parts, each like the whole thing except that lengths in the parts are only one-fifth of those in the whole thing. There is nothing magical about the number 5 used here; for any positive integer $n$ we could have divided the square into $n^2$ parts, each exactly like the whole thing with a scaling factor of $n$.

We have an intuitive feeling that for similarity dimension to be a meaningful concept, it should satisfy $d = 1$ for the line, 2 for the square, and 3 for the cube. We see that in each of these three cases

(5.1.1) $$N = s^d,$$

where $N$, $s$, and $d$ are as above. For example, $25 = 5^2$ for the square in Figure 5.5. We take (5.1.1) as the fundamental relationship between $N$, $s$, and $d$ for any self-similar object. In order to write this as an equation for $d$, we need to use logarithms.

We mentioned logarithms briefly in Section 3.1, in conjuction with the Prime Number Theorem. We recall briefly that $\log_b x$ is the number $y$ such that $b^y = x$. This satisfies

$$\log_b(x_1 x_2) = \log_b(x_1) + \log_b(x_2),$$
$$\log_b(y_1/y_2) = \log_b(y_1) - \log_b(y_2),$$
$$\text{and} \quad \log_b(x^a) = a \log_b(x).$$

Taking logarithms of both sides of (5.1.1) yields

$$\log_b(N) = d \log_b(s) \quad \text{or} \quad d = \frac{\log_b(N)}{\log_b(s)}.$$

It is easy to see (Exercise 2) that the ratio of logarithms on the right-hand side above does not depend on the value of the base $b$, and so the $b$ is usually omitted from the notation. This means that any base can be used, although in higher mathematics the base $e \approx 2.78128$ discussed in Section 3.1 is by far the most common base. On many calculators, LOG means $\log_{10}$, while LN means $\log_e$.

The above discussion leads us to the following definition.

DEFINITION 5.1.2. *If an object can be decomposed into N subobjects, each of which is exactly like the whole thing except that all lengths are divided by s, then the object is exactly self-similar, and the* **similarity dimension** *d of the object is defined by*

$$d = \frac{\log N}{\log s}.$$

For example, the similarity dimension of the Koch curve is $\log 4/\log 3 \approx 1.2619$. This number is a measure of its tendency to fill space. Although it is a curve, it is so crinkly that it is given a dimension greater than 1 according to a precise formula. Note that writing $d = \log 16/\log 9$ would have given the same answer, since $\log 16 = 2 \log 4$ and $\log 9 = 2 \log 3$.

This is our first example of a fractal, which, roughly speaking, Mandelbrot defined to be an object whose fractal dimension is a fraction that is not an integer. The term *fractal dimension* can be defined in several ways, with various degrees of generality and mathematical sophistication. The similarity dimension of 5.1.2 only applies to abstract objects having the very rigid self-similarity property stated in its hypothesis. In the next subsection, we will define a type of fractal dimension which we will call the *box dimension*, a concept which applies to objects that are only approximately self-similar.

We will omit a third, and still more complicated, definition called *Hausdorff dimension*. This concept was introduced by Felix Hausdorff in 1920 at a time when integration theory of calculus was being extended to more and more general scenarios. It has the advantage over the others in that it specifies not only the dimension but also the size, called *measure*. For example, a 4 × 5 square has Hausdorff dimension 2, and 2-dimensional Hausdorff measure 20. Hausdorff's definition fell out of favor for quite a few decades, until Mandelbrot revived interest in it. Mathematicians prove that these various definitions agree for certain kinds of objects. For example, one can prove that for objects that are exactly self-similar, the similarity dimension, the box dimension of the next subsection, and the Hausdorff dimension are all equal.

A fourth kind of dimension is the *topological dimension*. This is always an integer. For example, the topological dimension of the Koch curve is 1. We can now give a more precise statement of Mandelbrot's definition.

DEFINITION 5.1.3. *A fractal is an object whose Hausdorff dimension is greater than its topological dimension.*

We have to say it this way, rather than saying that the dimension is a fraction, because some curves are so crinkly that their Hausdorff dimension is 2. (See Exercise 6.) We want these curves to be considered as fractals since their Hausdorff dimension, although it is not a fraction, is greater than their topological dimension, which is 1 for all curves.

In his book, *The Fractal Geometry of Nature*, Mandelbrot stated that he considered 5.1.3 only as a tentative definition of "fractal," and he now apparently prefers the vaguer definition "A fractal is a shape made of parts similar to the whole in some way."[1]

### BOX DIMENSION AND THE CANTOR SET

In this subsection, we will discuss another type of fractal dimension called *box dimension*. Whereas similarity dimension was defined only for objects that are perfectly self-similar, box dimension can at least be contemplated for any subset of Euclidean space. This Euclidean

---

1. Feder 1988, p. 11.

space can be 1- or 2- or 3-dimensional (or higher). When we talk about a box in Euclidean space, we mean an interval, square, or cube, depending on whether the space is 1-, 2-, or 3-dimensional. For any positive number $s$, the Euclidean space can be covered by a grid of boxes of side length $s$. For example, we indicate in Figure 5.6 part of a grid of boxes of side length $1/6$ which covers 2-dimensional space.

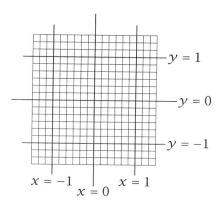

Figure 5.6. A grid of boxes in the plane.

DEFINITION 5.1.4. *If $X$ is a bounded subset of Euclidean space, and $g$ is a positive number, let $N(X,g)$ denote the minimal number of boxes in the grid of boxes of side length $g$ which are required to cover $X$. We say that $X$ has box dimension $D$ if the following limit exists and has value $D$:*

$$\lim_{g \to 0^+} \frac{\log(N(X,g))}{\log(1/g)}.$$

The expression $\log(N(X,g))$ which appears in this definition could be stated in words as "the logarithm of the number of boxes of side $g$ needed to cover $X$." One can prove that if the limit in 5.1.4 exists when we restrict attention to numbers $g$ of the form $1/q^n$ for some fixed number $q > 1$, then it will exist and have the same value when we consider all sufficiently small positive values of $g$.

For the reader who has not been exposed to limits before, informally we can say that the condition in the definition means that if $g$ is positive and very close to 0, then the ratio $\log(N(X,g))/\log(1/g)$ is very close to the value $D$. As established in Exercise 2, it does not matter what base we use for our logarithms.

We will illustrate first with a famous example called the *Cantor set*, and then with another old example called the *Sierpiński triangle*.

The Cantor set was introduced in 1883 by Georg Cantor, whom we mentioned in the "Aside" of Section 2.1 as the founder of modern set theory. It is the set obtained from a line segment by successively removing the middle third of all segments remaining at each step of the procedure. We illustrate the procedure in Figure 5.7. For convenience, suppose that the initial segment has length 1. We first remove the middle third, leaving two segments of length $1/3$. Then we remove the middle third of both, leaving four segments of length $1/9$. At the next step, we will have eight segments of length $1/27$, and the procedure continues forever. After five or six steps, the segments become so tiny that the subsequent divisions will be invisible on a computer screen or in the most carefully drawn depiction, but mathematically the later divisions are every bit as important as the first few.

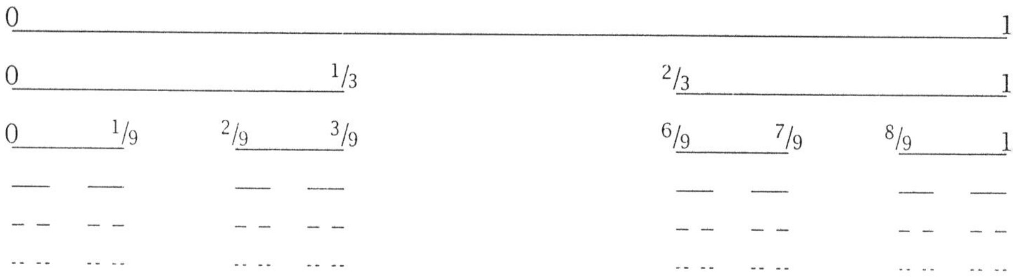

Figure 5.7. The first five steps in forming the Cantor set.

The Cantor set, which is the limiting set to which the infinitely many steps converge, is self-similar. It should be clear from the definition or from Figure 5.7 that it can be written as the union of two sets, namely the part between 0 and $1/3$ and the part between $2/3$ and 1, each of which is exactly like the whole with all lengths divided by 3. Thus the similarity dimension of the Cantor set is $\log 2 / \log 3 \approx .631$. The similarity dimension of the Cantor set could also have been determined by dividing it into four parts, each like the whole with all lengths divided by 9. This would yield

$$d = \frac{\log 4}{\log 9} = \frac{2 \log 2}{2 \log 3} \approx .631,$$

as before. Note how this dimension is between that of a line (dimension 1) and a set of isolated points (dimension 0).

On the other hand, we can apply Definition 5.1.4 with $g = 1/3^n$ to compute the box dimension of the Cantor set. It is easy to see from Figure 5.7 that it can be covered by two intervals of length $1/3$, or by $2^2$ intervals of length $1/3^2$, or in general by $2^n$ intervals of length $1/3^n$. Thus in Definition 5.1.4, $N(X, 1/3^n) = 2^n$, and the box dimension is

$$\lim_{n \to \infty} \frac{\log(2^n)}{\log(3^n)} = \lim_{n \to \infty} \frac{n \log 2}{n \log 3} = \frac{\log 2}{\log 3}.$$

As will happen with all self-similar objects, the similarity dimension and box dimension of the Cantor set are equal. The Cantor set contains no intervals, yet its infinite number of disjoint points are bunched in such a way that its dimension should be considered to be about .631. For a contrasting example of an infinite set of disjoint points with box dimension 0, see Exercise 7.

For the Koch curve, it is not easy to see exactly how many $1/3^n$-boxes are necessary to cover it. (It must be roughly $4^n$.) The way in which portions of it are turned at various angles complicates the analysis, and so we shall not pursue this here.

A pleasing fractal introduced by the Polish mathematician W. Sierpiński in 1915 is called the *Sierpiński triangle* or *gasket*. It is formed by starting with any solid triangle. One first removes the triangle formed by the midpoints of the sides of the triangle, yielding the picture on the left in Figure 5.8. Next, one removes a similar triangle from the middle of each of the three smaller triangles that are left after the first step. This leaves nine triangles, each having side lengths one-fourth of those of the original triangle. This procedure of removing triangles from the middle of remaining triangles is continued ad infinitum, and the Sierpiński triangle is the figure that remains at the end of this infinite sequence of removals. It looks something like Figure 5.8c.

Because the area remaining after each step is three-fourths of the area remaining at the beginning of the step, the area remaining at the end is 0. This is true because if $3/4$ is raised to a large enough power, one can obtain a number as close to 0 as desired. More succinctly,

$$\lim_{n \to \infty} \left( \tfrac{3}{4} \right)^n = 0.$$

The Sierpiński triangle can be broken up as the union of three subtriangles, each exactly like the whole thing except that all lengths are only half as large. Thus the similarity dimension of the Sierpiński triangle is $\log 3 / \log 2 \approx 1.585$. This is a precise measure of the extent to which it fills up space—more than a 1-dimensional curve, but less than a 2-dimensional solid region.

In order to calculate the box dimension of a Sierpiński triangle, it is convenient to start with the isosceles right triangle with vertices

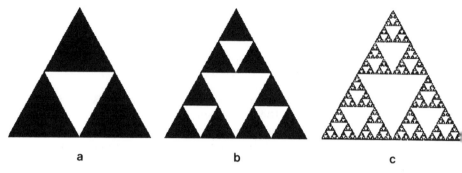

Figure 5.8. Forming the Sierpiński triangle.

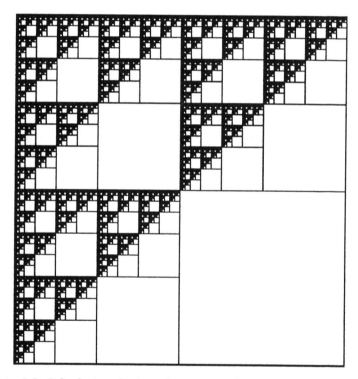

Figure 5.9. Calculating the box dimension of a Sierpiński triangle.

at (0,0), (0,1), and (1,1). Figure 5.9 makes it clear that it can be covered by three squares of side length $1/2$, or by nine squares of side length $1/4$, and in general by $3^n$ squares of side length $1/2^n$. Thus by

Definition 5.1.4, its box dimension is $\log 3 / \log 2$, similar to the above calculation for the Cantor set.

It is not obvious that the Sierpiński triangles in Figures 5.8 and 5.9 necessarily have the same box dimension, since one has to be bent to form the other. In Exercise 11, you will verify that the box dimension of the Sierpiński triangle in Figure 5.8 is also $\log 3 / \log 2$.

## FRACTALS IN NATURE

Until 1951, when Mandelbrot began a series of papers in which he showed that a variety of natural phenomena exhibited similar fractal properties, the earlier mathematical work such as that of Koch and Hausdorff was largely ignored. Fractals were viewed as being mathematical artifacts, whereas nature was viewed as being composed of Euclidean shapes such as circles and straight lines.

Mandelbrot's first papers showed that the noise[2] in data transmission lines came in bursts somewhat similar to those in the Cantor set. It was not until his 1967 paper, "How Long Is the Coast of Britain?" that more than a handful of people began to listen to what he was saying about fractals.[3]

Self-similarity in nature was not a brand new idea. The great English satirist Jonathan Swift (1667–1745) wrote in "On Poetry. A Rhapsody":

> So, naturalists observe, a flea
>
> Hath smaller fleas that on him prey;
>
> And these have smaller still to bite 'em;
>
> And so proceed *ad infinitum.*

But Mandelbrot made it mathematical. A rocky coastline is not perfectly self-similar, as were the mathematical sets described above. The coastline has statistical self-similarity. It will exhibit roughly the same amount of wiggliness regardless of whether it is viewed on a scale of thousands of meters, hundreds of meters, tens of meters, or meters. The wiggliness will be caused by global features (inlets) in the former case, and local features (big rocks) in the latter, but the outlines in the two cases will be quite similar.

The answer to the question posed in Mandelbrot's 1967 paper is: "It depends on the size of the measuring stick." Suppose you measured the coast by laying unmarked rods, 100 meters long, end to end

2. In the sense of "chance fluctuations or errors."
3. *Science* 155 (1967): 636–38.

Benoit Mandelbrot

as closely as possible to the coastline, and then did the same thing using 10-meter rods. The latter will pick up a lot more of the fine detail and give a longer estimate of the actual length. (See Fig. 5.10 for an illustration.) Here a portion of a map of the coast measured

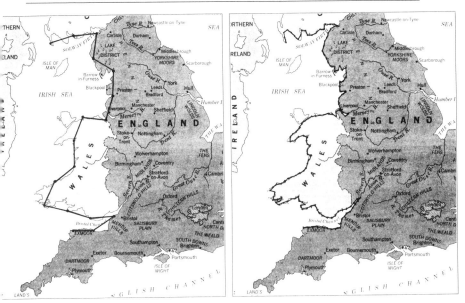

Figure 5.10. Two measurements of the coastline of Britain.

with $1/2$-inch rods requires 15.9 rods, and so seems to have length 7.95 inches, while if it is measured with $3/32$-inch rods, it requires 141 rods, and so it seems to have length 13.2 inches.

Mandelbrot's ideas about lengths of coastlines or borders were influenced by the earlier work of L. F. Richardson.[4] Among other things, Richardson pointed out that encyclopedias in Spain and Portugal listed values for the length of the common border of these two countries that differed by more than 20%. This was apparently due to their using different scales of measurement.

In Section 1.1, we discussed the definition of the length of a smooth curve such as a circle as a limit of the lengths of approximations of it by straight line segments. Here, also, the approximations to the length become greater as the measuring rods become smaller. However, in the case of a smooth curve, the approximations to the length of the curve approach a fixed limiting value as the measuring rods become smaller, while in the case of a fractal curve, the approximations to the length grow arbitrarily large as the measuring rods become very small.

A type of fractal dimension can be computed for an empirical curve such as a coastline that is approximately statistically self-similar. Al-

4. "The problem of contiguity: An appendix of statistics of deadly quarrels," *General Systems Yearbook* 6 (1961): 139–87.

though in principle this could be done using 5.1.4, an easier way is to note the way in which the measured length varies with the length of the measuring rod. A fractal curve has fractal (or divider) dimension $D$ if its length $L$ when measured with rods of length $\ell$ is given by

(5.1.5)                          $L = C \cdot \ell^{1-D}.$

Here $C$ is a constant that is a certain measure of the apparent length, and (5.1.5) must be true for several different values of $\ell$.

The divider dimension (5.1.5) and the box dimension 5.1.4 may differ slightly for a real-world curve such as a coastline, but they are equal for self-similar curves. We shall show this is true for the Koch curve. The Koch curve appears to have length $L = (4/3)^n$ if measured with rods of length $\ell = (1/3)^n$. Thus (5.1.5) says that its divider dimension $D$ should satisfy

$$\left(\tfrac{4}{3}\right)^n = C\left(\tfrac{1}{3}\right)^{n(1-D)},$$

for some number $C$. Taking the logarithm of both sides yields

$$n \log \tfrac{4}{3} = \log C + n(1 - D) \log \tfrac{1}{3},$$

hence, noting that $\log 1/3 = -\log 3$,

$$n \log 4 - n \log 3 = \log C + n(D - 1) \log 3.$$

The terms $-n \log 3$ cancel out, and in order for this equation to be true for all values of $n$, we must have $\log C = 0$ and $\log 4 = D \log 3$. Thus $C = 1$, and $D = \log 4 / \log 3$, compatible with our earlier result for its box dimension.

When (5.1.5) is applied to the maps of the west coast of Britain in Figure 5.10, a value of about 1.3 is obtained for the fractal dimension. You are asked to verify this in Exercise 13, using the data for the two measurements noted earlier in the text. Note that the fractal dimension of the coast of Britain is approximately equal to that of the Koch curve.

For physical objects, one does not require statistical self-similarity at all magnifications. It is generally agreed that a physical object may be considered to be a fractal if it has statistical self-similarity over a range of magnifications in which the largest is at least ten times the smallest.

In his book, Mandelbrot pointed out many other examples of fractals in nature. One group of these he called *trees*. This family includes not only actual trees, in which the fine-scale structure of the tiniest twigs is similar to that of the largest branches, but also rivers, with all their tributaries, and the lungs, whose bronchial tree has a fractal dimension of 3. The latter example has been adopted by some physiologists to help explain biological development, form, and function.[5]

Another idea of Mandelbrot's which has been followed up by specialists is that of the structure of the universe as a fractal. It has been observed that galaxies do not occur uniformly over the universe, but rather occur in clusters. The fractal dimension of the distribution of galaxies turns out to be about 1.2. This number is determined by noting that the number of galaxies in a region of radius $R$ is, on average, proportional to $R^{1.2}$. Cosmologists hope that an understanding of the cause of this fractal structure might lead to a better understanding of the origin of the universe.[6]

"Nuclear winter" is the term used for the expected conditions following a nuclear war. Sooty smoke would greatly decrease the amount of sunlight that can get to the surface of the earth, leading to year-round winterlike conditions. In a 1989 paper,[7] it was noted that sooty smoke particles cluster with a fractal dimension of 1.7 to 1.9. This observation implies that even less light will get through than had been computed under the former assumption that when small spheres merge, they form larger spheres. Thus the implication is that a nuclear winter would be more severe than had been previously thought.

An application where the notion of fractal dimension has been useful is in drilling for oil. Oil is forced out of the ground by pumping in water under high pressure. At the interface between the oil and water, the fluids often develop fractal fingers, which hinder the separation. Understanding how the fractal dimension is related to the ability to separate the two, and how to control the fractal dimension, is the topic of much current research. (See Feder's book for more information.)

As a final application of fractal dimension, we mention computer-generated fractal landscapes for use in movies. A computer can draw a very realistic mountain range, with a fractal dimension of about 2.2 to 2.4, with larger values corresponding to craggier mountains. A certain amount of randomness must be inserted into the computer program in order to create realistic scenes. (See Voss's contribution to Peitgen and Saupe 1988 for some of the important considerations in this procedure.) These landscapes have been used in movies such as *Star Trek II: The Wrath of Khan* and *The Return of the Jedi*.

5. B. J. West and A. L. Goldberger, "Physiology in fractal dimensions," *American Scientist*, 75 (1987): 354–65.

6. J. Maddox, "The universe as a fractal structure," *Nature* 329 (1987): 195.

7. J. Nelson, "Fractality of sooty smoke: Implications for the Severity of Nuclear Winter," *Nature* 339 (1989): 611–13.

ASIDE: FRACTAL FERNS AND IMAGE COMPRESSION

Note: A reader with minimal background in computers may want to postpone reading this "Aside" until after reading Section 5.2. In this "Aside" we will jump quite quickly into BASIC programming, while Section 5.2 gives a more leisurely introduction to programming.

In the late 1980s, drawing on ideas of John Hutchinson,[8] Michael Barnsley developed a method of generating many elaborate and realistic fractals by a very simple method. Moreover, these complicated forms can be described by just a few numbers, so that this method has tremendous promise as a method of image compression. "Compression" of an image refers to the amount of computer space required to store a description of a picture, or to the number of bits of data required to transmit the picture electronically. The traditional method is to tell the color of every little square, or pixel, in the grid on which the picture is displayed. Barnsley claims that he can reduce by a factor of 500 the required amount of storage.[9] We will illustrate Barnsley's method with one of his favorite examples—a fern. He has adopted the fern as the logo of his company, Iterated Systems, which is performing research toward the goal of making this a practical method of image compression. He and his colleague Alan Sloan had been mathematics professors at Georgia Institute of Technology before they realized the commercial potential of this idea.

The fern in Figure 5.11 is generated by four affine transformations, each describable by six numbers. Thus only twenty-four numbers are required to describe this fern completely.

The parallelograms in Figure 5.11a illustrate three of the four affine transformations. An *affine transformation* is a method of transforming figures in space by rotating, stretching, reflecting, and translating (moving). The portions of the fern in each of the three parallelograms in Figure 5.11a are images of the entire fern under affine transformations. We will elaborate on this later.

The computer draws the fern one point at a time. Each time it marks a point, it then randomly chooses one of the four affine transformations the programmer has given to it, and applies this transformation to the point just marked to obtain the next point. This procedure goes on for as long as the programmer stipulates. The fern in Figure 5.11a was obtained by plotting 20,000 points, which took several minutes on our 1987-model personal computer, but just a few

8. J. Hutchinson, "Fractals and self-similarity," *Indiana University Journal of Mathematics* 30 (1981): 713–47.
9. *Scientific American* 262 (1990): 77.

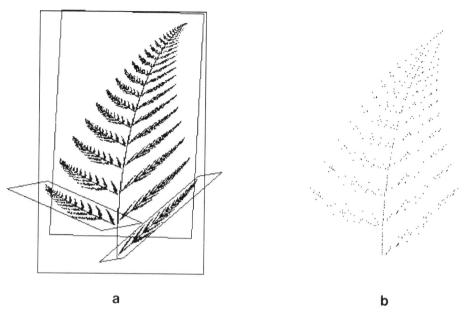

a                                                    b

Figure 5.11. Generating a fern.

seconds on our new (1991) 486 machine. Plotting 10,000 points, which takes only half the time, gives nearly as good an image. It is fun to watch the fern take shape, as one point at a time is added to the image. The image after the first 500 points is displayed in Figure 5.11b.

The screen on which the fern first appeared (before PRINT SCREEN was applied) is a grid of many little rectangles, or pixels. "Plotting a point" means coloring, or filling in, a pixel. During the course of plotting 20,000 points, many pixels will be filled in many times. This doesn't change them. If they have been filled in once, they stay filled in regardless of how many times they have been hit.

A difficulty with trying to teach graphics programming in a textbook is that it varies more from system to system than purely numerical programming does. One source of difficulty is that in many computers a pixel will not appear on the screen as a perfect square, which causes images to appear distorted. For example, what the computer thinks is a circle may look like an ellipse to us. The programmer can adjust for this in various ways, such as changing the formulas in the programs, adjusting knobs on the monitor, or changing the "aspect ratio," but here we will not worry about this advanced topic in computer programming.

The formulas we will use to generate the fern involve describing points of the fern as pairs $(x, y)$ of numbers, with both $x$ and $y$ being between 0 and 1. For our first example, we will use a grid of 200 by 200 pixels for the image. When the computer calculates the coordinates of a point to be a pair $(x, y)$ of numbers between 0 and 1, it rounds off the pair $(200x, 200y)$ to the nearest integer, and fills in the pixel corresponding to this pair of whole numbers, using the usual Cartesian coordinates. Ordinarily, the computer makes the vertical coordinate increase as you move from top to bottom, rather than bottom to top, as is usually done in mathematics. To avoid this complication, we will use a WINDOW statement that makes the computer's $y$-axis agree with the mathematician's. For example, with this WINDOW statement in effect, if $x = .77863$ and $y = .3165$, the computer calculates $200 \cdot .77863 = 155.72$ and $200 \cdot .3165 = 63.3$, and so turns on the pixel 156 over from the left side, and 63 up from the bottom. (See Fig. 5.12.) Note that the whole screen has a grid of 320 by 200 pixels in medium-resolution graphics.

We denote the coordinates of the $n$th point by $(x_n, y_n)$. The $n$th point determines the $(n + 1)$st according to the following formula, which lists the result of applying the four affine transformations to the point $(x_n, y_n)$:

(5.1.6)
$$(x_{n+1}, y_{n+1}) =$$
$$\begin{cases} (.486, .216y_n + .065) & \text{with probability } .05 \\ \begin{aligned}(-.144x_n + .39y_n + .527, \\ -.026x_n + .33y_n + .081)\end{aligned} & \text{with probability } .1 \\ \begin{aligned}(.244x_n - .385y_n + .393, \\ .043x_n + .156y_n + .171)\end{aligned} & \text{with probability } .1 \\ \begin{aligned}(.856x_n + .0414y_n + .07, \\ -.017x_n + .857y_n + .146)\end{aligned} & \text{with probability } .75 \end{cases}$$

The way $(x, y)$ takes on four values with probabilities .05, .1, .1, and .75, respectively, is by the computer choosing a random number $t$ between 0 and 1, and if $t$ is between 0 and .05, then $(x, y)$ takes the first value; if $t$ is between .05 and .15, then $(x, y)$ takes the second value, etc. For example, suppose the computer has just computed its one-hundredth pair of numbers $(x_{100}, y_{100})$ to be $(.32, .45)$. It chooses a random number, which we suppose turns out to be .4762. Since this is greater than .25, the computer uses the fourth formula in (5.1.6), and calculates

$$x_{101} = .856 \cdot .32 + .0414 \cdot .45 + .07 = .3625$$
$$y_{101} = -.017 \cdot .32 + .857 \cdot .45 + .146 = .5262.$$

Because of the limited resolution of the computer screen, we are not seeing a perfect image of the fern. It can be proved that the fern that an ideal screen with perfect resolution would depict, based on (5.1.6), would be perfectly affinely self-similar, by which we mean that at any magnification it can be divided into portions that are exact affine images of the whole object. We have illustrated this by showing in Figure 5.13b a magnified version of the portion of Fig-

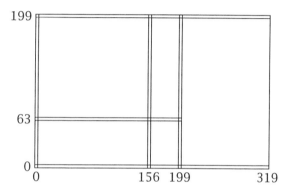

Figure 5.12. Turning on a pixel.

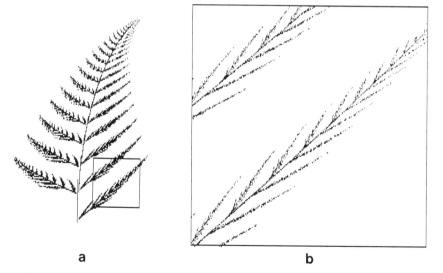

a                                    b

Figure 5.13. Enlarging a portion of the fern.

ure 5.13a in the box. This enlargement is not done by just adjusting the magnification on a copy machine, but rather by a separate calculation. In this separate calculation for the enlargement, the computer does not plot a point $(x, y)$ unless $.55 < x < .75$ and $.1 < y < .3$, which corresponds to the coordinates of the box being enlarged. Because it is only depicting one-fifth of the range in each direction, it can depict it five times as finely. Each pixel in the box in Figure 5.13a corresponds to a 5 by 5 grid of pixels in Figure 5.13b. In effect, the numbers $x$ and $y$ are not being rounded off or truncated as crudely in Figure 5.13b as in 5.13a.

We list a BASIC computer program which will draw the fern. There are variations among different versions of BASIC, and so it is unlikely that this very program will work for the reader; but with a little help even a computer novice should be able to modify this program into one that runs. In Section 5.2 we will start with simpler programs and explain more carefully what the statements mean. A reader who has no programming experience should perhaps look at some of those programs before tackling this one.

```
CLS: KEY OFF: SCREEN 1
WINDOW (0,0)-(319,199)
X=0: Y=0
FOR I=1 TO 20000
  T=RND
  XT=X
  IF T<.05 THEN
    X=.486
    Y=.216*Y+.065
  ELSEIF T<.15 THEN
    X=-.144*X+.39*Y+.527
    Y=-.026*XT+.33*Y+.081
  ELSEIF T<.25 THEN
    X=.244*X-.385*Y+.393
    Y=.043*XT+.156*Y+.171
  ELSE
    X=.856*X+.0414*Y+.07
    Y=-.017*XT+.857*Y+.146
  END IF
  IF I>20 THEN PSET (200*X,200*Y)
NEXT I
END
```

The first line clears the screen, turns off a menu that usually appears at the bottom of the screen, and puts the screen in medium-resolution graphics mode. The three statements here could have been written on separate lines, but for short related statements, colons separating them on the same line are fine. The second line makes the lower left-hand corner have coordinates (0,0) and the upper right-hand corner have coordinates (319,199). As discussed earlier, this is done to make the computer label its vertical axis in the way that mathematicians usually do it. The third statement sets initial values of $X$ and $Y$. It doesn't matter much what values are chosen, as long as they are between 0 and 1. The points will quickly converge to points on the fern. We do not display the first twenty points in order to allow the moving point to make its way to the fern. Once it is on the fern, it will stay on the fern.

The FOR I ... NEXT I statements cause the lines between them to be iterated 20,000 times. The line T=RND gives $T$ a random value between 0 and 1. The next line stores the value of $X$ under the name $XT$. This is done because some of the next lines change the value of $X$, but we still need the old value of $X$ in order to compute $Y$ in lines that follow them. These lines compute the appropriate expression on the right-hand side, and then store them as the value of the variable on the left hand side. The PSET statement turns on the pixel with the indicated coordinates. We remind the reader who feels overwhelmed by this program that we will have a more leisurely introduction to programming in Section 5.2, after which a second look at this program might be more fruitful.

Now that we have seen how the fern is generated, we can approach the big question: Why does it work, and where did the numbers come from? The reason that it works is that the fern can be written as the union of four sets, each of which is an affine image of the entire set. In Figure 5.11a, it is fairly clear how the two skinny parallelograms are rotated and squeezed versions of the whole thing. If the contents of these two parallelograms and the stem are removed from the fern, then the remaining part $P$ lies in the slightly tilted large parallelogram. This $P$ is also an affine image of the whole fern, under the transformation that pushes each major branch up by one. Under this transformation, the lowest major branch on each side, which is part of the whole fern but not part of $P$, is mapped to the lowest branch on its side in $P$, which is the second lowest major branch on its side of the whole fern. Finally, the stem may be considered to be an affine image of the whole fern, under the transformation that squeezes the whole fern into a little straight line.

The four affine transformations have been chosen so as to map the entire rectangle onto (i) the stem, (ii) the skinniest parallelogram on the right, (iii) the skinny parallelogram on the left, and (iv) the large parallelogram in the middle. The probabilities have been chosen so as to be roughly proportional to the areas of the parallelograms. This way we won't waste the computer's time by, for example, mapping onto the stem one-fourth of the time. Suppose $(x_n, y_n)$ lies on the fern, and the random number is between .15 and .25, which specifies the frond on the left. Then $(x_{n+1}, y_{n+1})$ will be the point on the left frond that lies in the same relative position on the left frond as $(x_n, y_n)$ did on the whole fern.

To determine the six numbers that will give an affine transformation mapping the whole rectangle onto a prescribed parallelogram, it suffices to specify where you want three points to go. Standard methods of solving equations then allow you to determine the desired six coefficients. We will investigate this further in Exercise 14. The interested reader who has more than a minimal mathematical background is urged to consult Barnsley's delightful book, *Fractals Everywhere*, which includes many more details.

The Cantor set, Koch curve, and Sierpiński triangle can all be easily obtained by Barnsley's method. The Cantor set, being a subset of the 1-dimensional line, is particularly easy. It can be written as $A \cup B$, where $A$ is obtained from the whole set by multiplying all numbers in it by $1/3$, and $B$ is obtained from the whole set by multiplying all numbers in it by $1/3$ and then adding $2/3$. Thus if you start with a number such as $1/3$, and at each step replace your current number $x$ by a random choice of either $x/3$ or $x/3 + 2/3$, then the Cantor set will be obtained.[10] In Exercise 17, you are asked to investigate this procedure for the Sierpiński triangle.

Most pictures are not so clearly representable as the union of affine images of themselves. Barnsley's method for a fractal approximation of a photograph involves breaking it up into many subpictures, each of which can be attacked by methods similar to the one outlined above. The extent to which it will become a viable method of image compression is dependent on further research.

### Exercises

1.* Show that the area of a Koch Snowflake is exactly 1.6 times the area of the equilateral triangle with which it begins. [Hint: The

---

10. If you start with a number that is not in the Cantor set, your points will rapidly approach points in the Cantor set, but it is nicer to start with a point that is actually in it.

area of an equilateral triangle of side length $s$ is $s^2\sqrt{3}/4$. How many triangles are added at the $n$th step? You will get an infinite series for the sum of the areas of the triangles added at all the steps. You will need to use the fact that if $r$ satisfies $-1 < r < 1$, then

$$1 + r + r^2 + r^3 + r^4 + \cdots = \frac{1}{1-r}.$$

This is a consequence of (3.1.9).]

2. Prove that, for any positive numbers $b$ and $c$,

$$\frac{\log_b N}{\log_b g} = \frac{\log_c N}{\log_c g}.$$

(Hint: Each side is the solution $x$ of the equation $g^x = N$.)

3. A fractal similar to the Koch curve can be obtained by changing, at each step, every segment into five segments, each one-fourth as long as the original. The first step and fourth steps are drawn in Figure 5.14. Draw the second step, and determine the similarity dimension of the limit curve.

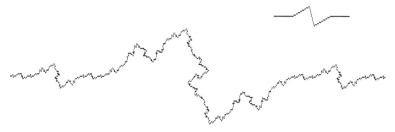

Figure 5.14. A curve for Exercise 3.

4. A fractal similar to the Koch curve can be obtained by changing at each step every segment into eight segments, each one-fourth as long as the original. The first and fifth steps are drawn in Figure 5.15. Draw the second step, and determine the similarity dimension of the limit curve. One can form an analogue of a Koch island by performing this construction on each of the four sides of a square. Do the first two steps in this procedure, and color in your "islands."

5. Describe how to make a fractal curve with similarity dimension $\log 6/\log 4 \approx 1.29$. Draw the first two steps in the formation of this curve. (There is more than one way to do this.)

6. The first two stages of a space-filling curve due to Sierpiński are drawn in Figure 5.16. The limit curve will pass through every

Figure 5.15. A curve for Exercise 4.

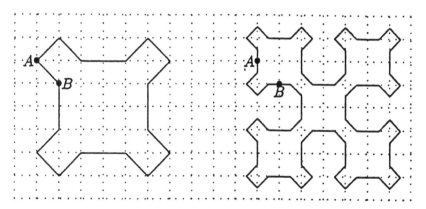

Figure 5.16. A space-filling curve for Exercise 6.

point of the square, although it is still a curve in the sense that its coordinates are given by continuous functions $x(t)$ and $y(t)$, as $t$ runs from 0 to 1. This limit curve has fractal dimension 2. Draw part of the third stage of the construction of this curve by mimicking the pattern of the first two stages. You will need a $16 \times 16$ grid rather than the $8 \times 8$ used in Figure 5.16. Note that diagonal lines are replaced by five shorter lines as illustrated between points $A$ and $B$ in Figure 5.16.

7. Let $X$ denote the infinite set of points $\{1/2, 1/4, 1/8, 1/16, \ldots\}$ on the real line. Show that $X$ can be covered by $n$ intervals of length $1/2^n$. Deduce that its box dimension is 0. Here you need to know that $\lim\limits_{n \to \infty} (\log n)/n = 0$.

8.* Let $Y$ denote the infinite set of points $\{1/1, 1/2, 1/3, 1/4, 1/5, \ldots\}$ on the real line. Show that for any positive integer $k$, the number of intervals of length $1/k^2$ required to cover $Y$ is $2k - 1$. (Hint: The distance between $1/i$ and $1/i+1$ is less than $1/k^2$ if and only if $i \geq k$.) Deduce that the box dimension of $Y$ is $1/2$. [Hint: You may approximate $\log(2k - 1)$ by $\log(2k) = \log(2) + \log(k)$.] This result is a bad feature of Definition 5.1.4. We would prefer to think of the fractal dimension of this set $Y$ as being 0. For example, its Hausdorff dimension of 0. (Hausdorff dimension is the notion of fractal dimension preferred by many mathematicians, but was omitted from our discussion because of its complicated definition.)

9. Use Definition 5.1.4 to show that the box dimension of a square is 2.

10. The *Sierpiński carpet* is formed from a solid square by first removing the middle $1/3 \times 1/3$ subsquare as in the diagram on the left in Figure 5.17. Next, one removes the middle $1/9 \times 1/9$ subsquare from each of the eight remaining $1/3 \times 1/3$ subsquares, yielding the right side of Figure 5.17. The Sierpiński carpet is what remains after this procedure has been iterated infinitely often.

Figure 5.17. First two steps in the Sierpiński carpet.

    a. Draw the picture of the next stage in the formation of the Sierpiński carpet.

    b. Show that the area of the Sierpiński carpet is 0.

    c. Calculate the similarity dimension of the Sierpiński carpet.

    d. Calculate the box dimension of the Sierpiński carpet.

11. Suppose that the Sierpiński triangle in Figure 5.8c has base 2 and height 2. Show that if this triangle is evenly covered by a grid of

squares of side length $1/2^n$ for $n = 0, 1$, or 2, then the number of squares needed to cover the triangle is $4 \cdot 3^n$. For example, with $n = 0$, when the square whose horizontal and vertical sides go from 0 to 2 is divided into four equal subsquares, all four squares contain part of the Sierpiński triangle. The $4 \cdot 3^n$ formula works for all values of $n$. Show that this yields $\log 3 / \log 2$ as the box dimension of the Sierpiński triangle. You need to use the fact that

$$\lim_{n \to \infty} \frac{4 + n \log 3}{n \log 2} = \frac{\log 3}{\log 2}.$$

To see why this last step is true, compare $(4 + 1,000,000 \log 3) / (1,000,000 \log 2)$ and $\log 3 / \log 2$.

12. Mandelbrot presents data of L. C. Richardson regarding the measured length $L$ of the west coast of Britian if it is measured with rods of length $\ell$. If $\ell = 10$ km, then $L = 3020$ km, while if $\ell = 100$ km, then $L = 1700$ km. Use (5.1.5) to determine the fractal (divider) dimension $D$ from this information. [Hint: Substitute the information into (5.1.5), divide the equations to eliminate $C$, and then take log of both sides of the resulting equation to solve for $D$. Alternatively, substitute the information into (5.1.5), take logs of both equations, and subtract the resulting equations to eliminate $C$.]

13. Use (5.1.5) and the data discussed in the text on the measurements of the maps in Figure 5.10 to determine the approximate fractal (divider) dimension of the coastline measured in Figure 5.10. (Hint: Use the method of Exercise 12.)

14.* The image in Figure 5.18a is the union of three affine images of itself. One is the stem, which we want to go from the point $(.5, 0)$ to $(.5, .5)$. The other two are equal to the whole thing shrunk by one-half, rotated by $45°$ either way, and situated with base at the point $(.5, .5)$. It turns out that this will make the largest $y$-value in the picture equal to approximately .9. (See Exercise 15.)

a. Show that $(x_{n+1}, y_{n+1}) =$

$$\begin{cases} (.5, .55 y_n) & \text{for the trunk} \\ \left( \frac{\sqrt{2}}{4} x_n - \frac{\sqrt{2}}{4} y_n + \frac{1}{2}\left(1 - \frac{\sqrt{2}}{4}\right), \right. \\ \left. \qquad \frac{\sqrt{2}}{4} x_n + \frac{\sqrt{2}}{4} y_n + \frac{1}{2}\left(1 - \frac{\sqrt{2}}{4}\right) \right) & \text{for one branch} \\ \left( \frac{\sqrt{2}}{4} x_n + \frac{\sqrt{2}}{4} y_n + \frac{1}{2}\left(1 - \frac{\sqrt{2}}{4}\right), \right. \\ \left. \qquad -\frac{\sqrt{2}}{4} x_n + \frac{\sqrt{2}}{4} y_n + \frac{1}{2}\left(1 + \frac{\sqrt{2}}{4}\right) \right) & \text{for the other} \end{cases}$$

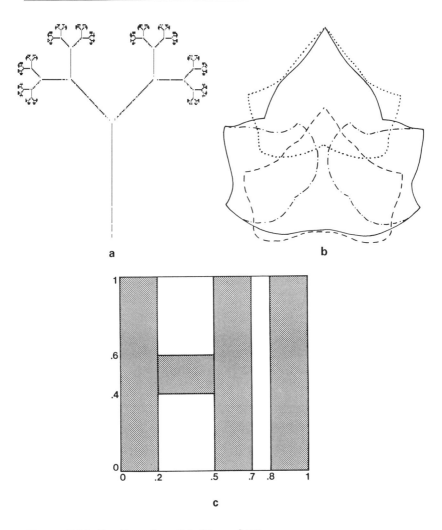

Figure 5.18. For Exercises 14, 22, and 23.

will generate this tree. The easiest way to discover this is to note that sending $(x, y)$ to $(\sqrt{2}/2x \pm \sqrt{2}/2y, - \pm \sqrt{2}/2x + \sqrt{2}/2y)$ causes a 45 degree rotation one way or the other. You want to multiply this by one-half, and then arrange things so that $(.5, 0)$ goes to $(.5, .5)$.

b. Write and run a program to generate this tree.

15. Evaluate the infinite series of the vertical components of all the trunks in Exercise 14, as follows. Since each trunk is half as long

as its predecessor, and rotated by $45°$ from its predecessor, the vertical components are $1/2, 1/4(\sqrt{2}/2), 1/8, \ldots$ Write the next few terms and explain. Show that the sum is $2/3 + \sqrt{2}/6 \approx .902$ by using the formula, derived from (3.1.9), that if $-1 < x < 1$, then

$$1 + x + x^2 + x^3 + \cdots = \frac{1}{1 - x}.$$

16.* An affine transformation that sends $(x, y)$ to $(ax + by + e, \pm(bx - ay) + f)$ is called a *similitude*. A similitude has the property that it shrinks distances between all points by the same amount. This amount is $\sqrt{a^2 + b^2}$, called the *scaling factor* of the similitude. It can be proved that for an object formed by Barnsley's method discussed in the "Aside," if the affine transformations are similitudes, and the affine images of the object are disjoint, then the box dimension of the object is the unique number $D$ satisfying

(5.1.7)     $$s_1^D + s_2^D + \cdots + s_n^D = 1,$$

where the numbers $s_i$ are the scaling factors associated to the various affine transformations.

a. Let $(x_{n+1}, y_{n+1})$ be determined from $(x_n, y_n)$ by one of the following four formulas, each with probability $1/4$.

$$(x_{n+1}, y_{n+1}) = \begin{cases} (.5x_n, .5y_n) \\ (.5x_n + .5, .5y_n) \\ (.5x_n, .5y_n + .5) \\ (.4x_n + .6, .4y_n + .5) \end{cases}$$

Modify the program that produced the fern to use these transformations, and run it. You should observe that the four affine images of the whole object are disjoint.

b. Use (5.1.7) to determine the box dimension of the object produced by this program. You must solve (5.1.7) with $s_1 = s_2 = s_3 = .5$ and $s_4 = .4$. You will have to use a calculator or computer to determine the number $D$ experimentally.

17. For the Sierpiński triangle with vertices at $(0,0)$, $(0,1)$, and $(1,1)$, as in Figure 5.9, write three affine transformations that send the whole triangle onto the three subtriangles, each of which is half as large as the whole. Incorporate these into the computer program in the "Aside," yielding a program that traces the Sierpiński triangle one random point at a time.

18.* Draw the first three steps in the construction of a modified Koch curve, which at each step replaces the middle fifth of every straight line by two sides of an equilateral triangle above it. Explain how this curve could be produced by a Barnsley-type program with four affine transformations, which are, in fact, similtudes. Use Exercise 16 to show that the box dimension of the limiting curve is the number $D$ which satisfies

$$.4^D + .2^D + .2^D + .4^D = 1.$$

Use a calculator to find the approximate value of this number $D$. You can simplify the work slightly by changing the equation to $.4^D + .2^D = 0.5$

19. a. Number systems to base $b$, such as discussed in the "Aside" of Section 3.3, can be used for decimal places as well as integers. For example,

$$(.1021)_3 = \frac{1}{3} + \frac{0}{3^2} + \frac{2}{3^3} + \frac{1}{3^4} = \frac{27 + 6 + 1}{81} = \frac{34}{81}.$$

Evaluate $(.21011)_3$ as an ordinary fraction.

b. Show that the Cantor set consists of all real numbers between 0 and 1 whose base-3 expansion consists entirely of 0's and 2's. (There is a subtle difference between saying it this way and saying that it is those whose base-3 expansion has no 1's. For example, $1/3$ can be written as $(.1)_3$ or as $(.02222\ldots)_3$. We want $1/3$ to be included as part of the Cantor set.)

20. A fractal can be formed starting with a solid square, and at each stage dividing all remaining squares into a $3 \times 3$ grid of subsquares, and removing the four subsquares that are not in the center or in a corner. Draw the first two stages of this construction, and part of the third stage. What is the similarity dimension of this fractal?

21. Experiment with some changes in the coefficients in the program in the "Aside" which produces the fern. Try to find images that are significantly different from the original, but still "fernlike."

22. A maple leaf can be approximated by four affine images of itself, as in Figure 5.18b. Affine transformations accomplishing this are given by

$$(x_{n+1}, y_{n+1}) = \begin{cases} (.6x_n + .18, .6y_n + .36) \\ (.6x_n + .18, .6y_n + .12) \\ (.4x_n + .3y_n + .27, -.3x_n + .4y_n + .36) \\ (.4x_n - .3y_n + .3, .3x_n + .4y_n + .09), \end{cases}$$

each with probability $1/4$.

> a. Modify the fern program of the "Aside" so as to produce this leaf.
>
> b. In a square whose $x$-values go from 0 to 1 and whose $y$-values go from 0 to 1, draw the four parallelograms that are the images of the whole square under the four transformations above. You should apply each transformation to the points (0,0), (1,0), (0,1), and (1,1), and connect the dots.

23. Make a fractal HI, using transformations that send the whole square onto each of the four shaded rectangles in Figure 5.18c. For the three vertical rectangles, you want the $x$-axis to be mapped to a vertical edge. For example, $x_{n+1} = .2y_n$ and $y_{n+1} = x_n$ accomplishes this for the first rectangle. For the crossbar of the H, make horizontal go to horizontal.

## 5.2   Iteration and Computers

In this section, we will study the very different things that can happen when a simple formula is iterated many times. We will see how a tiny change in the formula can cause tremendous differences in the behavior after many iterations. We will use simple computer programs to perform this study, and hope that the reader will at least be able to run our programs, or, preferably, to write modifications of them.

### POPULATION GROWTH

The main example we will study in this section is a formula that gives a simple model for population growth. We will be concerned with the population at a regular interval of time, such as once a month or once a year. In order to normalize the various sorts of populations that we might study, such as humans or insects or fish, we represent the population by a number $p$, which represents the percentage that the population has out of some theoretical maximum value. Thus $p$ might denote the population of the United States divided by 300,000,000, or it might represent the number of fish in Lake Erie divided by 1 billion. We let $p_n$ denote the value of $p$ at the $n$th time interval, and will call it "population," with the understanding that it is really this normalized population, and has a value between 0 and 1.

The formula that has been found to model certain kinds of population growth reasonably well is

(5.2.1) $$p_{n+1} = kp_n(1 - p_n),$$

where $k$ is a constant that has to do with how fast the population grows. This constant $k$ will probably be different for the population of the United States or the number of fish in Lake Erie. The formula gives the population at the next time interval in terms of its current value. It is a reasonable formula because the $p_n$-factor reflects the tendency of a larger population to replenish itself more quickly, while the $(1 - p_n)$-factor reflects the tendency of the growth to level off when the population gets too large.

We illustrate (5.2.1) using $k = 2$ and $p_0 = 0.1$. The reader can use a calculator to verify the following calculations of $p_1$ through $p_5$. The value of $p_1$ is computed to be $2 \cdot .1 \cdot .9 = .18$. Then $p_2 = 2 \cdot .18 \cdot .82 = .295$ is computed. The next few values are $p_3 = 2 \cdot .295 \cdot .705 = .416$, $p_4 = .486$, and $p_5 = .499$. A few more calculations should convince the reader that in this case the values of $p_n$ approach .5.

The surprising behavior that can result from the simple formula (5.2.1) was brought to the attention of the scientific community in two papers written by biologists in 1976.[11] The word "dynamics," which appears in the title of both articles, refers to long-term behavior of systems for which the way in which the system changes is related to the current state of the system.

We begin by describing the simple behavior of the values of $p_n$ that occurs when $k \le 1$ in (5.2.1).

PROPOSITION 5.2.2. *If $0 \le k \le 1$, then the values of $p_n$ in (5.2.1) will approach 0 as n gets large, regardless of the initial value $p_0$.*

This initial value $p_0$ is always assumed to be some number between 0 and 1. If $p_0$ equals 0 or 1, then $p_n = 0$ for all $n \ge 1$, for any value of $k$. (Verify this.) Since the behavior when $p_0 = 0$ or 1 is so simple, we will always assume $0 < p_0 < 1$.

### COMPUTER EXPERIMENTS

Before explaining why Proposition 5.2.2 is true, we want to show the reader how to do computer experiments. The BASIC program below

11. R. M. May, "Simple mathematical models with very complicated dynamics," *Nature* 261 (1976): 459–67; R. M. May and G. F. Oster, "Bifurcations and dynamic complexity in simple ecological models," *American Naturalist* 110 (1976): 573–99.

will allow the user to vary the values of $k$, $p_0$, and the number of iterations, and to quickly see the successive values of $p_n$. The reader is strongly encouraged to perform the computer experiments in the text and exercises of this section. The programs here are simple, yet powerful. Now here is our first program, which we shall name ITERATE. (When you save a program, you have to give it a name.)

```
INPUT "What is k"; K
INPUT "Initial value"; P
INPUT "Number of iterations"; N
FOR I=1 TO N
    PRINT P
    P=K*P*(1-P)
NEXT I
END
```

Some versions of BASIC require you to give line numbers in front of each line of the program. If that is the case on your system, it is recommended that you number the lines 10, 20, ..., 80. These gaps make it easier for you to insert additional lines that you may think of later. We will assume that we are using a version of BASIC such as TURBOBASIC or QBASIC which does not require line numbers. As different versions of BASIC and different types of computer require slightly different ways of getting started, we will not discuss that here. You may need a manual for your computer or help from an experienced user to get going the first time.

When the computer runs this program, it goes through the statements one at a time, starting with the first. When the computer executes the first line of this program, it displays on the screen the phrase

What is k?

and awaits the response of the user. Note that the quotation marks that appeared in the first line of the program will not appear on the screen, but there will be a question mark that was not part of the program. The INPUT statement is used to prompt the user to enter a value, which will be assigned as the value of the variable whose name appears after the semicolon. It is essential that the programmer use a semicolon, rather than a colon or a comma, to separate the prompting statement from the name of the variable. The computer is very intolerant of deviations from its syntax.

After seeing the computer type "What is k?" the user of the program should respond by typing a value such as 0.8 or 1.0 or 1.1. After typing the desired number, the user presses the Enter or Return key.

Then the first statement will have been executed, and the variable $K$ will have the value (say) 0.8.

The second and third lines will be executed similarly. The computer will prompt the user to enter values the computer will assign to the variables $P$ and $N$. Note the roles of the programmer, computer, and user here. The programmer tells the computer to ask the user to input the appropriate numbers, and the computer carries out this task whenever the program is used. Note also that the programmer has the freedom to choose the names of the variables used in the program. It is good practice to use names that are suggestive of the role that the variable plays in the program, such as $P$ for population, and $N$ for number. Variables can have names that are several letters long. They are the way in which the programmer communicates with the computer about locations in the computer where numbers are stored. When the value of a variable is changed, the number in the appropriate location in the computer is changed, and the old value is lost.

The next four lines constitute a loop. Each time it gets to the NEXT I statement, it will add 1 to the value of $I$ and go back to the FOR I= statement. It will continue doing this until the value of $I$ exceeds $N$, which has the value the user gave to it in the third line. Each time the computer goes from the FOR statement to the NEXT statement, it will execute the intervening steps one at a time. Thus it will first print the current value of the variable $P$, which, as we will see, will usually change from one printing to the next. Then it executes the crucial statement,

$$(5.2.3) \qquad\qquad P = K * P * (1 - P).$$

To do this, the computer substitutes the current values of the variables into the expression on the right-hand side, computes the value, and assigns this value to the variable on the left-hand side. The equal sign in BASIC (and in many other computer languages) is somewhat different than the equal sign in mathematics. It means "is assigned the value." Thus (5.2.3) says: "Take the current values of $K$ and $P$, compute $KP(1-P)$ (* means multiply), and give to the variable $P$ the value which is the result of this calculation." The statement K*P*(1−P)=P does not make sense in BASIC. It would cause the program to stop running.

Suppose the user responded to the three prompts by saying $K = 0.8$, the initial value equals 0.6, and the number of iterations is 2. So $P = 0.6$ and $N = 2$. After setting $I = 1$, the computer prints 0.6 and then computes the value $.8 \times .6(1 - .6) = .192$ and stores this as the new value of the variable $P$. When this is done, the previous value $P = .6$ is erased from the computer's memory. The NEXT I statement causes the value of $I$ to be increased from 1 to 2. The computer then goes back to the FOR I=1 TO N statement and notes that the current value of $I$, namely 2, does not exceed the value of $N$, which is also 2, and so it goes on. It next prints the current value of $P$, which is .192. By "print," we mean to display on the screen. If instead the programmer had wanted the printing to be done on a printer, then LPRINT should have been written in the program instead of PRINT. Next, the computer calculates $.8 \times .192(1 - .192) = .124$, and stores this as the new value of $P$. Now the computer increments $I$ to 3, and when it notes that this is larger than $N$, it goes on to the next statement after the NEXT I statement, namely END. Note that it never printed the value .124. It might have been better if the programmer had included another PRINT P statement between the NEXT I and END statements, in order that the last value of $P$ be printed.

The results obtained if the program is run with $k = 0.8$, $p_0 = 0.6$, and $N = 10$ are .6, .192, .124, .087, .064, .048, .036, .028, .022, and .017. It seems pretty clear that these numbers are approaching 0, and if ten iterations is not enough to convince you, then you can run the program again with more iterations. The numbers at each step are being multiplied by a number less than .8, and since a sufficiently large power of .8 can be made as close to 0 as desired, this provides a proof that the population is approaching 0 if $k = .8$.

If $k$ is precisely equal to 1, the population still approaches 0, but it does so more slowly. You might want to have the computer compute many values of $P$, but not print them all out, so that you are not overwhelmed with a plethora of numbers. One way to accomplish this would be to have it only print out only every tenth value of $P$. This can be accomplished by modifying the PRINT P statement to say

$$\text{IF I MOD 10 = 0 THEN PRINT P.}$$

Similar to our study of congruence arithmetic in Section 3.3, I MOD 10 gives the remainder when $I$ is divided by 10. This will be 0 only when $I = 10, 20, 30$, etc., and so the program prints out the tenth, twentieth, etc., values of $P$. The results of every tenth iteration out of one hundred iterations when $k = 1.0$ and $p_0 = .6$ are .074, .042, .029, .022, .018, .015, .0133, .0117, .0105, .0095. Thus, for example, after ninety iterations $P$ will have the value .0105. Printing every tenth number is not a good thing to do if the behavior is oscillatory, as will be the case when $k > 3$.

Note that the role of the "=" sign in the IF statement above is similar to its use in ordinary mathematics, and different from its role in a statement such as (5.2.3). After IF or WHILE, the statement A=B has the value TRUE or FALSE, depending on whether or not the variables $A$ and $B$ have the same value. On the other hand, the statement A=B by itself says, "Give to the variable $A$ the value which is currently held by the variable $B$."

Next we study the behavior when $1 < k \leq 3$.

PROPOSITION 5.2.4. *If* $1 < k \leq 3$ *and the initial value* $p_0$ *satisfies* $0 < p_0 < 1$, *then the numbers* $p_n$ *in (5.2.1) approach* $(k-1)/k$ *as n gets large.*

For example, with $k = 1.5$, the numbers $p_n$ defined by $p_{n+1} = 1.5p_n(1-p_n)$ will approach $(1.5-1)/1.5 = \frac{1}{3}$ for any initial value $p_0$ strictly between 0 and 1. Although we will not give a rigorous proof of Proposition 5.2.4, we will show graphically why it is true. The reader is encouraged in Exercise 3 to run some computer experiments to verify that the proposition works, and to get some idea of how fast the convergence is.

One fine point of this kind of computer experimentation is the notion of how many decimal places of the numbers are accurate. In BASIC, ordinary numbers are only accurate in the first six or seven decimal places, yet the computer may print sixteen. The user should just ignore the digits beyond the sixth. If more precision is required, the program can be changed to make the variables into "double precision" variables, which are accurate to fifteen or sixteen decimal places. This is done by putting a # after every occurrence of the variable. Then (5.2.3) would become P#=K#\*P#\*(1.0-P#). The reader may have to become comfortable with the scientific notation that some versions of BASIC use to print the values of numbers that are close to 0. They print $2.75643E - 002$ for $2.75643 \cdot 10^{-2} = .0275643$. Many calculators use this notation, too.

*FIXED POINTS*

Next we show how the number $(k-1)/k$ in 5.2.4 is obtained. It is a fixed point of (5.2.3). A *fixed point* of a function $f$ is a number $x$ such that $f(x) = x$. The formula (5.2.3) can be thought of as a function,

$$(5.2.5) \qquad f(x) = kx(1-x),$$

which gives the population at the next time interval in terms of the current population. Thus a fixed point of this function is a value of the population which, if once obtained, will stay the same. If $x$ is a fixed point of (5.2.5) and $p_n = x$, then $p_{n+1}, p_{n+2}$, and all subsequent values of $p$ will equal $x$. To find a fixed point of (5.2.5), or equivalently a value of $p$ for which $p_{n+1} = p_n$, we must solve the equation

$$kx(1 - x) = x \quad \text{or} \quad kp(1 - p) = p.$$

The solution $x = (k - 1)/k$ or $p = (k - 1)/k$ is easily obtained. (Do it.)

A more difficult question is why the population should approach this fixed point from any initial value $p_0$ satisfying $0 < p_0 < 1$. We remind the reader that we are still assuming that $k$ is between 1 and 3.

To see why the values of $p_n$ converge toward the fixed point, we consider the graph of the functions $y = kx(1 - x)$ and $y = x$ drawn on the same axes. The value of $x$ where these graphs intersect is the fixed point of (5.2.5). The graphs in Figure 5.19 have $k = 2.5$, for which the fixed point occurs at $p = {}^{1.5}\!/_{2.5} = 0.6$. In the graph on the right, we have shown the succession of values of $p_n$ if $p_0 = 0.1$.

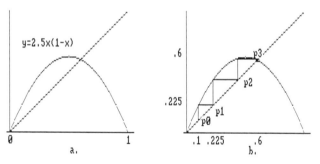

Figure 5.19. Convergence to a fixed point.

This is done by moving vertically from the point $(p_0, p_0) = (.1, .1)$ to a point on the graph of $y = f(x)$, then horizontally from there to a point on the graph of $y = x$. These points will be $(p_0, p_1) = (.1, f(.1)) = (.1, .225)$ and $(p_1, p_1) = (.225, .225)$, respectively. Next, we move vertically from $(.225, .225)$ to the point on the graph of $y = f(x)$ directly above it, whose $y$-value is $p_2$, and then horizontally to the line $y = x$ to change the $x$-value to $p_2$. This continues for as many steps as we tell the computer to perform, and demonstrates the way in which the values of $p_n$ are approaching the fixed point. The graph displays only about five steps because subsequent changes in the value of $p_n$ are too small to show up on the graph. The computer program that drew this graph is given in Exercise 4.

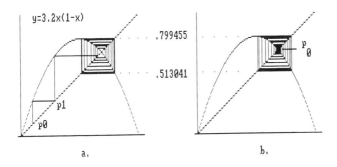

Figure 5.20. $k = 3.2$ in both. $p_0 = 0.1$ on left; $p_0 = 0.69$ on right.

Now we consider a value of $k$ greater than 3. If we run this program with $k = 3.2$ and $p_0 = 0.1$ or $p_0 = 0.69$, we obtain the graphs in Figure 5.20. These graphs look different than the one in Figure 5.19b. Here the values of $p_n$ oscillate—after approximately a dozen initial steps required to get there—back and forth between values very close to .799455 and values very close to .513041. We list the first twenty-five values of $p_n$ when $p_0 = 0.1$:

.1, .288, .656, .722, .642, .735, .623, .751, .598, .770,

.567, .785, .539, .795, .521, .7985, .515, .7993, .5133,

.7994, .5131, .79945, .51305, .799455, .51304

Note how each step of the oscillation brings them a little bit closer to the two limiting values.

The reason for the choice $p_0 = 0.69$ as the other initial value is that it is close to the fixed point corresponding to $k = 3.2$, which is $p = {}^{2.2}/_{3.2} = .6875$. There is a big difference between this fixed point and the one we had in the previous case ($k = 2.5$ with fixed point ${}^{1.5}/_{2.5} = 0.6$). That fixed point was *attractive*, because if you start with any initial value in the general vicinity of 0.6, the values of $p_n$ approach 0.6. On the other hand, the fixed point when $k = 3.2$ is *repulsive*. This means that unless the initial value is *exactly* equal to the fixed point ${}^{2.2}/_{3.2} = 0.6875$, the values of $p_n$ will move away from .6875, in fact approaching the alternating values .799455 and .513041. The first five values of $p_n$ after .69 are .683, .692, .682, .694, and .680. These values appear as a blob in Figure 5.20b. The values continue moving farther away from .6875 on either side, until from $p_{22}$ to $p_{31}$ we have

.789, .534, .796, .519, .7988, .514, .7993, .5132, .7994, .51307.

The following result generalizes the above observation.

PROPOSITION 5.2.6. *The fixed point of (5.2.1) at $p = (k - 1)/k$ is attractive if $1 \leq k \leq 3$, and repulsive if $k > 3$.*

To understand why Proposition 5.2.6 is true requires some ideas of calculus. The reader who has not been introduced to calculus may find this difficult. The slope of a straight line is defined to be the change in $y$ divided by the change in $x$ between any two points on the line. For a straight line, this ratio will be the same regardless of which two points you choose. The slope will be negative if the line drops as you move from left to right, as in Figure 5.21a.

In a small region around a point on it, a smooth curve can be closely approximated by a straight line. This line is called the *tangent to the curve* at the point, and the slope of the curve at that point is defined to be the slope of the line. See Figure 5.21b, which shows the tangent to a curve. In calculus, formulas are derived that tell the slope of a curve at any point, and one of the first formulas that you learn tells you that the slope of the graph of $y = kx(1 - x)$ at the point $(x, kx(1 - x))$ is $k - 2kx$. In Exercise 7, you are asked to show that when $x = (k - 1)/k$, which is the fixed point of $f$, the slope of the curve at this point is less than $-1$ if $k > 3$, and is greater than $-1$ if $k < 3$.

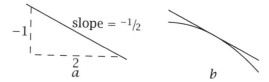

Figure 5.21. Slope of a line, and tangent to a curve.

By considering the graphical approach to iteration, the attracting or repelling nature of a fixed point is seen to be the same as that of a straight line passing through the origin and having the same slope as the slope of the curve at the fixed point. See Figure 5.22, which illustrates iteration around a curve, and iteration around a straight line with the same slope.

Note that if $f(x) = -0.9x$, then after $n$ iterations an initial value of $x_0$ will go to $\pm 0.9^n x_0$, which will be close to 0 if $n$ is very large. This could be restated to say that 0 is an attractive fixed point of $f(x) = -0.9x$, and similarly for any other straight line passing through the origin with negative slope which is greater than $-1$. (Note that, for a negative number, being greater than $-1$ means being closer to 0.) Combining this with the remarks of the two preceding paragraphs,

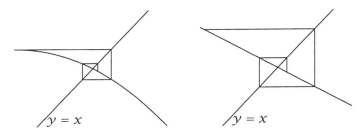

Figure 5.22. Iteration around a curve is approximated by iteration around a straight line with the same slope.

we see that if $k < 3$ in (5.2.1), then the fixed point has negative slope greater than $-1$, hence it has the same behavior as a straight line through the origin with this slope, and hence it is attractive.

On the other hand, if $f(x) = -1.1x$, then after $n$ iterations an initial value of $x_0$ will go to $\pm 1.1^n x_0$, which becomes very far from 0 if $n$ is large. From this, an argument similar to that of the previous paragraph shows that if $k > 3$ in (5.2.1), so that the slope at the fixed point is less than $-1$, then the fixed point is repulsive.

This completes a sketch of proof of Proposition 5.2.6, except for the case $k = 3$, which is more delicate, and will be omitted.

Although the following result can be proved rigorously, we will let our computer experiments suffice.

PROPOSITION 5.2.7. *For any value of k satisfying* $3.0 < k \leq 3.44$*, the limiting behavior of (5.2.1) is to approach a cycle of period 2, if the initial value* $p_0$ *satisfies* $0 < p_0 < 1$ *and* $p_0 \neq (k-1)/k$.

This was illustrated for $k = 3.2$ in Figure 5.20, and for $k = 3.3$ in Exercise 5. A cycle of period $d$ means $d$ values of $p$ such that if $p_0$ equals one of them, then the values of $p_n$ will cycle through them, always in the same order. To approach a cycle of period $d$ means that after sufficiently many iterations, the values of $p_n$ will be extremely close to the values in a cycle. For example, if $k = 3.2$, then the cycle of period 2 which the numbers $p_n$ approach consists of two numbers whose first six decimal places equal 0.513041 and 0.799455.

For some value of $k$ between 3.44 and 3.45, the period of the cycle that is approached by almost any initial value jumps from 2 to 4. For example, if $k = 3.45$, after two thousand iterations from virtually any initial value of $p_0$, the values of $p_n$ to four decimal places will repeat the values .4459, .8524, .4339, .8474, over and over again. By doing delicate enough computer experiments, you can pin down the value where the jump occurs as finely as you like. (See Exercise 8.)

By now, we hope that the reader might guess that if $k$ is increased far enough beyond the value of 3.45, the period of the cycle to which almost every initial value $p_0$ converges jumps from 4 to 8. This value of $k$ is between 3.54 and 3.545. In Exercise 9, the reader is asked to do computer experiments to determine with more accuracy the value of $k$ where this jump occurs. Similarly, the period jumps from 8 to 16 for some value of $k$ between 3.564 and 3.566.

Note how the values of $k$ at which the period of the cycle doubles get closer together. The switch from period 1 to 2 occurred for $k =$ 3.0, from 2 to 4 near 3.44, from 4 to 8 near 3.54, and from 8 to 16 near 3.564. The differences of these values of $k$ are approximately .44, .10, and .024. Alternatively, these are the lengths of the intervals of values of $k$ over which the periods are 2, 4, and 8, respectively. In Figure 5.23 we illustrate the behavior of (5.2.1) over these intervals of values of $k$.

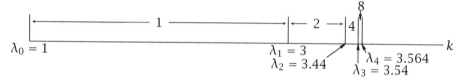

Figure 5.23. Period of cycle approached by (5.2.1).

Note that the ratios of these lengths, $.44/.10$ and $.10/.024$, are both about 4. This was noticed by Mitchell Feigenbaum, a physicist at Los Alamos National Laboratories, in 1976. Feigenbaum did much more than just this. We summarize some of the things he proved, along with some of the things we have mentioned, in the following proposition.

PROPOSITION 5.2.8. *There are numbers $\lambda_0, \lambda_1, \lambda_2, \ldots$, such that if $\lambda_n < k < \lambda_{n+1}$, then for almost any initial value $p_0$, the numbers $p$ in (5.2.1) approach a cycle of length $2^n$. The values of $\lambda_n$ satisfy $\lambda_0 = 1$, $\lambda_1 = 3.00$, $3.44 < \lambda_2 < 3.45$, $3.54 < \lambda_3 < 3.545$, $3.564 < \lambda_4 < 3.566$, and, for all $n$, $\lambda_n < 3.56995$. As $n$ gets large, the ratio $(\lambda_n - \lambda_{n-1})/(\lambda_{n+1} - \lambda_n)$ approaches a constant value, which is approximately 4.6692.*

The ratios $.44/.10$ and $.10/.024$ noted above are the cases $n = 2$ and $n = 3$ of the ratios in the last sentence of this proposition. The limiting value, 4.6692..., which is known to at least twenty decimal places, is called *Feigenbaum's constant*. The proposition does not say that the ratio of differences of $\lambda$'s always equals Feigenbaum's constant, but only that as $n$ gets large, they get very close to this number. Feigenbaum also showed that for many other families of functions that are determined by one free parameter, similarly to the way in which the functions $f(x) = kx(1 - x)$ are determined by the parameter $k$, many of the same things that happen for the family $f(x) = kx(1 - x)$ also happen. Two examples of such families are $f(x) = k\sin(x)$ and $f(x) = x^2 + k$, the latter of which will be studied later in this section. There will be an interval of values of $k$ for which the function with this value of $k$ has an attractive fixed point. Next to it will be an interval of values of $k$ for which the function has an attractive cycle of period 2. Similarly there will be, for every value of $n$, an interval of values of $k$ for which the function has an attractive cycle of period $2^n$. Moreover, the ratio of the lengths of these intervals will approach Feigenbaum's constant. This is the amazing part—that the same number occurs as the limit of ratios of lengths for many different families of functions. Thus this recently discovered number seems to be an important universal constant. Much recent research has been concerned with learning more about it.

CHAOS

Proposition 5.2.8 says that, for all positive values of $k$ that are less than 3.56995, the numbers $p_n$ determined by (5.2.1) will, from almost every initial value $p_0$, converge to a cycle of period $2^n$ for some number $n$. What happens if $k$ is greater than 3.56995 is considerably more complicated. For most such values of $k$, the values of $p_n$ appear to oscillate randomly and wildly. This is called *chaos*. The numbers defined by $p_{n+1} = 3.58p_n(1 - p_n)$ and $p_0 = 0.5$ (the initial value doesn't matter much, so we choose this one for concreteness) are every bit as well defined as those defined by $p_{n+1} = 3.56p_n(1 - p_n)$ and $p_0 = 0.5$, yet the latter approach a nice cycle of period 8, while the former appear random. A mathematical definition of chaos would say something like "deterministic behavior that appears to be random."

You can see the chaotic range of values by running ITERATE with $k = 3.58$. A more dramatic way of viewing the change from periodic behavior to chaos is to run a program that draws a graph of the values of $p_n$. The program turns on pixel $(n, 200p_n)$ for values of $n$ from 0 to 639. The reason for the 200 in the second component is that the values of $p_n$ will range from 0 to 1, and there are 200 pixels vertically, so that multiplying the value by 200 will fill in a pixel at the appropriate height. We use SCREEN 2, which is high-resolution graphics on a CGA monitor and gives a grid of 640 by 200 pixels. Both TURBOBASIC and QBASIC have SCREEN numbers from 0 to 12. Larger numbers give better resolution, if you have a monitor of sufficiently high quality.

The program that draws these graphs is listed below. We call it GRAPH.

```
CLS: KEY OFF: SCREEN 2
WINDOW (0,0)-(639,199)
INPUT "What is k"; K
INPUT "Initial value"; P
FOR I=0 TO 639
    PSET (I, 200*P)
    P=K*P*(1-P)
NEXT I
END
```

In the "Aside" to Section 5.1, we discussed briefly the graphics statements used in this program. The first line clears the screen, the second line compensates for the computer's way of numbering the vertical axis oppositely from the usual way in mathematics, and PSET turns on the indicated pixel.

Figure 5.24 shows part of the output of this program for $p_0 = 0.1$ and $k = 3.5$ (period 4), $k = 3.56$ (period 8), and $k = 3.58$ (chaos). You can see that, up to the accuracy displayed by these graphs, the periodic behavior is achieved after a very small number of iterations. This is largely because in drawing these graphs the computer is rounding values to the nearest $1/200 = .005$, which is not as fine a distinction as we were making when we looked at values printed by ITERATE.

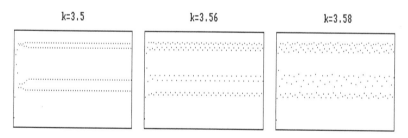

Figure 5.24. Graph of values of $p_n$ for three values of $k$.

*BIFURCATION DIAGRAMS*

For most values of $k > 3.56995$ and most initial values $p_0$, chaotic behavior will occur. However, there are some very short intervals of values of $k$ between 3.56995 and 4.0 on which periodic behavior occurs for all initial values. Moreover, the periods here are not usually powers of 2, as they were for $k < 3.56995$. To see this, it is useful to make what is now called a bifurcation diagram, which we have pictured in Figure 5.25.

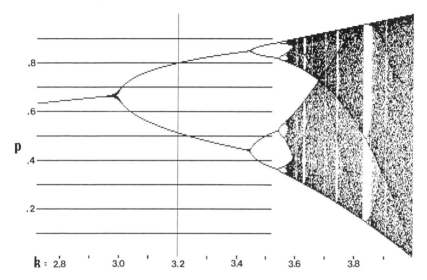

Figure 5.25. Values obtained by (5.2.1) for various values of $k$.

This diagram displays the values of $p_{100}$ through $p_{250}$ if $p_0 = 0.5$ and $k = 2.72, 2.722, 2.724, \ldots, 3.996, 3.998$. This gives 640 values of $k$, every .002 from 2.72 to 4.0. The number 640 was chosen because that is the number of horizontal pixels of many computer screens. We don't display the first one hundred values of $p_n$ in order to give the values plenty of time to settle into their pattern.

Note that for $k < 2.98$, only one value of $p$ is obtained. This is not precisely true, but true as far as the computer display is concerned. As in Figure 5.24, the $p$-values are rounded to the nearest $1/200 =$ .005, and the fluctuations in the value of $p_n$ after $p_{100}$ will be much smaller than .005. For $2.98 < k \leq 3.00$, there is a slight thickening of the graph, which indicates that the initial undisplayed one hundred iterations were not enough to cause the value to be within .005 of the value to which it is converging.

A *bifurcation* is a splitting in two, and the first bifurcation occurs at $k = 3.0$. As observed earlier, for $3.0 < k < 3.44$, the numbers $p_n$ approach a cycle of period 2. This is consistent with the two branches in Figure 5.25 that lie above each of these values of $k$. We have drawn in the line at $k = 3.2$ to show that, as noted earlier, the values of $p_n$ when $k = 3.2$ are eventually very close to .513 or .799. Figure 5.25 shows clearly that the second bifurcation occurs around $k = 3.44$, as also noted earlier. For values of $k$ between roughly 3.44 and 3.54, there are four branches of the curve. This shows not only that for such values of $k$ the values of $p_n$ approach a cycle of period 4, but also it shows roughly what are the four values of $p$ corresponding to each value of $k$. An example in an earlier subsection showed that if $k = 3.45$, then the values of $p_n$ approach the four numbers .4459, .8524, .4339, and .8474. These can be seen in Figure 5.25 by noting where a vertical line drawn up from 3.45 intersects the curve. The bifurcation to eight branches around 3.54 is also clear in Figure 5.25, but after that it gets a bit murky. The resolution of the screen is not fine enough to display the later bifurcations at the scale presented here.

We could make minor changes in the program that produced Figure 5.25 and produce a magnification of the region with $k$ between 3.535 and 3.575 and the vertical coordinate between .34 and .39. This would display the next few bifurcations in finer detail. (See Exercise 12, which lists the program and suggests how to modify it.) The bifurcation diagram is a fractal, inasmuch as portions of it are self-similar at all magnifications.

The right side of the bifurcation diagram shows that for most values of $k$ between 3.57 and 4.0, $p_n$ will at least come close to all values in appropriate intervals. For example, if $k = 3.58$, we showed in Figure 5.24 the chaotic behavior of the values of $p_n$ and how they lie in two bands, each of which is subdivided into two sub-bands. This can also be seen in Figure 5.25 by running a vertical line up from 3.58, and noting that it intersects four black portions of the bifurcation diagram. Each vertical slice of the bifurcation diagram contains the information of one of the graphs of Figure 5.24; it just doesn't show how the values $p_n$ vary with $n$, but this is either extremely regular or chaotic.

The white bands in the dark part of the bifurcation diagram show short intervals of values of $k$ where periodic behavior recurs. For example, if $3.627 \le k \le 3.629$, then running GRAPH will show that the values of $p_n$ appear to have period 6. Here, as in our earlier periodic cases, for most values of $p_0$, the values of $p_n$ are not strictly periodic, but rather approach values that have period 6. For $k = 3.631$ or 3.632, it appears from the output of GRAPH that at least some of the six periodic values have bifurcated. It might be reasonable to guess that the other values have bifurcated as well, but that the two periodic values into which they have split are so close together that they show

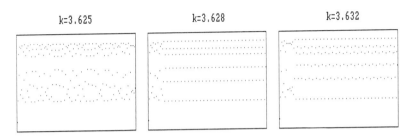

Figure 5.26. Graphs of (5.2.1) for $k = 3.625$, 3.628, and 3.632.

up as the same pixel in our graph. We list in Figure 5.26 the output of GRAPH for $p_0 = 0.5$ and $k = 3.625$ (chaotic), 3.628 (period 6), and 3.632 (probably period 12). The reader is asked in Exercise 13 to do similar experiments in the broad white band around 3.83.

The most important lesson to be learned from this example is the extreme sensitivity of the qualitative properties of iterated behavior to some small changes in the functions being iterated. Why should it happen that iterating $f(x) = 3.628x(1 - x)$ converges from almost any initial value to a cycle of period 6, while iterating $f(x) = 3.625x(1 - x)$ has chaotic behavior for almost any initial value? This is a question that has important implications for real world science. We shall say more about it in the "Aside" at the end of this section.

### A SIMILAR FAMILY OF FUNCTIONS

We mentioned earlier that the most important part of Feigenbaum's discovery was that the same sort of behavior occurs for many families of functions. We briefly consider one other example, which will be important to us in Section 5.3. We iterate the function $f(x) = x^2 - c$. The number $c$ plays the role that $k$ did in the functions $f(x) = kx(1 - x)$ considered above. One can easily modify the programs ITERATE, GRAPH, or that of Exercise 12 to deal with this family.

We shall suppose that the initial value $x_0$ equals 0, and that numbers $x_n$ are defined by

$$(5.2.9) \qquad x_{n+1} = x_n^2 - c.$$

We shall ask how does the limiting behavior change as $c$ is varied. One could also ask about how variations in the initial value $x_0$ affect the limiting behavior, but we will not worry about that now. We can form a bifurcation diagram similar to Figure 5.25. To maximize similarity with the bifurcation diagram obtained earlier, we let the

$x$-values run opposite of the usual direction, so that $x$ increases as we move from top to bottom. Figure 5.27 is the bifurcation diagram for (5.2.9) with $x_0 = 0$. It displays the values $x_{100}$ through $x_{250}$ for each value of $c$ from 0.67 to 1.95 in increments of 0.002. Note the extreme similarity with Figure 5.25. Indeed, as you may investigate in Exercise 16, one of these diagrams can be directly transformed into the other, but the transformation is not linear.

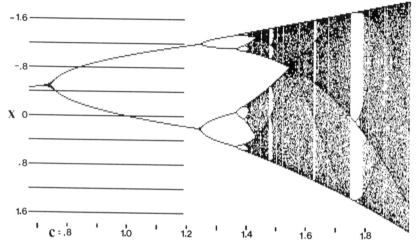

Figure 5.27. Values of $x_n$ in (5.2.9) for various values of $c$.

Figure 5.27 shows that for $c$ between .67 and .75 the values of $x_n$ converge to a number between $-0.4$ and $-0.5$. It shows that for $c$ between 0.75 and some number around 1.25 the values of $x_n$ approach a cycle of period 2. As $c$ increases above 1.25, the periods of the cycles start doubling faster and faster until chaos sets in around 1.4. Figure 5.27 also shows how there are, amidst the chaos, short intervals of values of $c$ for which (5.2.9) approaches a periodic cycle. You are asked in Exercise 14 to investigate some of these more carefully by writing and running a computer program that accepts as input a value of $c$, and then draws a graph of the values of $x_n$.

ASIDE: LORENZ'S WEATHER FORECAST

In 1963 Edward Lorenz, a meteorologist at Massachusetts Institute of Technology, discovered chaotic behavior while using a computer to simulate the weather. By considerably oversimplifying the situation for actual weather, he had reduced his consideration to three parameters that described the aspects of the weather he wished to study. He had derived some simple differential equations that described the way these variables changed. The computer could plot the solutions of the differential equations—the way that the variables would vary with time. He noticed that the tiniest changes in the initial values could cause huge changes after a long enough period of time.

Lorenz called this the "butterfly effect." By this he meant that a butterfly flapping its wings in Brazil today can cause a tornado in Texas a week from now. Of course, the conditions would have to be just right, but it can happen that the differential equations the wind satisfies are so sensitive to a small change in its initial value (today in Brazil) that a huge difference can be caused in its value a week from now in a distant place (Texas). Even more important than this is the implication for long-range weather forecasting. Even if the differential equations that govern the weather are perfectly understood, one will never be able to take perfectly accurate initial readings. The slight errors in the initial reading can cause tremendous differences between the forecasted weather and the actual weather in the distant future.

Several factors make Lorenz's equations more complicated than those we considered earlier in this section. One is that they are differential equations (equations involving the derivative, from calculus) rather than iterative equations. However, this is not terribly significant because the way in which we approximate their solution on the computer is to replace them by iterative equations. The bigger difference is that Lorenz's equations involve three variables, rather than the single variable we had before. The solution of his equations would be a curve in 3-dimensional space. As this is quite difficult to envision, one frequently ignores one of the variables ($y$), plotting the curve in 2-dimensional space traced by the values of other two variables ($x$ and $z$). This is like taking the projection on the $xz$-plane of the actual curve in 3-dimensional space.

Lorenz's variables are

$x$ = rate of convective overturning
$y$ = horizontal temperature variation
$z$ = vertical temperature variation.

We let $\dot{x}$ denote the derivative of $x$ with respect to $t$. This is a measure of how fast $x$ is changing as time passes. Similarly, $\dot{y}$ and $\dot{z}$ denote the derivatives of $y$ and $z$, respectively, with respect to time. Then

Lorenz's differential equations are

$$\dot{x} = 10(y - x)$$
$$\dot{y} = x(28 - z) - y$$
$$\dot{z} = xy - 8z/3.$$

For small changes in $t$, the change in the value of $x$ is approximately equal to $\dot{x}$ times the change in $t$. We will use 0.01 as our small value for the change in $t$ from one step to the next. The differential equations become the iterative equations,

$$x_{n+1} = x_n + 0.01 \cdot 10(y_n - x_n)$$
$$y_{n+1} = y_n + 0.01(x_n(28 - z_n) - y_n)$$
$$z_{n+1} = z_n + 0.01(x_n y_n - 8z_n/3).$$

We choose $x_0 = 1$, $y_0 = 1$, and $z_0 = 1$, and ask the computer to plot and connect the points $(x_n, z_n)$ for all values of $n$ through a range (we use 5000). We do not plot the $y$-values because the screen can only display 2-dimensional data conveniently. But we must keep track of the $y$-values, as they affect the changes in $x$ and $z$. The computer output is listed in Figure 5.28. In Exercise 15, the program which produced this is given, along with some computer experiments.

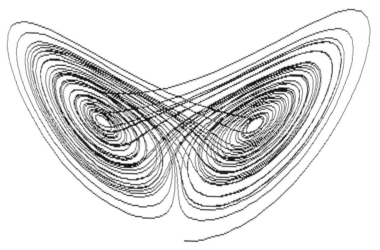

Figure 5.28. Lorenz's butterfly.

This striking figure is called *Lorenz's strange attractor*, or *Lorenz's butterfly*. For almost any initial values of $x$, $y$, and $z$, the values of the $x$ and $z$ variables will quickly be attracted to points on this figure, and then they will cycle around the two loops in apparently chaotic patterns. It is fun to watch the image being formed on the computer screen as the path is traced out. If your computer is so fast that the figure seems to be produced instantly, you might be able to modify the program as suggested in Exercise 15, so that bands of color are displayed as the point moves. Though we know that the values of $x$ and $z$ soon lie on (or extremely close to) this figure, one can hardly predict at what point on the attractor they will lie at any given time. This is because points that start close together will stay close together for a little while, but all of a sudden they will split apart, and from that time on, there seems to be no relationship between the position of one and the position of the other. This has significance for long-range weather forecasting. The weather conditions will certainly lie very close to the attractor, and this restricts the possibilities for the weather; however, we don't know where on the attractor they will lie, and so we can't make accurate long-range weather forecasts.

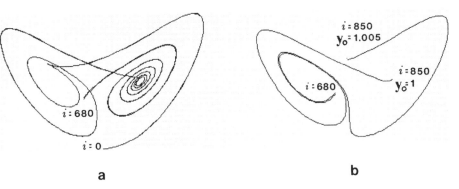

Figure 5.29. Sequences based on nearby initial values stay close together for about 730 iterations, and then split apart.

For example, in Figure 5.29a we show how the curves determined by the initial values $x_0 = y_0 = z_0 = 1$ and by the initial values $x_0 = 1$, $y_0 = 1.005$, $z_0 = 1$ stay indistinguishably close together for the first 680 iterations, while Figure 5.29b shows them splitting apart between iterations 680 and 850. You have to look very closely to see that there are two curves plotted in Figure 5.29a, one for each set of initial values, but they occasionally separate enough so that you can distinguish them. This pattern continues during the first part of Figure 5.29b, up to about iteration 730, but once they split, they follow completely independent paths.

There are many other aspects of these equations that could be investigated. For example, one could magnify a small portion of the strange attractor to reveal its fractal properties. Or, one could study how changes in the coefficients of the equations (10, 28, and $^8/_3$) change the appearance of the attractor. But the main lesson to be learned is how you can know that a moving point lies on an attractor without having any control over where on the attractor it lies.

## Exercises

1. List the output obtained when $p_{n+1} = .7p_n(1 - p_n)$ is iterated twenty times starting with $p_0 = .5$.
2. List every twentieth value of $p_n$ when $p_{n+1} = p_n(1 - p_n)$ is iterated two hundred times starting with $p_0 = .5$.
3. Run computer programs that check Proposition 5.2.4 for various values of $k$, $p_0$, and number of iterations. Summarize the highlights of your conclusions.
4. A computer program that produces something like Figure 5.19 is written below.

```
CLS: KEY OFF: SCREEN 1
WINDOW (0,0)-(319,199)
FOR I=0 TO 199
    PSET (I, I)
    X=.005*I
    Y=2.5*X*(1-X)
    PSET (I,200*Y)
NEXT I
X=0.1
FOR I=1 TO 30
    Y=2.5*X*(1-X)
    LINE (200*X,200*X)-(200*X,200*Y)
    LINE (200*X,200*Y)-(200*Y,200*Y)
    X=Y
NEXT I
END
```

a. Using the "Aside" to Section 5.1 as a guide, explain what is happening in the various parts of this program. The LINE statement draws a straight line between the indicated points.
b. Revise the program so that it just prints the left part of Figure 5.19.
c. Change the program so that it has $k = 2.9$, $p_0 = .2$, and runs for twenty iterations. Run your program and, if you have a

printer connected to your computer, press the PRT SC key, which prints the screen. You may also have to hold down the SHIFT key with the PRT SC key. It is also possible that you might have to type some instruction such as LASERJET A at the beginning of your session in order to activate the capability of printing the screen.

d. Modify the program so that it divides the $x$-axis into four hundred points rather than just two hundred. You will have to change SCREEN 1 to SCREEN 2, which is high-resolution graphics on a CGA monitor. It gives a 640 by 200 grid of pixels. The $y$-axis can still only be divided into two hundred intervals, and so a pair $(x, y)$ of numbers between 0 and 1 should be displayed as pixel (400*X,200*Y). You will also have to change the WINDOW statement and the statement X=.005*I.

5. a. By running the program ITERATE, find, to four decimal places, the two values of $p$ between which the population will eventually alternate if $k = 3.3$ in (5.2.1).

b. If $f(x) = 3.3x(1 - x)$, write a formula for $f(f(x))$. Explain why the two values of $p$ that you found experimentally in part (a) should be fixed points of the function $f(f(x))$.

c. Use a calculator to verify directly that the two numbers found experimentally in part (a) satisfy $f(f(x)) = x$, using the formula for $f(f(x))$ that you found in part (b).

6. If $k = 3.0$ in (5.2.1) then the values of $p_n$ approach 2/3, but if the initial value $p_0$ is not precisely equal to 2/3, thousands of iterations are required before the approximation is very close. If you want to run the program ITERATE for thousands of iterations but don't want to see thousands of numbers flash by on the screen, you can change the PRINT statement to read

IF I>N-10 THEN PRINT P

which will print just the last ten values. By running ITERATE several times with $k = 3.0$ and $p_0 = .68$, possibly modifying the program as just suggested, write to four decimal places the values of $p_{1000}$ and $p_{1001}$, $p_{5000}$ and $p_{5001}$, $p_{10000}$ and $p_{10001}$, and $p_{20000}$ and $p_{20001}$. You should use double precision for the variables $K$ and $P$ in your program. As explained in the text, this is done by putting # after every occurrence of the variable in your program.

7. Show that when $x = (k - 1)/k$, the value of $k - 2kx$ is less than $-1$ if $k > 3$, and is greater than $-1$ if $k < 3$. This was used in the proof of Proposition 5.2.6.

8. Try to determine the third decimal place of the number between $k = 3.44$ and 3.45 where the period of the cycle of (5.2.1) jumps

from 2 to 4. You should let ɪᴛᴇʀᴀᴛᴇ run to *at least* five thousand iterations, and you should incorporate the modifications suggested in Exercise 6. For each value of $k$ that you try, write out to four decimal places the values of $p$ that are involved in the cycle. The value of $p_0$ that you use is pretty much irrelevant.

9. Similarly to Exercise 8, pin down more accurately the place between $k = 3.54$ and $k = 3.545$ where the period of the cycle of (5.2.1) jumps from 4 to 8. For each value of $k$ that you try, write out to five decimal places the values of $p$ that are involved in the cycle. You may need to go up to 15,000 iterations to clarify the period.

10. Similarly to Exercise 9, find to greater accuracy the value of $k$ between 3.564 and 3.566 where the period jumps from 8 to 16.

11. Using Proposition 5.2.8 as a guide, determine with as much accuracy as you can the value $\lambda_5$ of $k$ where the cycle jumps from 16 to 32. Using your results of Exercises 8 through 11, compute the ratios of differences of $\lambda_n$'s in Proposition 5.2.8, and compare with Feigenbaum's limiting value.

12.* The program that produced Figure 5.25 is listed below.

```
CLS: KEY OFF: SCREEN 2
WINDOW (0,0)-(639,199)
FOR I=0 TO 639
    K=2.72+.002*I
    P=0.5
    FOR J=1 TO 250
        IF J>100 THEN PSET (I,200*P)
        P=K*P*(1-P)
    NEXT J
NEXT I
END
```

Revise this program so that it will produce a magnified version of the portion of Figure 5.25 with $3.535 \le k \le 3.575$ and vertical coordinate between .34 and .39. Run your program. Some tips follow. You will want to turn on a pixel only if $P$ lies between .34 and .39. If $P$ is not in this range, don't turn on any pixel. If it lies in this range, then you should turn on pixel $(I, 200*(P-.34)/.05)$. The value of $K$ in terms of $I$ could be chosen to be $3.535+.04*I/640$.

13. Run ɢʀᴀᴘʜ for $p_0 = 0.5$ and various values of $k$ between 3.82 and 3.87. Show that the behavior is chaotic for 3.82 and 3.87. Find a

value of $k$ where there appears to be period 3 and another value of $k$ where there appears to be period 6.

14. a. Modify GRAPH to run for $x_{n+1} = x_n^2 - c$ with $x_0 = 0$. One thing that you will have to adjust is the correspondence between $x$-values and pixel numbers. Instead of using $200*P$ as the pixel number, you should use $50(X + 2)$. This will make the values of $x$, which range from $-2$ to $2$, expand over the range of pixels from 0 to 199. If you do it this way, you should not use a WINDOW statement, since you want the vertical axis to be numbered in the opposite of the usual direction.

   b. Run it for values of $c$ close to 1.25 to try to determine as accurately as you can where the period shifts from 2 to 4.

   c. Run it for values of $c$ close to 1.4 to try to determine as accurately as you can where chaos begins.

   d. Run it for values of $c$ slightly greater than 1.62 to see the periodic behavior in the second of the three major white bands in Figure 5.27. What is the period for most values of $c$ in this band? Find a value of $c$ in the band where the period appears to be twice as large.

15. The computer program that drew Figure 5.28 is given below.

```
CLS: KEY OFF: SCREEN 2
WINDOW (0,0)-(639-199)
X=1: Y=1: Z=1
FOR I=1 TO 5000
    DX=0.01*10*(Y-X)
    DY=0.01*(X*(28-Z)-Y)
    DZ=0.01*(X*Y-8*Z/3)
    LINE (15*(X+20),3.7*Z)-(15*(X+DX+20),3.7*(Z+DZ))
    X=X+DX: Y=Y+DY: Z=Z+DZ
NEXT I
END
```

The numbers in the LINE statement are chosen so as to display values of $x$ between $-20$ and 20 on pixels 0 to 600, and values of $z$ between 0 and 55 on pixels 0 to 200. These ranges of values for $x$ and $z$ must be obtained by trial and error (or from books).

   a. Run the program above, and then modify it so that it just prints out points rather than lines between the points. Run this latter program, too.

   b. Modify the program so as to use changes in $t$ of 0.001 instead of 0.01. Observe the effect that this has on the output.

c. See how changing the initial values $X = Y = Z = 1$ affects the output.

d. Try changing some of the coefficients in the equations for $DX, DY$, and $DZ$, and see how it affects the output.

e.* Figure 5.29b was made by making an outer loop that runs the main loop of the program twice, once with $Y = 1$ and once with $Y = 1.005$. Both times it lets $I$ go to 850, but only executes the LINE statement if $I > 680$. Experiment with other small changes in the initial values of the variables, printing out portions of the curves until you can see a portion where they diverge dramatically.

f. If you have a color monitor, you can print bands of the Lorenz butterfly in different colors. This is one way of seeing the butterfly develop. For example, SCREEN 7 or 8 with an EGA monitor and SCREEN 12 with a VGA monitor will display sixteen colors, numbered from 0 to 15. The statement

$$\text{LINE } (a,b)-(c,d),j$$

colors the line from the point (a,b) to the point (c,d) color j. Change the main loop of the above program to a double loop, the outer part of which lets j go from 0 to 15, and the inner part lets i go from 1 to 1000. Put ,j at the end of the LINE statement. This will follow the butterfly in fifteen colored bands of 1000 points each.

16.* a. Show that the equations

$$p_{n+1} = kp_n(1 - p_n) \quad \text{and} \quad x_{n+1} = x_n^2 - c$$

of (5.2.1) and (5.2.9) are equivalent under the correspondence

$$p_n \leftrightarrow \frac{1}{2} - \frac{x_n}{k} \quad \text{and} \quad c \leftrightarrow \frac{k^2}{4} - \frac{k}{2}.$$

b. List the values of $k$ and $c$ where bifurcations occur in (5.2.1) and (5.2.9). These should either be approximate values mentioned in the text or more accurate values found in Exercises 8, 9, and 14. Show that these agree under the correspondence verified in part (a).

c. We saw in the text that if $k = 3.2$, (5.2.1) eventually cycles between .513041 and .799455. What does this imply about (5.2.9)?

## 5.3   Mandelbrot and Julia Sets

We saw in Section 5.2 how the behavior of the function $f(x) = x^2 + c$ under iteration changes as $c$ is changed. In 1980 Benoit Mandelbrot decided to investigate the same question if $x$ and $c$ are complex numbers. The result of his computer investigation was a remarkably complicated and beautiful subset of the plane, which was later named after its discoverer.[12] In this section, we will discuss this Mandelbrot set, and show how to generate it with a simple computer program.

### COMPLEX NUMBERS

Complex numbers are numbers of the form $a + bi$, where $a$ and $b$ are real numbers, and $i^2 = -1$. These numbers were introduced in the 1500s in order to write solutions of equations. For example, the equation $x^2 + 1 = 0$ has no solution $x$ among the real numbers, because the square of any real number is nonnegative. So we artificially introduce a number $i$ that is a solution of this equation. The number $i$ is sometimes called *imaginary*, but, with the graphical representation of complex numbers we will describe below, the imaginary numbers are every bit as real as the real numbers.

Addition, subtraction, and multiplication of complex numbers follows naturally from the usual rules of arithmetic. For example,

$$(3 + 4i) + (5 - 2i) = (3 + 5) + (4 - 2)i = 8 + 2i$$

$$(3 + 4i) - (5 - 2i) = (3 - 5) + (4 - (-2))i = -2 + 6i$$

$$(3 + 4i)(5 - 2i) = 3 \cdot 5 + 3(-2i) + 4i \cdot 5 + 4i(-2i)$$
$$= 15 - 6i + 20i - 8i^2 = 15 + (20 - 6)i - 8(-1)$$
$$= (15 + 8) + 14i = 23 + 14i.$$

Real numbers can be considered to be complex numbers in which the imaginary part is 0. Dividing a complex number by a real number poses no problem. For example, $(3+4i)/2 = 3/2 + 4/2 i = 3/2 + 2i$. Division

---

12. An earlier computer drawing of the same set was given in 1979 by R. Brooks and J. P. Matelski. (See *Ann. of Math. Studies* 97 (1980): 65-71.) Theirs lacked the attention-grabbing detail of those done by Mandelbrot.

by a nonreal complex number is slightly trickier. One should note that $(a + bi)(a - bi) = a^2 + b^2$ is a real number, as the imaginary part $(-ab + ba)i$ equals 0. If a quotient has $a + bi$ in the denominator, the numerator and denominator should both be multiplied by $a - bi$. This does not change the value of the fraction, but it turns the denominator into a nonzero real number, from which point the division can be completed as above. For example,

$$\frac{3 + 4i}{5 - 2i} = \frac{(3 + 4i)(5 + 2i)}{(5 - 2i)(5 + 2i)} = \frac{15 + 6i + 20i + 8i^2}{25 + 4}$$

$$= \frac{7 + 26i}{29} = \frac{7}{29} + \frac{26}{29}i.$$

Complex numbers occur naturally as solutions of equations. For example, the solutions of the equation $ax^2 + bx + c = 0$, which are given by the quadratic formula

$$x = \frac{-b \pm \sqrt{b^2 - 4ac}}{2a},$$

require complex numbers if $b^2 - 4ac < 0$. The Fundamental Theorem of Algebra states that any polynomial $x^n + \cdots + a_1 x^1 + a_0$ whose coefficients $a_i$ are real or complex numbers can be factored into linear factors over the complex numbers. This means that there are complex numbers $w_1, \ldots, w_n$ so that

(5.3.1)   $x^n + \cdots + a_1 x^1 + a_0 = (x - w_1) \cdots (x - w_n).$

This can also be stated to say that if $p(x)$ is a polynomial of degree $n$ with real or complex coefficients, then the equation $p(x) = 0$ has $n$ complex solutions, if the solutions are counted with their multiplicity as factors of (5.3.1). Note that this ability to factor and find roots does not work if we restrict ourselves to real numbers. For example, $x^4 + 4x^2 + 4$ has no real roots but can be factored as $(x^2 + 2)^2 = (x - i\sqrt{2})^2(x + i\sqrt{2})^2$, and so the equation

$$x^4 + 4x^2 + 4 = 0$$

has solutions $i\sqrt{2}$ and $-i\sqrt{2}$, each of multiplicity 2.

The Fundamental Theorem of Algebra was proved by C. F. Gauss in 1799. We have already read of many other important contributions to mathematics by this great German mathematician. His proofs of this theorem were facilitated by the geometrical interpretation of complex numbers, which he developed at about the same time as several other mathematicians.[13] The idea is to think of the complex number $a + bi$ as corresponding to the point $(a, b)$ in the plane. The nice thing about this correspondence is that both addition and multiplication have geometric interpretations. The sum of two complex numbers corresponds to the point in the plane obtained by drawing the line from the point $(0,0)$ to the points corresponding to the numbers being added, and then forming the parallelogram having these lines as two sides, and taking the fourth vertex of that parallelogram as the sum. (See Fig. 5.30a.) The product of two complex numbers corresponds to the point in the plane whose distance from $(0,0)$ equals the product of the distances from $(0,0)$ of the two points, and whose counterclockwise angle from the positive $x$-axis equals the sum of the angles of the two points. See Exercise 2 for an explanation of why this works, and see Figure 5.30b for an illustration of how it works. The plane of pairs of numbers is called the *complex plane* when the pairs of numbers are being interpreted as complex numbers.

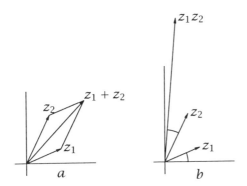

Figure 5.30. Geometric representation of complex addition and multiplication.

Complex numbers are not just an artifact of abstract mathematics. Because of this geometric interpretation, they have had many important applications in physics and electrical engineering.

### THE MANDELBROT SET

As we did for real numbers in the subsection "Another Family of Functions" in Section 5.2, we can ask, for any complex number $c$,

---

13. Jean-Robert Argand and Caspar Wessel published the same idea at about the same time. John Wallis had come close one hundred years earlier.

what is the limiting behavior of the sequence of numbers defined by

(5.3.2) $$z_0 = 0, \quad z_{n+1} = z_n^2 + c.$$

The letter $z$ is the letter most commonly used to denote complex numbers, which is why we use it rather than the letter $x$, preferring to reserve that letter for the real part of $z$. That is, we will sometimes write $z = x + yi$. We used $x^2 - c$ rather than $x^2 + c$ in Section 5.2 so that the relevant values of $c$ would be positive, but for the Mandelbrot set the function is usually written as $z^2 + c$.

Another way of writing (5.3.2) is to let

$$f_c(z) = z^2 + c,$$

and $f_c^n(z) = f_c(f_c(\cdots(f_c(z))\cdots))$ with the function $f_c$ being iterated $n$ times. Note that the subscript $c$ describes the function being iterated, while the superscript $n$ tells how many times it is iterated. For example,

$$f_2(0) = 0^2 + 2 = 2$$
$$f_2^2(0) = f_2(f_2(0)) = f_2(2) = 2^2 + 2 = 6$$
$$f_2^3(0) = f_2(f_2(f_2(0))) = f_2(f_2^2(0)) = 6^2 + 2 = 38$$

Then the numbers $z_n$ in (5.3.2) satisfy $z_n = f_c^n(0)$. A nice feature of the $f_c$-notation is that the role of $c$ is made explicit in it.

The main question is: For a given complex number $c$, does the sequence of numbers

(5.3.3) $$f_c(0), \ f_c^2(0), \ f_c^3(0), \ f_c^4(0), \ldots$$

eventually give very large numbers or does it stay moderately small? The size of a complex number is measured by its absolute value,

$$|a + bi| = \sqrt{a^2 + b^2},$$

which just gives its distance from $(0,0)$ in the geometric interpretation. One can prove (Exercise 4) that if, in the sequence (5.3.3), the absolute value of any of the numbers exceeds 2, then the sequence will approach $\infty$, and in fact it will do so very rapidly.

We now introduce the definition of a very interesting set of complex numbers.

DEFINITION 5.3.4. *The Mandelbrot set M is the set of complex numbers $c$ such that the sequence $\langle f_c^n(0) \rangle$ does not approach $\infty$ as $n$ gets large.*

Another way to say this is that $c$ is in $M$ if the sequence

$$(5.3.5) \qquad c, \; c^2 + c, \; (c^2 + c)^2 + c, \; ((c^2 + c)^2 + c)^2 + c, \ldots$$

stays bounded. By Exercise 4, if $c$ is in $M$, then each number in the infinite sequence (5.3.3) will have absolute value $\leq 2$. For example, for $c = 0$, we have $f_0(0) = 0^2 + 0 = 0$, and hence $f_0^n(0) = 0$ for all $n$, and so 0 is in the Mandelbrot set. On the other hand, for $c = 1$, the sequence (5.3.3) becomes $1$, $1^2 + 1 = 2$, $2^2 + 1 = 5$, $5^2 + 1 = 26$, which is clearly becoming large very quickly. Thus 1 is not in $M$.

We consider two other examples, $c = -1$ and $c = 1/4$. If $c = -1$, the sequence (5.3.3) is $-1$, $(-1)^2 + (-1) = 0$, $0^2 + (-1) = -1$, $(-1)^2 + (-1) = 0, \ldots$. It is clear that this sequence will alternate between $-1$ and 0 forever, and so $-1$ is in $M$. If $c = 1/4$, then the sequence begins $1/4$, $1/16 + 1/4 = 5/16$, $25/256 + 1/4 = 89/256$. The numbers in this sequence seem to be staying below $1/2$, and you can prove that they will stay below $1/2$ by noting that if $z$ is a positive real number which is less than $1/2$, then $z^2 + 1/4 < 1/2$. Thus $1/4$ is in $M$. In Exercise 3, you show that $-2$ and $i$ are both in $M$, as well.

The Mandelbrot set, as defined in 5.3.4, is a precisely defined subset of the complex plane. Every complex number $c$ is either in it or not. In theory, one could picture the Mandelbrot set by darkening every point $c$ of the complex plane that lies in it. In practice, there are infinitely many points $c$ to check, and for some of them it may be very difficult to tell whether or not they lie in $M$. This would be the case if the numbers $|f_c^n(0)|$ don't seem to be getting large, but you have no good reason to know that they might not eventually get large. It was *a priori* conceivable that the set defined by 5.3.4 might turn out to have some simple description, such as a combination of disks and lines, and therefore turn out to be perfectly describable and drawable. That turned out not to be the case. Indeed, the Mandelbrot set has been described as "the most complex object mathematics has ever seen."[14]

The set defined by 5.3.4 had been considered before 1920 by the French mathematicians Pierre Fatou and Gaston Julia. In the next subsection, we will learn about some of the things they proved. But long before computers were invented, they had no idea what the set looked like. Mandelbrot, a research fellow at IBM, had, by 1980, already spent two decades investigating fractal phenomena in nature, as we discussed in Section 5.1. He was a visiting professor at Harvard in 1980 and decided to let a computer sketch the set defined by

14. Peitgen and Saupe 1988, p. 177.

5.3.4.[15] Surprised by what he saw, he went back to IBM's better computer equipment, and found that the set appeared to be a striking fractal, with baby copies of the whole set appearing at all magnifications.

No computer picture is going to give a perfect picture of the Mandelbrot set. We divide the appropriate portion of the complex plane into a grid of tiny squares corresponding to a grid of pixels on our computer screen. We leave a pixel uncolored if the sequence (5.3.3) with $c$ equal to the complex number in the lower left corner of the pixel exceeds 2 in absolute value before a prescribed number of iterations has been performed.[16] This guarantees that this number $c$ is not in $M$, and we leave the entire pixel uncolored to reflect this. If the sequence (5.3.3) does not become larger than 2 before this prescribed number of iterations has been attempted, then the computer guesses that this number $c$ is probably in $M$ and colors this pixel black. The programmer has to tell the computer in advance how many times to iterate (5.3.3) before declaring that $c$ is in $M$ (provided, of course, that the numbers in the sequence never exceed 2 in absolute value). There is a trade-off here. If you choose this maximum number of iterations too small (e.g., 20), then you will miscolor many of the pixels, because there are many numbers $c$ for which the sequence (5.3.3) stays small for at least the first twenty numbers, but later gets large. On the other hand, if you choose this maximum number too large (e.g., 5000), then you will waste a lot of computer time running through the five thousand iterations on points in $M$, for it is only a very sparse set of points close to the boundary of $M$ where more than five hundred iterations are required for the sequence (5.3.3) to get large.

It is known that $M$ lies in the square whose $x$-values are between $-2$ and $0.5$, and whose $y$-values are between $-1.25$ and $1.25$. Suppose we wanted to divide this into a 200 by 200 grid, which is the largest square that could be displayed on many computer screens. The length of side of each square in the grid would be $2.5/200 = .0125$. Then we would be testing 40,000 values of $c$, namely $c = a + bi$, where $a$ takes the values $-2$, $-2 + .0125$, $-2+.025, \ldots, -2+199 \cdot .0125 = .4875$, and $b$ takes the values $-1.25$, $-1.25 + .0125$, $-1.25 + .025, \ldots, -1.25 + 199 \cdot .0125 = 1.2375$. For each of these 40,000 values of $c$, it will compute the sequence (5.3.3), stopping if any number exceeds 2 in absolute value or if the maximum prescribed number of iterations (e.g., 130) is obtained. If the

---

15. He now holds a professorship at Yale along with his position at IBM.
16. Many books use the point in the center of the pixel rather than the corner point.

maximum number of iterations is obtained, then the pixel will be colored in.

You should understand why this will miss a lot of fine detail. Along the boundary of the Mandelbrot set, there are tiny "warts," little eruptions whose shapes are very similar to that of the entire Mandelbrot set. (See Fig. 5.32 for a crude depiction, and Color Plate 1 for one with better resolution.) Many of these warts are much smaller than .0125 by .0125 (or whatever the size of the square is in the complex plane corresponding to one pixel). These will not be seen or displayed by the computer unless a corner point of one of the squares of the grid happens to lie within the baby wart, in which case a whole square bigger than the baby will be turned on.

In Figure 5.31 we illustrate with a portion of the grid described above. For example, in the second column the three middle pixels are turned on because if $c = -1.3875$ or $-1.3875 \pm .0125i$, then the sequence (5.3.3) is still small after 130 iterations, while if $c = -1.3875 \pm .025i$, $-1.3875 \pm .0375i$, or $-1.3875 \pm .05i$, the sequence (5.3.3) gets large before 130 iterations.

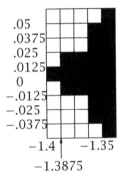

Figure 5.31. A grid for a small portion of the Mandelbrot set.

The following computer program will allow the reader to display $M$ and to enlarge portions of it. We call this program MANDEL.

```
DEFDBL C,D,X,Y: DEFINT H,I,J,M,N,V
INPUT "Enter number of pixels horiz, vert"; H,V
INPUT "Enter GAP--same for horiz and vert"; DEL
INPUT "Enter coordinates of low point"; CR,CINIT
INPUT "Maximum number of interations"; MAX
SCREEN 1: CLS: KEY OFF
WINDOW (0,0)-(319,199)
FOR I=1 TO H
  CI=CINIT
```

```
        FOR J=1 TO V
        NUM=1: X=0: Y=0
        WHILE X*X+Y*Y<=4 AND NUM < MAX
          X1=X*X-Y*Y+CR
          Y=2*X*Y+CI
          X=X1
          NUM=NUM+1
        WEND
        IF NUM>=MAX THEN PSET (I,J)
        CI=CI+DEL
      NEXT J
      CR=CR+DEL
    NEXT I
    END
```

This program contains a number of features that were not present in our earlier programs. The first line stipulates that any variable whose name begins with the letters C, D, X, or Y is a double precision variable, while any variable whose name begins with H, I, J, M, N, or V is an integer variable. This is easier than writing a # after every occurrence of a double precision variable.

The user is asked to input four parameters each time the program is run. We require that the gap size be the same in the horizontal and vertical directions so that the displayed picture is not distorted. In the example discussed prior to the listing of the program, the user would enter 200, 200 for H and V, then .0125 for the gap size, then −2, −1.25 for the coordinates of the lower left-hand corner, and finally 130 for the maximum number of iterations that the computer will try before concluding that the point is probably in the Mandelbrot set.

The program uses a double loop. It will start with $I = 1$ and let $J$ run through all values from 1 to $V$. Then it will let $I = 2$ and again let $J$ run from 1 to $V$. It will do this for each value of $I$ from 1 to $H$. So this double loop will be gone through for every pair of values of $I$ and $J$ corresponding to each pixel to be filled in. The complex number $c$ that is being tested as to whether or not it is in the Mandelbrot set is represented in the computer by its real part $CR$ and its imaginary part $CI$. That is, $c = $ CR+CIi. The variables start off with the values given by the user in response to the third INPUT statement. Each time the value of $J$ changes, the value of CI is increased by DEL, which is the gap-size given by the user, and each time the value of $I$ changes, the value of $CR$ is increased by DEL. Since each change of $I$ causes the values of $J$ to start over again, it is important that the values of CI start over again, too. This is accomplished by the statement CI=CINIT. This initial value, CINIT, is not changed throughout the running of the program.

For each such value of $I$ and $J$, CR and CI will satisfy $c$ = CR+CIi, and the computer will try to decide whether or not $c$ is in $M$. This is done by letting $z = X + Yi$ start with the value 0, and repeat the process of replacing $z$ by $z^2 + c$. Since the computer doesn't automatically know how to work with complex numbers,[17] we must tell it that

$$(X + Yi)^2 + (CR + CIi) = (X^2 - Y^2 + CR) + (2XY + CI)i.$$

This is accomplished in the first two lines after the WHILE statement. The variable $X1$ is used to store the new value of $X$ temporarily, so that the old value of $X$ can be used to compute $Y$.

The statements between WHILE and WEND are repeated as long as the conditions following the WHILE statement are satisfied. The WEND statement acts like a NEXT I statement, inasmuch as its role is to send the program back to the WHILE statement. The two conditions that must be satisfied in order for the computer to keep changing the value of $z$ are that $X^2 + Y^2 \le 4$ (which is equivalent to saying $|z| \le 2$) and that the prescribed maximum number of iterations has not yet been obtained. Note how BASIC uses < = and > = for $\le$ and $\ge$. When this WHILE loop is exited, the computer colors the appropriate pixel if the maximum number of iterations was attained, since in this case the point $c$ is probably in the Mandelbrot set.

When this program is run with the initial values discussed earlier, an image that crudely approximates that of Figure 5.32 is obtained. Whether this picture is produced in a few seconds, a few minutes, or a few hours depends on your computer. If you have a fast computer and good monitor, such as a 386 machine with VGA monitor, you can get an image like that of Figure 5.32 by using a 400 by 400 grid to display the same region of the complex plane. Your gap size will be half as large, so that you see detail that is twice as fine as you were seeing before. You must use SCREEN 12 to instruct the computer to display this fine detail.

If you are plotting the entire Mandelbrot set, you can take advantage of its symmetry around the horizontal axis to cut the computing time in half. This symmetry, which is proved in Exercise 9, implies that the pixel corresponding to the complex numbers $c = a + bi$ and $\bar{c} = a - bi$ will be filled in exactly the same way. Therefore, if

17. Some newer versions of BASIC have arithmetic of complex numbers built in.

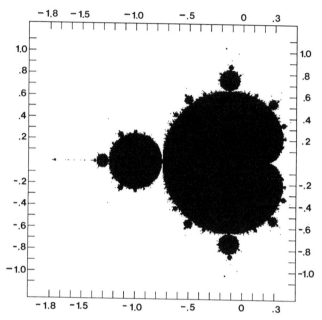

Figure 5.32. A depiction of the Mandelbrot set.

your program computes one of these points, it need not compute the other. The program MANDEL can easily be modified to incorporate this, but it won't be helpful for sketching subportions of the Mandelbrot set such as that in Figure 5.34.

This picture is frequently called the Gingerbread Man or the Snowman. We will call it a Bug, and refer to the baby bugs that appear around its boundary. We will want to take a better look at some of these baby bugs, and also to investigate the stray dots that seem to be separated from the main body of the bug.

Before we begin this, we mention a fine point about the proboscis of the Mandelbrot set, which sticks out to the left in Figure 5.32, where it looks as if it contains several disconnected parts. One can prove (see Exercise 5) that the intersection of $M$ with the $x$-axis consists precisely of all numbers between $-2$ and $0.25$, inclusive. If this were to be reflected by Figure 5.32, the proboscis would have to be a solid line out to $-2$ at the left side of the picture. The reason that it didn't come out this way is due to a limitation on the accuracy of the numbers used by the computer. The points in question should correspond to values of $c$ whose imaginary part is 0. Instead, the author's computer uses complex numbers whose imaginary part is

$6.9 \cdot 10^{-17}$. So the proboscis in Figure 5.32 displays the intersection of the actual proboscis of $M$ with a line ever so slightly above the $x$-axis.

One way of seeing more of the structure of the Mandelbrot set is by having the computer assign colors to the points outside it. The choice of color is determined by how many iterations were required before the sequence (5.3.3) exceeded 2. In Figure 5.33, the gray points are those points outside $M$ which required thirteen or more iterations to get large,[18] while the white points which fill up the exterior region are those values of $c$ for which (5.3.3) got large in less than thirteen iterations.

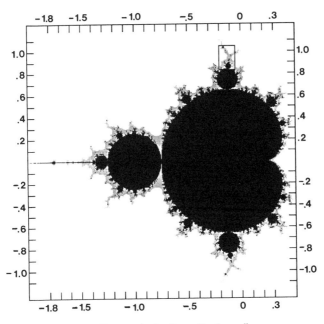

Figure 5.33. The Mandelbrot set in four "colors."

Although, strictly speaking, this gray area does not give any specific information about the set $M$ itself, it suggests that there might be thin strands of $M$ embedded in these gray arms, so thin that they escape detection by the computer's grid of points tested. That this is true is based on experience—it usually happens. The existence of values of $c$ that require many iterations for (5.3.3) to get large sug-

18. From now on, when referring to the sequence (5.3.3), we will often use the phrase "get large" to mean "exceed 2 in absolute value."

gests the likelihood of nearby values of $c$ for which (5.3.3) will never get large.

More gradations of color or tone can be displayed, depending on one's computer. The pictures look much more spectacular when bright colors are used, as we shall see in the Color Plates. In Figure 5.33, there is actually a second shade of gray fringing the Mandelbrot set, but it doesn't show up well in the picture; it was much more apparent when this picture was displayed on a color terminal. It corresponds to values of $c$ for which the number of iterations for (5.3.3) to get large was greater than thirty.

The choice of the number of iterations at which to change the color is made by the user of the program. The program MANDEL must be modified to give the user this flexibility. The fourth INPUT statement is changed to read

```
INPUT "Three iteration thresholds"; N1,N2,N3
```

The word MAX in the WHILE statement should be changed to N3. The PSET statement near the end of the program is changed to

```
IF NUM>=N3 THEN
    PSET (I,J),3
ELSEIF NUM>=N2 THEN
    PSET (I,J),2
ELSEIF NUM>=N1 THEN
    PSET (I,J),1
ELSE
    PSET (I,J),0
END IF
```

The format of this compound IF statement works in TURBO BASIC and QBASIC. If you are using a different version of BASIC, you may have to modify it. The numbers at the end of the PSET statements refer to the four different colors that can be used when SCREEN 1 is in effect.[19] Thus if, when the Mandelbrot algorithm is run for a complex number $c$ corresponding to a certain pair $(I, J)$ of values, the number of iterations required was greater than or equal to the inputted value N3, then the pixel at position $(I, J)$ is colored using color 3. When we ran the program, modified in this way, and the computer asked us

```
Three iteration thresholds?
```

we typed 13, 30, 130. Using a number smaller than 13 will give more

---

19. SCREEN 12 on a VGA monitor allows colors numbered from 0 to 15, and these may be chosen from a palette of sixty-four colors, using a PALETTE statement.

gray, but begins to lose some of the detail of the tentacles present in Figure 5.33.

A better way to see more detail of the Mandelbrot set is to rerun the program with the parameters chosen so as to display a small portion of the set in finer detail. For example, suppose we want to take a closer look at the little box near the top of Figure 5.33. This region consists of numbers $c = a + bi$ such that $-0.2 \leq a \leq -0.05$ and $0.83 \leq b \leq 1.05$. If we want to display this on a screen with two hundred vertical pixels, we can use a much smaller gap size. This means that the values of $c$ being tested are much closer together, and so the picture will contain fine detail that was missed by Figure 5.33. We choose as the gap size $(1.05 - 0.83)/200 = 0.0011$. This is more than ten times smaller than was the gap size $(0.0125)$ that led to the 200 by 200 analogue of Figure 5.33. Therefore a .0125 by .0125 region of the complex plane, which was colored in one color in Figure 5.33 according to whether or not its lower left point was in $M$, will be divided into a grid of more than 100 little squares in our new calculation, each of which will be given individual attention for its choice of color.

As the region in the box that we are going to expand is not exactly square, we will not want to use 200 horizontal pixels. Remember that it is essential that the horizontal and vertical gap sizes be the same so that the picture not be distorted. The number of horizontal pixels will be $(-0.05 - (-0.2))/0.0011 = 136$. Thus the user should tell the computer that $H = 136$, $V = 200$, gap size $= 0.0011$, and the coordinates of the low point are $-0.2$, $0.83$. We run the modified version of the program that displays four colors, and use as our thresholds 20, 50, and 150. These are somewhat larger than those used in Figure 5.33 because we are dealing with a region close to the boundary of the Mandelbrot set, so that a more stringent requirement should be made for a point to be considered to be in a colored region. The output appears in Figure 5.34. This picture shows the fourth color slightly better than did Figure 5.33, inside the gray closer to the black.

Note how at this magnification the tiny bug in the bottom of the box in Figure 5.33 appears in Figure 5.34 as a big bug with lots of tiny bugs on it. The reader should certainly not be surprised now to be told that, if a magnification of the little bug on top of the big bug in Figure 5.34 is performed by running the program again with a smaller gap size, then a very similar picture will be obtained. The Mandelbrot set is not perfectly self-similar, because most of these tiny bugs are slightly different; the way in which tentacles extend out is slightly different for the different babies. Also, as illustrated dramatically in Color Plate 4, many of them are not perfectly sym-

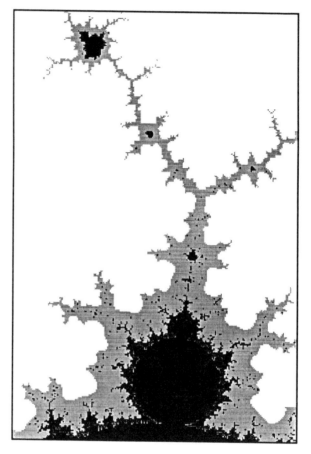

Figure 5.34. A magnification of the region in the box in Figure 5.33.

metrical. Much recent research (e.g., Milnor 1987) has studied the extent of this self-similarity. One proven fact is that in any tiny disk around any point on the boundary of $M$, there are infinitely many baby bugs.

The stray dot near the top of Figure 5.32 was present in Figure 5.33, but was sort of lost amid the gray. In Figure 5.34, it has been given much sharper focus, where it has become the bug near the top. What appear to be tinier bugs along the way from the main body toward the big baby are apparent in Figure 5.34. These did not show up at all in Figure 5.32 or 5.33 because the coarser grid completely skirted these points. When Mandelbrot first saw these stray dots or bugs in 1980, he wondered whether they were connected to the main body.

The gray strands suggest that they might be, but those strands are not part of $M$. They are just a reminder that for these values of $c$, the sequence (5.3.3) takes a long time to get large. It is extremely difficult to make your computer find points that are actually in $M$ inside these gray strands.

It was proved in 1982 by Adrien Douady of Ecole Normale Supérieure in Paris and John Hubbard of Cornell University that the Mandelbrot set is a connected set. This means that it cannot be split into two completely separated parts. The proof involves a very delicate argument in complex analysis.[20] Their analysis shows that there are extremely fine filaments connecting all the baby bugs. This theorem illustrates the power of mathematical proof—computer programs of the type we have been discussing will never see all these filaments, but the proof guarantees that they are there.

There are several unsolved mathematical questions about the Mandelbrot set which have attracted some of the most famous mathematicians in the world to study them. Most of them are very technical, but one that we can explain was unsolved at the time of the writing of the first draft of this book. This is the determination of the Hausdorff dimension of the boundary of the Mandelbrot set. It was shown to equal 2 in July 1991, by Mitsuhiro Shishikura while working at the State University of New York at Stony Brook. It says that the boundary of $M$ is so wiggly that it fills space near it as much as would something with thickness.

In Plates 1 to 11, we show a new sequence of blowups of the Mandelbrot set, produced by Ken Monks of the University of Scranton. Each plate is a blowup of the region inside the little box pictured in the preceding plate.[21] All the detail of Plates 2 to 11 is taking place in what appears in the original picture as a black area near what might be considered the anus of the bug. Plate 2 shows that this crevice is much sharper than one would expect from looking at Plate 1. Plate 4 shows a highly asymmetrical bug, which, although we cannot see the filament, is apparently attached to the larger bug of Plate 3. This asymmetry is real; there is no distortion in these photos.

From the little box in Plate 6, it appears as if Plate 7 will be an enlargement of a region (relatively) far away from the Mandelbrot set. This is not the case, as we shall see, and if one then inspects Plate 6 more carefully, a wiggly golden thread coming up from the

20. Douady and Hubbard, "Étude dynamique des polynomes complexes, I, II," *Publ. Math. Orsay*, 1984, 1985.
21. Actually, the small box in Plate 1 is several times larger than it should be. A box in Plate 1 depicting the actual region being blown up into Plate 2 would have been too small to see clearly.

brighter yellow streak will be seen. Embedded in that thread must be a filament of points of $M$, for by the time we get to Plate 11, we see clearly that there is a baby bug that was too small to be visible in Plates 7, 8, and 9, but nevertheless is present in the appropriate parts of them, when they are viewed with sufficient care. Remember that no physical picture can show all the detail of the Mandelbrot set.

In Table 5.1, we list the coordinates $(x_0, y_0)$ of the lower left corner of each of Plates 1 through 11. This is for the use of the reader who would like to create some of these pictures. Be forewarned that for all but the first few, the requisite number of digits of precision demands a great deal of care. The number $\Delta y$ in Table 5.1 is the vertical length of the region of the complex plane being displayed. The horizontal length is always $4/3$ times this amount. This is true because each picture depicts a screen of 800 by 600 pixels. For example, Plate 2 depicts the region of the complex plane bounded on bottom by $y = -0.0032$, on top by $y = -0.0032 + 0.0060 = 0.0028$, on the left by $x = 0.2500$, and on the right by $x = 0.2500 + 4/3 \cdot 0.0060 = 0.258$.

TABLE 5.1. Parameters for Plates 1 to 11.

| No. | $x_0$ | $y_0$ | $\Delta y$ |
|---|---|---|---|
| 1 | −2.6666666666667 | −2.0 | 4.0 |
| 2 | 0.2500 | −0.00319935 | $6.0 \times 10^{-3}$ |
| 3 | 0.2540330216288567 | −0.0005096439272165 | $1.4 \times 10^{-4}$ |
| 4 | 0.2541128750890493 | −0.000427307561039 | $1.3 \times 10^{-5}$ |
| 5 | 0.2541177679989487 | −0.0004275961574166 | $2.6 \times 10^{-6}$ |
| 6 | 0.2541190116705356 | −0.000425963230244 | $2.6 \times 10^{-7}$ |
| 7 | 0.254119281284086 | −0.000425800640350 | $1.3 \times 10^{-8}$ |
| 8 | 0.2541192884101678 | −0.000425800088364 | $6.5 \times 10^{-10}$ |
| 9 | 0.2541192888594526 | −0.000425799745035 | $3.2 \times 10^{-11}$ |
| 10 | 0.2541192888800564 | −0.000425799728823 | $2.8 \times 10^{-12}$ |
| 11 | ·0.2541192888817539 | −0.0004257997273911 | $1.4 \times 10^{-13}$ |

The magnification factor for any of the pictures can be determined as the quotient of 4 divided by its value of $\Delta y$. The number 4 occurs here as the value of $\Delta y$ for Plate 1. Thus Plate 11 is being magnified by a factor of $4/(1.4 \times 10^{-13}) = 2.8 \times 10^{13}$, as compared to Plate 1. If all of Plate 1 were magnified by this amount, its vertical side would be about 1 billion ($10^9$) miles, which is approximately equal to the diameter of the orbit of Jupiter around the sun. In other words, if the Mandelbrot set were enlarged so as to fill up a screen the size of the orbit of Jupiter, then Plate 11 would be the same size it is in your book. See Exercise 10 for a verification of these size estimates.

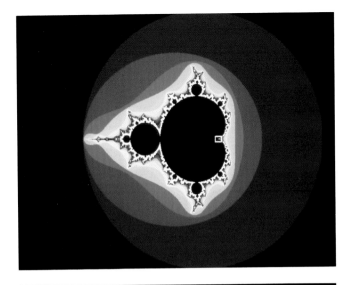

PLATE **2**

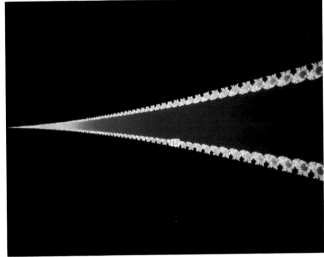

PLATE **1**

PLATE **3**

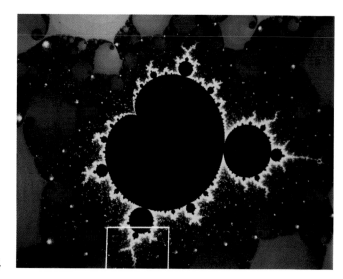

PLATE **4**

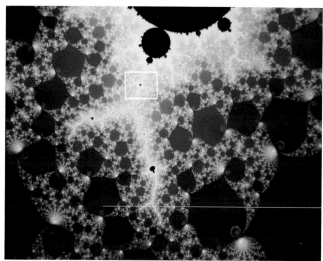

PLATE **5**

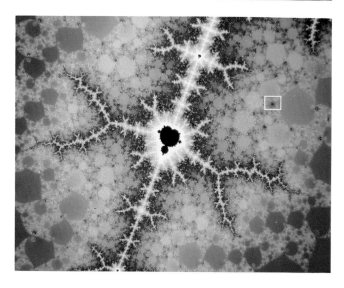

PLATE **6**

PLATE 7

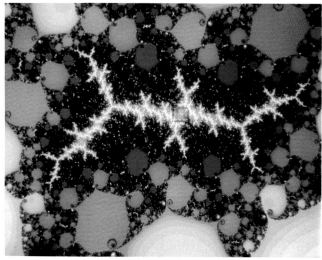

PLATE 8

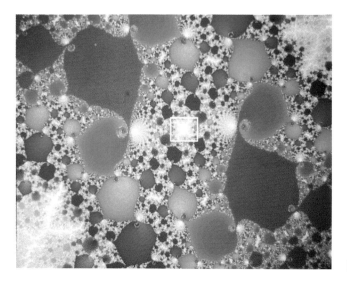

PLATE 9

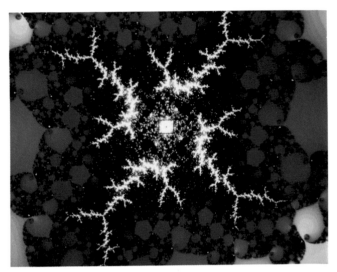

PLATE **10**

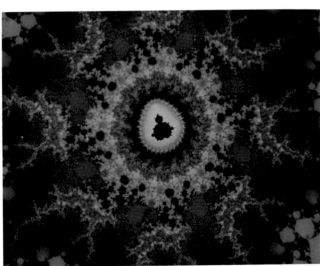

PLATE **11**

In order to minimize the number of points that were misclassified as being in $M$, the computer went to 30,000 iterations before declaring a point to be in $M$. This requires a lot of computer time (six days on a 386 machine for Plate 2), but for points as close to $M$ as those being depicted here, it is necessary, for some of them take thousands of iterations before the sequence gets large.

*JULIA SETS*

We have been considering what happens to the sequence defined by

$$(5.3.6) \qquad\qquad z_{n+1} = z_n^2 + c$$

when $z_0$ is fixed (at 0) and $c$ is varied. One could instead fix $c$ and vary $z_0$. This is perhaps a more natural thing to do, because it can be done for any function $f$. That is, for any function $f$ and any number $z$, you can ask what is the limiting behavior of the sequence $\langle f^n(z) \rangle$ as $n$ goes to infinity. As usual, $f^n(z)$ means $f(f(\cdots(f(f(z))\cdots))$, with $f$ occurring $n$ times, and $\langle f^n(z) \rangle$ means the sequence as $n$ runs through all positive integers. The possibilities include converging to a fixed value, converging to a cycle, diverging to $\infty$, behaving chaotically but staying bounded, or getting unbounded chaotic behavior. You could color the complex $z$-plane according to how this sequence behaves.

DEFINITION 5.3.7. *If $f$ is a polynomial, the* **filled-in Julia set** *of $f$ is the set of complex numbers $z$ such that the sequence $\langle |f^n(z)| \rangle$ is bounded. The* **Julia set** *$J_f$ of $f$ is the boundary of the filled-in Julia set. If $f = f_c$ is the function defined by $f_c(z) = z^2 + c$ for some complex number $c$, then we abbreviate the Julia set of $f_c$ as $J_c$.*

The term *boundary* of a set is a technical term that agrees with one's intuition of boundaries. A point is in the boundary of a set $S$ if every little disk around the point contains some points in $S$ and some points outside $S$. (See Exercise 12 for an example.) Thus $J_c$ is the set of complex numbers $z$ such that every little disk around $z$ contains some points $w$ such that the sequence $\langle |f_c^n(w)| \rangle$ approaches infinity, and some points $v$ such that $\langle |f_c^n(v)| \rangle$ stays bounded.

The above description suggests a way of having a computer sketch the Julia set $J_c$. Make the pixels correspond to a grid of tiny squares that covers the region where you think the Julia set should lie. Each of these little squares has four corners. Color a pixel if, of the four complex numbers $z$ corresponding to its corners, the sequence $\langle |f_c^n(z)| \rangle$ approaches infinity at some but not all of these four complex numbers. This is the closest the computer can come to easily implementing the condition for a point to be in the boundary of the filled-in Julia set.

There is, at least in principle, a different Julia set $J_c$ for each complex number $c$. We shall write and run a program to examine some of these. The program is quite similar to MANDEL. We shall call this one JULIA.

```
DEFDBL C,D,X,Y,Z
DEFINT H,I,J,K,L,M,N,S,T,V

DEF FNTEST(X,Y)
  SHARED CR, CI, MAX
  LOCAL X1, N
  N=1
  WHILE X*X+Y*Y<1000 AND N<MAX
    X1=X*X-Y*Y+CR
    Y=2*X*Y+CI
    X=X1
    N=N+1
  WEND
  IF N=MAX THEN FNTEST=1 ELSE FNTEST=0
END DEF

INPUT "Enter number of pixels horiz, vert"; H,V
INPUT "Enter GAP--same for horiz and vert"; DEL
INPUT "Enter coordinates of low point"; X0,Y0
INPUT "Maximum number of interations"; MAX
INPUT "Real and imaginary parts of c"; CR,CI
SCREEN 1: CLS: KEY OFF
WINDOW (0,0)-(319,199)
DIM N1(200), N2(200)
X=X0
Y=Y0
FOR J=0 TO V
  N1(J)=FNTEST(X0,Y)
  Y=Y+DEL
NEXT J
FOR I=1 TO H
  X=X+DEL
  Y=Y0
  FOR J=0 TO V
    N2(J)=FNTEST(X,Y)
    Y=Y+DEL
```

```
        NEXT J
        FOR J=0 TO V-1
          S=N1(J)+N1(J+1)+N2(J)+N2(J+1)
          IF S=1 OR S=2 OR S=3 THEN PSET(I,J)
        NEXT J
        FOR J=0 TO V
          N1(J)=N2(J)
        NEXT J
      NEXT I
      END
```

This program incorporates two major features not present in our earlier programs—functions and arrays. The lines beginning DEF FN and ending END DEF define a function named FNTEST which is called many times during the running of the program. These statements are not executed at the very beginning of the program; they are executed whenever the main program says $= $ FNTEST$(-,-)$. Each time this happens, the two blanks will be filled in with numbers, which are called the *arguments* of the function.

For example, suppose the program is being run, and the values $X0 = -2$ and $Y0 = -1$ are inputted. Then one of the first statements the computer will execute is

(5.3.8)          N1(J)=FNTEST(X0,Y),

with $J = 0$, $X0 = -2$, and $Y = -1$. At this point, the computer jumps back to the statement DEF FNTEST(X,Y).[22] The arguments X and Y will take on the values $-2$ and $-1$. This illustrates the crucial role played by the arguments of the function; the values of the variables $X0$ and $Y$ when the function is called are transferred to the variables $X$ and $Y$ in the body of the definition of the function.

The next two lines say that the variables CR, CI, and MAX are shared with the main part of the program, while the variables X1 and N have nothing to do with any variables in the main part of the program. These lines are necessary in TURBOBASIC, but will be unrecognizable to QBASIC. The function FNTEST takes a value of 1 if the formula (5.3.6) starting with $z_0 = X + Yi$ can be iterated MAX times without getting large. Otherwise, FNTEST takes the value 0. This is accomplished by the eight lines in the definition of FNTEST beginning with N=1. After this iteration has been performed, the computer returns to the statement (5.3.8) and assigns the value 0 or 1 to the variable N1(0). It then proceeds to the next line in the program.

22. The format of functions may vary in different forms of BASIC.

This variable N1(0) is part of an array. This is the computer's way of working with a subscripted family of variables. The statement DIM N1(200) tells the computer to reserve 201 words of storage for variables named N1(0), N1(1), ..., N1(200). Every time the variable N1 appears in the program it will have a number in parentheses after it, telling which of these 201 variables is being used. The first little 4-line loop, for which (5.3.8) is the central line, gives to the variables N1(0), N1(1), ..., N1(V) the values 0 or 1 depending upon whether or not the iteration (5.3.6) starting at the 0th, 1st, ..., Vth points along the left side of the vertical rectangle being investigated stays small or becomes large.

We keep track of these values for all the points in a column because every point (except those on the boundary of the whole region) will be relevant for determining whether four different pixels should be turned on. For example, the big dot in the middle of Figure 5.35 affects whether the four pixels surrounding it are turned on.

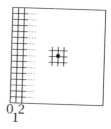

Figure 5.35. Grid illustrating aspects of calculating Julia sets.

Since the bulk of the computer time is spent iterating (5.3.6), we don't want to do this more than once for each point. Therefore we store this value (0 or 1) until the point has been used as a corner point of all four pixels. After storing the values along column 0 in the array N1, the computer computes the values along column 1, and stores them in the array N2. Then for each pixel in the first column of pixels, it turns the pixel on if 1, 2, or 3 of the corner points has the value 1. This would say that for some but not all corner points of the pixel (5.3.6) gets large, which is our criterion for the pixel to be in the Julia set. This is accomplished in the program by the lines that define $S$ as a certain sum, and ask whether it equals 1, 2, or 3. It is illustrated along the left side of Figure 5.35. When the computer has finished filling in the first column of pixels, it puts the N2-values in the N1-array, and forgets the old N1-values, which it will no longer need. It now applies FNTEST to column 2 of points, and puts the 0's or 1's in the array N2. Now it colors the second column of pixels using the new values in N1 and N2. This procedure is continued one column at a time.

Calculating a Julia set on a 200 by 200 grid in this manner takes about as long as calculating a portion of the Mandelbrot set on a 200 by 200 grid—anywhere from seconds to hours, depending on your computer. You can take advantage of the symmetry discussed in Exercise 14 to cut your computing time in half if you are plotting the whole Julia set.

For values of $c$ that lie inside the Mandelbrot set, the Julia sets calculated by the above program are generally quite spectacular. We include several of these as Figures 5.36 and 5.37a. Those in Figure 5.36 were drawn using a 200 by 200 grid to fill the square whose $x$- and $y$-values go from $-2$ to 2. Thus the gap size is $4/200 = .02$. We used 100 as the maximum number of iterations to try. Figure 5.37 uses a 400 by 400 grid, using SCREEN 12 on an EGA monitor, and goes up to 150 iterations. Figure 5.37b shows a magnification of the indicated subsquare of Figure 5.37a. It illustrates the fractal property of the Julia sets. For example, Figure 5.37b shows holes and sea-horse tails that were invisible in Figure 5.37a, but have the same character as other larger ones that were visible in Figure 5.37a. It can be proved that these Julia sets can be covered by copies of themselves, but these copies are obtained by nonlinear transformations, which can affect the appearance significantly.

Before 1920, Fatou and Julia independently proved the following theorem, which has played a central role in current work regarding mathematical properties of the Mandelbrot set and Julia sets. It is

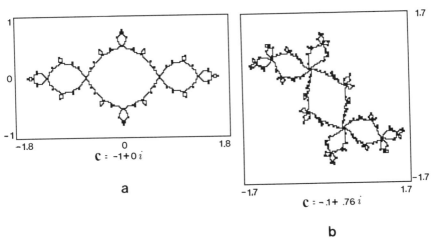

Figure 5.36. Julia sets for $c = -1 + 0i$ and $c = -0.1 + 0.76i$.

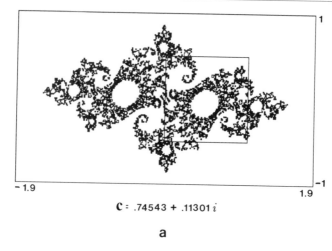

C = .74543 + .11301 $i$

a

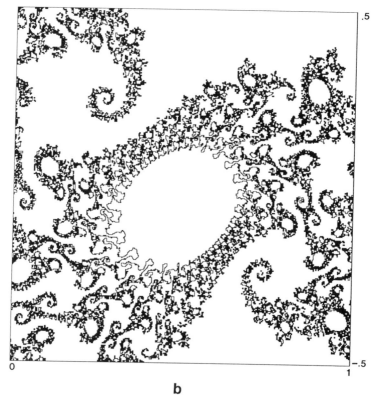

b

Figure 5.37. Julia set for $c = -0.74543 + 0.11301i$ and an enlargement.

remarkable that they could discover a theorem such as this without computers to show them what some of these sets really looked like.

THEOREM 5.3.9. (Fatou and Julia). The Julia set $J_f$ of any polynomial $f$ is nonempty and is either connected or else totally disconnected. It is connected if and only if, for all critical points $z$ of $f$, the sequence of numbers $\langle |f^n(z)| \rangle$ is bounded.

We have already explained that a connected set is one that cannot be split into two separated parts.[23] Any set can be broken down into its connected components. A set is *totally disconnected* if its connected components consist of single points. Thus any two points in a totally disconnected set can be separated from each other. The Cantor set (Fig. 5.7) is a typical example of a totally disconnected set.

A *critical point* of a polynomial $f$ is a number $z$ such that $f'(z) = 0$, where $f'$ denotes the derivative of $f$. If $f$ is the polynomial defined by $f(z) = z^2 + c$ for some (complex) number $c$, then the only critical point of $f$ is the point $z = 0$. Thus applying Theorem 5.3.9 to the polynomial $f(z) = z^2 + c$ immediately yields the following corollary. The fact that $z = 0$ is the only critical point of $z^2 + c$ is the reason why it is important to start at 0 in the sequence that defines the Mandelbrot set.

COROLLARY 5.3.10. *A complex number $c$ is in the Mandelbrot set if and only if the Julia set $J_c$ of $z^2 + c$ is a connected set. If $c$ does not lie in the Mandelbrot set, then $J_c$ is totally disconnected.*

Thus the Mandelbrot set displays in one picture certain information about all the Julia sets $J_c$—it tells which ones are connected. It actually does more than this. It has been observed that, for complex numbers $c$ that are very close to the boundary of the Mandelbrot set, the portion of the Mandelbrot set very close to $c$ bears a striking resemblance to the portion of the Julia set $J_c$ very close to the value $c$. A rigorous formulation for certain points $c$ on the boundary of the Mandelbrot set was given in Lei (1990), and current research is attempting to extend those results. Discussion and examples are given on pages 199–206 of Peitgen and Saupe (1988). For example, it is shown on page 206 of that book how the portion of the Mandelbrot set in a small region near the value $c = -.74543 + .11301i$ resembles a portion of our Figure 5.37. Extreme magnifications are required in order to see this.

The values of $c$ that gave rise to Figure 5.36 are deep inside the Mandelbrot set. As you can see from Figure 5.32, $c = -1$ lies in the

23. This is not the most general definition of connected set, but works for the compact sets with which we are dealing here.

large disk attached to the left of the main body of the bug, while $c = -0.1 + .76i$ lies in the largest disk above the main body. The connected appearance of Figures 5.36a and 5.36b is thus in agreement with Corollary 5.3.10. The number $c = -.74543 + .11301i$ lies very close to the boundary of $M$, in the crevice between the main body and the large disk to the left. It is, in fact, slightly outside $M$, as the sequence (5.3.2) with $c = .74543 + .11301i$ becomes large after 295 iterations. Thus the Julia set approximated in Figure 5.37 is totally disconnected, despite the fact that it looks as if it might be connected. Although the notions of connected set and totally disconnected set are diametrically opposed, there is actually a fine line between them. A minuscule change in $c$ as it crosses the boundary of $M$ causes just a minuscule change in the Julia set $J_c$, but that change can be the difference between being connected and totally disconnected. (See, for example, Handler, Kauffman, and Sandin 1987, for many computer-generated pictures illustrating this fact dramatically.)

As $c$ moves farther away from $M$, the Julia set $J_c$ becomes more clearly totally disconnected. For example, Figure 5.38 is the Julia set corresponding to $c = -0.1 + 1.11i$. The program JULIA is unable to draw this Julia set because for all $z$-values outside it, the sequence (5.3.6) with $z_0 = z$ becomes infinite. Because the $z$-values in this $J_c$ are so sparse, it is extremely likely that all corner points in our grid will be attracted to $\infty$. Hence no squares in our grid satisfy the criterion our program requires for being in the Julia set, namely that some but not all of its corner points are attracted to $\infty$. In our earlier cases (Figs. 5.36 and 5.37), there was a large region enclosed by the Julia set consisting of points that were not attracted to $\infty$.

For values of $c$ that are away from $M$ and have sparse Julia sets, there is a completely different method of having a computer sketch the Julia set $J_c$. This is discussed, with a computer program, in Peit-

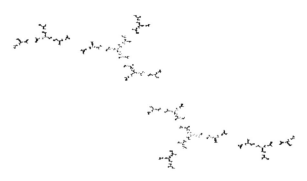

Figure 5.38. Julia set for $c = -.1 + 1.11i$.

gen and Saupe (1988: 153-54). It is based on another theorem of Fatou and Julia, which gives a completely different characterization of Julia sets.

### NEWTON'S METHOD

Frequently in applied mathematics one needs to find an approximate value of a solution of an equation. If the equation is written as $f(x) = 0$, this can be interpreted as saying that one wants to find, to a specified degree of accuracy, a value of $x$ at which the graph of $y = f(x)$ crosses the $x$-axis. This value of $x$ is sometimes called a root of $f(x)$. A popular method of doing this is Newton's method, which utilizes elementary calculus. It was developed by Isaac Newton, upon whom we shall focus in the final subsection.

The method is easily envisioned when $x$ and $y$ are real variables (as opposed to complex). The idea is to make an initial guess, $x_0$, and then to iteratively try to move from $x_n$ closer to the root. This is done by letting $x_{n+1}$ be the $x$-value where the tangent line to the curve $y = f(x)$ at the point $(x_n, f(x_n))$ meets the $x$-axis. Elementary calculus shows that

(5.3.11)
$$x_{n+1} = x_n - \frac{f(x_n)}{f'(x_n)},$$

where $f'$ denotes the derivative. (See Exercise 17.) This is pictured in Figure 5.39 for a certain curve.

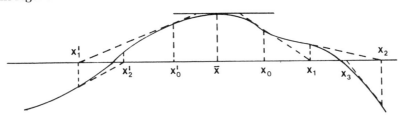

Figure 5.39. Newton's method for real variables.

In Figure 5.39, we have chosen two different initial values $x_0$ and $x_0'$. The indicated points $x_0$, $x_1$, $x_2$, and $x_3$ show how Newton's method starting at $x_0$ converges to the root on the right hand side of the graph, while $x_0'$, $x_1'$, and $x_2'$ show that starting at $x_0'$ will lead us to the other root. Note also that if the initial value had been the point $\bar{x}$ where the tangent is horizontal, so that $f'(\bar{x}) = 0$, then Newton's method wouldn't converge to either root. Given an equation $f(x) = 0$, one could color each point on the $x$-axis a color that depends on the root of $f(x)$ to which (5.3.11) converges if $x_0$ equals the $x$-value being colored. In the case of Figure 5.39, it appears that all points to the right of $\overline{x}$ will be one color, and those to the left will have the other color. This procedure is very similar to studying the Julia set of the function $x - f(x)/f'(x)$.

Newton's method also works for complex variables. We write the variable as $z$, and iterate

(5.3.12) $$z_{n+1} = z_n - f(z_n)/f'(z_n)$$

to find a root of $f(z)$. Now we will color the point $(x, y)$ in the plane a color depending upon to which root of $f(z)$ the sequence (5.3.12) converges if $z_0 = x + yi$. This question was already considered by the great English mathematician Arthur Cayley in the 1870s. His work came to a halt when he was unable to visualize the complicated answer that occurs when $f$ is a cubic polynomial.

One might naïvely think that Newton's method would just converge to the root closest to $z_0$, but it is much more complicated than that. In fact, suppose that $f(z)$ is a polynomial with roots $r_1, \ldots, r_n$. For each root $r_i$, let $A_i$ denote the set of points $z_0$ for which (5.3.12) converges to $r_i$, and let $B_i$ denote the boundary of the region $A_i$. Then an amazing fact, proved by Julia before 1920, is that all the sets $B_i$ are the same. In other words, the different colored sets corresponding to the different roots all have the same boundary! We state this amazing fact in a third way. Suppose $z$ is a complex number such that every little disk around $z$ contains some points from which (5.3.12) converges to the first root of $f$ and some points from which (5.3.12) does not converge to this root. Then every little disk around $z$ contains points converging to each root of $f$.

We illustrate by applying Newton's method to the equation $z^3 - 1 = 0$. The three roots of this equation are at $1$, $-1/2 + i\sqrt{3}/2$, and $-1/2 - i\sqrt{3}/2$. In Exercise 18, we list the program that colors pixels according to which of the three roots (or none) Newton's method approaches when starting at the complex number corresponding to the lower left corner point of the pixel. The result is given in Figure 5.40. The three roots are marked with little boxes. The white region is the set of points from which (5.3.12) converges to the root $z = 1$, near the right edge of the pictured square. The gray region is the set of points which, if used as $z_0$ in (5.3.12), lead to the root $-1/2 + i\sqrt{3}/2$ near the top of the picture. Similarly, the black region contains the points from which Newton's method leads to the root in the black region. One can think of the three roots as competing for the attraction. The remarkably intricate diagram displays the result of this competition.

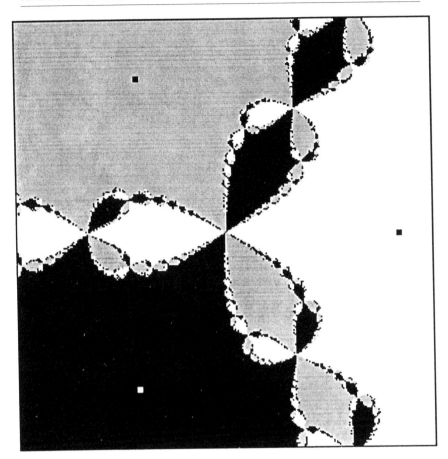

Figure 5.40. Newton's method for $z^3 - 1$.

Along the common boundary of the three regions are points that lead to none of the three roots. They form the Julia set of the function $z - f(z)/f'(z)$.[24] These don't show up well in Figure 5.40 because they are so sparse. A fourth color is required for the points that are not attracted to any of the roots. Such points are hardly visible until portions very close to the boundary are magnified, but upon magnification they take on familiar forms of Julia sets.

Analogous to the definition of the Mandelbrot set, one can define a family of cubic polynomials determined by a complex parameter $c$, and color the point $c$ depending on which of the three roots attracts

24. The Julia set of a function that is not a polynomial is defined slightly differently than Definition 5.3.7.

the initial value 0. This is illustrated in Plates 10 to 12 after page 86 of Devaney and Keen (1989). It shows how, after great magnification, the set of complex numbers $c$ for which 0 is not attracted to any of the roots forms a set that looks like the Mandelbrot set. It can be proved that the Mandelbrot set must appear in many such contexts.

## FOCUS: ISAAC NEWTON

Isaac Newton (1642–1727) is generally ranked along with Archimedes and Gauss as one of the three greatest mathematicians of all time. All three of these men were also physicists. We mentioned Archimedes' practical work in Section 1.1. Gauss made some fundamental advances to our knowledge of electricity and magnetism; he was co-inventor of the electromagnetic telegraph. Newton's discoveries about motion, gravity, and light revolutionized physics, and indeed all of scientific thought.

Newton's life got off to a bad start. He was born tiny and sickly,[25] his father died before he was born, and his mother soon remarried and left him to be raised by his grandmother. He did well in school, and enrolled in Trinity College, Cambridge, in 1661. The university was closed from 1665 to 1667 because of the bubonic plague. Newton spent this time at his family home in Woolsthorpe, England, and it was during this time, the *anni mirabiles* (miraculous years), that most of his famous theories were developed. When Cambridge reopened, Isaac Barrow, the Lucasian Professor of Mathematics, recognized Newton's genius and stepped down to allow Newton to occupy his professorship.[26]

Newton's discoveries owe as much to his diligence as to his genius. He let nothing interfere with his work. For example, in order to investigate how the curvature of the retina affected what the eye saw, he put a stick "betwixt my eye & ye bone as neare to ye backside of my eye as I could."[27] His depiction is reproduced in Figure 5.41.

Many of Newton's ideas were disseminated in lectures, although these were not usually well attended. In 1671 he spoke to the Royal Society in London about his invention of the reflecting telescope, which was highly lauded. Shortly thereafter, he spoke about his theory that white light was a composite of all the colors of the rainbow. This talk, which also included other topics in optics, aroused much controversy, which Newton detested, causing him to go into isolation for the next decade.

25. It is said that he could fit in a quart pot.
26. Some historians feel this rendering overromanticizes Barrow's motives.
27. Westfall 1980, p. 95.

Sir Isaac Newton, analyzing a ray of light.

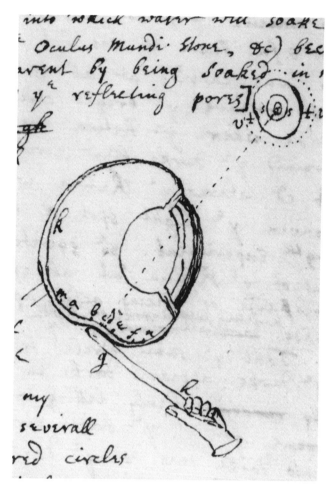

Figure 5.41. Newton's drawing of his eyeball experiment.

Newton's main fame as a mathematician is as the "inventor" of calculus. Calculus is involved with finding rates of change, and with finding things such as area under a curve. The Fundamental Theorem of Calculus expresses the relationship between these two problems. Many people prior to Newton had been toying with these ideas and applying them in special cases. Newton's main advance was to consolidate various disparate techniques into a general theory that could be applied to any function. Closely linked with Newton's insights in calculus were his ideas about infinite series. For example, he showed that the binomial theorem works with noninteger exponents. (See Exercise 19.) He also made important advances in the theory of equations, such as Newton's method discussed in the previous subsection.

In 1687, under the prodding and financial subsidy of astronomer Edmund Halley, after whom the famous comet is named, Newton published his major work, *Philosophiae Naturalis Principia Mathematica*. This book, whose title translates to *Mathematical Principles of Natural Philosophy*, is considered by many to be the most important single work in the history of modern science. This book is primarily about laws of motion and gravity. Newton's three laws of motion are the following:

1. An object remains at rest or with constant velocity unless a force is exerted on it (Law of Inertia).
2. The change in velocity is proportional to the force ($F = ma$).
3. To every action there is an equal and opposite reaction.

He showed that, assuming these laws of motion, Kepler's laws of planetary motion, which we discussed in Section 1.3, are equivalent to the existence of a gravitational force that is inversely proportional to the square of the distance between the bodies. Although Newton's book did not give a thorough treatment of calculus, some ideas from his calculus were used.

Newton's mathematical approach to motion was quite an innovation, although Galileo had introduced this general idea eighty years earlier, and his idea of gravity as a force acting on distant bodies was controversial. Subsequent to the publication of the *Principia*, Newton's scientific work was rather limited. Much of his time during these later years was spent in theological study. He wrote an interpretation of the prophecies of Daniel and St. John, and studied chronology in the Bible.

He became very famous. A poem by Alexander Pope includes these lines:

Nature and Nature's laws lay hid in night;
God said, 'Let Newton be,' and all was light.

He was named Warden of the Mint in 1696, a lucrative position he held for the rest of his life. He was named president of the Royal Society in 1703, a position he also held until his death, and in 1705 became the first scientist ever to be knighted. He never married.

Newton was involved in several bitter disputes with rival scientists. The most prominent of these was with Gottfried Leibniz, a German, over who invented calculus, which created a major British-Continental rift. It is now believed that their discoveries were independent, and that, although Leibniz published it first, Newton's discovery was about ten years earlier than that of Leibniz. Newton was unusually vigorous as Warden of the Mint, sending several counterfeiters to their death on the gallows. Some scholars attribute his disputative nature to his traumatic childhood.

### Exercises

1. Suppose $\alpha = 6 + 2i$ and $\beta = 5 - 3i$. Compute $\alpha + \beta$, $\alpha - \beta$, $\alpha \cdot \beta$, $\alpha / \beta$, and $\beta / \alpha$.
2. a. Show that any complex number $a + bi$ can be written as $r \cos \theta + (r \sin \theta)i$, where $r = \sqrt{a^2 + b^2}$ is the distance from $(0,0)$ to the point $(a, b)$, and $\theta$ is the counterclockwise angle from the positive $x$-axis to the line from $(0,0)$ to $(a, b)$.
   b. Use the formulas from trigonometry for $\sin(\theta_1 + \theta_2)$ and $\cos(\theta_1 + \theta_2)$ to show that the geometric interpretation of the product of complex numbers is as stated in the text. [Hint: Write the complex numbers in the form given in (a).]
   c.* The complex exponential function is defined by

   $$e^{a+ib} = e^a \cos b + ie^a \sin b,$$

   where $b$ is the radian measure of an angle. Show that this satisfies $e^{z_1 + z_2} = e^{z_1} e^{z_2}$. Recall that $\pi$ radians corresponds to $180°$. Show that

   $$e^{i\pi} + 1 = 0.$$

   This is the theorem that was voted to be the most beautiful in all of mathematics in the article cited preceding Theorem 3.1.4. Its appeal is largely due to the fact that it shows a relationship among what are probably the five most important numbers in mathematics.
3. a. Calculate $f^n_{2+i}(0)$ for $n = 1, 2,$ and $3$.
   b. Calculate $f^n_c(0)$ for all positive values of $n$ if $c = -2$. Do the same thing if $c = i$. Explain why your answer shows that $-2$ and $i$ are in $M$.
4.* a. Prove the triangle inequality $|z_1 + z_2| \leq |z_1| + |z_2|$ for any complex numbers $z_1$ and $z_2$. (Hint: Use the geometric interpretation of addition.)
   b. Suppose $|z| > |c|$. Explain the reason for each step in

   $$|z^2 + c| \geq |z^2| - |c| > |z^2| - |z| = |z|(|z| - 1).$$

   Use this to show that if $|z| > |c|$ and $|z| > 2$, then the numbers

$|f_c^n(z)|$ approach $\infty$ as $n$ gets large. (Hint: If $|z| - 2 = d$, then the ratio of successive numbers in the sequence

$$|f_c(z)|, |f_c^2(z)|, |f_c^3(z)|, \ldots$$

is greater than $1 + d$. [Why?])

c. Show that if $|c| > 2$, then $c$ is not in the Mandelbrot set. (Hint: Show directly that $|f_c^2(0)| > |c|$. Then apply [b].)

d. Show that if $|c| \leq 2$ and $|f_c^n(0)| > 2$ for some $n$, then $c$ does not lie in the Mandelbrot set. This is the fact, referred to frequently in the text, that it suffices to test whether the numbers in the sequence (5.3.3) ever exceed 2.

5. A real number $c$ is in $M$ if and only if $-2 \leq c \leq 1/4$. Carry out the following steps that investigate why this is true.

a. Explain how Exercise 4 shows that $c$ is not in $M$ if $c < -2$.

b. Adapt the argument in the text concerning the number $1/4$ to show that $c$ is in $M$ if $0 \leq c \leq 1/4$.

c.* Write a computer program to determine whether a real number $c$ is in $M$. It will be essentially the same as the program of Exercise 14 in Section 5.2. Use it to show that some numbers such as .2501 and .25001 are not in $M$. For each of these numbers, tell how many iterations were required before (5.3.3) exceeded 2.

d.* Show that if $c$ and $x$ are real numbers satisfying $-2 \leq c < 0$ and $|x| \leq |c|$, then $|x^2 + c| \leq |c|$. Deduce that a real number $c$ is in $M$ if $-2 \leq c < 0$. (Hint: Apply Exercise 4a to $|x^2 + c|$.)

e.** Use your computer program of (c) to convince yourself that when $c = 1/4$ the sequence (5.3.3) approaches $1/2$. Assuming this, write out details of the following argument, which shows that if $c > 1/4$, then the sequence (5.3.3) becomes arbitrarily large. Write $c = 1/4 + \epsilon_1$. Explain why the sequence (5.3.3) with this value of $c$ must eventually exceed $1/2$. Show that

$$\left(\frac{1}{2} + \epsilon_2\right)^2 + \frac{1}{4} + \epsilon_1 > \frac{1}{2} + \epsilon_2 + \epsilon_1.$$

Explain how this implies the result.

6. Run the program MANDEL, preferably modified so as to display four colors, to display the square consisting of values of $c = a + bi$ such that $-1.8 \leq a \leq -1.7$ and $-0.05 \leq b \leq 0.05$ on a 200 by 200 grid. Tell what you used as your input values. Describe the portion of Figure 5.33 that is being enlarged. If you have a fast computer and good monitor, you might try it on a 400 by 400 grid.

7. Run the program MANDEL, preferably modified so as to display four colors, to display the rectangle consisting of values of $c = a + bi$ such that $-0.78 \le a \le -0.73$ and $0 \le b \le 0.2$ on a grid with 200 vertical pixels. Tell what you used as your input values. Describe the portion of Figure 5.33 that is being enlarged. If you have a fast computer and good monitor, you might try it on a grid with 400 vertical pixels.

8. The complex number $c = -.106 + .924i$ lies inside the little bug slightly below the center in Figure 5.34. Run the program MANDEL, preferably modified so as to display four colors, to display a small square containing this bug on a 200 by 200 grid. Tell what you used as your input values. Because this region is so close to $M$ you will want to use threshold values slightly larger than we have been using.

9.* Prove that the Mandelbrot set is symmetrical around the $x$-axis, and explain why it is not symmetrical around the $y$-axis. The following comments and questions are meant to help you give the required proof and explanation.

   a. If $c = a + bi$ is a complex number, then its conjugate, written as $\overline{c}$, is the complex number $a - bi$. What is $\overline{4 + 3i}$? What is $\overline{2 - 5i}$?

   b. Show that $|\overline{c}| = |c|$ and that $\overline{c^2} = (\overline{c})^2$.

   c. Show that $-\overline{c}$ is the complex number obtained from $c$ by changing the sign of the real part of $c$.

   d. Show that each number in the sequence (5.3.2) based on $\overline{c}$ is the conjugate of the corresponding number in (5.3.2) based on $c$. Another way of saying the same thing is, in the notation of (5.3.3), $f_{\overline{c}}^n(0) = \overline{f_c^n(0)}$. The best way to prove this is by Mathematical Induction. Show that if it is true for $n$, then it is true for $n + 1$. You might use (5.3.5) to help you get started.

   e. Deduce from (d) that $\overline{c}$ is in $M$ if and only if $c$ is in $M$. Explain why this implies that $M$ is symmetrical around the $x$-axis.

   f. Explain why it is not true that each number in the sequence (5.3.2) based on $-\overline{c}$ is the negative of the conjugate of the corresponding number in (5.3.2) based on $c$.

   g. Explain, using (c) and (f), why $M$ is not symmetrical around the $y$-axis.

10. Show that the ratio of the sizes of the portions of the complex plane depicted by Plates 1 and 11 is roughly the same as the ratio of the size of the orbit of Jupiter to the actual size of Plate 11. The diameter of the orbit of Jupiter is about 1 billion ($10^9$) miles, and the side of Plate 11 is about 2 inches.

11. A variant of the Mandelbrot set could be obtained by using some

other complex number instead of 0 as the initial number $z_0$ of the iteration (5.3.2). It may then no longer be the case that if some number in the sequence exceeds 2 in absolute value then the numbers will approach infinity, so your test should be that $|z_n|$ exceed some large number such as 100. Modify the program MANDEL so that it accepts as input an additional complex number, which will be the number used as $z_0$ for each value of $c$. Run this program for a few values of $z_0$. Some interesting examples are $z_0 = 0.5i$, then $z_0 = 0.8i$, and then $z_0 = i$.

12. Let $D$ denote the disk that consists of all points whose distance from (0,0) is equal to or less than 1. Let $C$ denote the circle that consists of all points whose distance from (0,0) equals 1. Referring to the definition of "boundary" after Definition 5.3.7, show that $C$ is the boundary of $D$. In particular, explain why a point whose distance from (0,0) does not equal 1 is not in the boundary. For example, how about a point whose distance from (0,0) is .999?

13. What is the Julia set of the function $f(z) = z^2$? To answer this, you will need to determine for which complex numbers $z$ will repeated squaring eventually give very large numbers. (Hint: What does squaring do to the absolute value of a complex number?)

14. Let $J_c$ denote the Julia set of the function $f(z) = z^2 + c$. Prove that $z$ is in $J_c$ if and only if $-z$ is in $J_c$. Show how this explains the symmetry in Figure 5.36.

15. Run the program JULIA to plot the Julia set $J_c$ if $c = .32 + .43i$. For all input parameters except $c$, use the same values as were used in producing Figure 5.36.

16. Magnify an interesting area of either Figure 5.36a or 5.36b or 5.37b by running JULIA, similarly to the way that 5.37b was obtained from 5.37a. Tell what values you used as your input parameters.

17. Prove (5.3.11). All you need to know is that $f'(x)$ measures the slope of the tangent line to the curve $y = f(x)$ at the point $(x, f(x))$.

18. We list below the program that produced Figure 5.40, which displayed the results of applying Newton's method to $z^3 - 1$ on a grid of 300 by 300 pixels representing the square whose sides go from $-1.2$ to 1.2 in both the horizontal and vertical directions. It allowed the program to run up to one hundred iterations, and required that $z_n$ be within .02 of one of the roots.

    a. Show that the part of the program that changes the values of $x$ and $y$ accurately reflects (5.3.12). (Hint: The derivative of $z^3 - 1$ is $3z^2$.)

b. Run the program using parameters that magnify interesting portions of Figure 5.40. Tell what parameters you used. If you do not have a VGA monitor, you may have to change SCREEN 12 to SCREEN 1, and settle for a 200 by 200 grid.

```
DEFDBL X,Y,D,E,S: DEFINT I,J,H,V,N
INPUT "Coordinates of lower left corner"; X0,Y0
INPUT "gap"; DEL
INPUT "Number of pixels--horizontal, vertical"; H,V
INPUT "Maximum number of iterations"; MAX
INPUT "Distance-squared to roots"; EPS
CLS: SCREEN 12
WINDOW (0,0)-(319,199)
ST=SQR(3)/2
SF=2/3
FOR I=1 TO H
  FOR J=1 TO V
    N=0
    X=X0+(I-1)*DEL
    Y=Y0+(J-1)*DEL
    X1=(X-1)*(X-1)
    X2=(X+.5)*(X+.5)
    Y1=Y*Y
    Y2=(Y-ST)*(Y-ST)
    Y3=(Y+ST)*(Y+ST)
    WHILE N<MAX AND X1+Y1>EPS AND X2+Y2>EPS AND X2+Y3>EPS
    AND X*X+Y1<>0
      XT=SF*X+(X*X-Y1)/(3*(X*X+Y1)*(X*X+Y1))
      Y=SF*Y-(SF*X*Y)/((X*X+Y1)*(X*X+Y1))
      X=XT
      X1=(X-1)*(X-1)
      X2=(X+.5)*(X+.5)
      Y1=Y*Y
      Y2=(Y-ST)*(Y-ST)
      Y3=(Y+ST)*(Y+ST)
      N=N+1
    WEND
    IF N=MAX OR X*X+Y1=0 THEN
      PSET(I,J),3
    ELSEIF X1+Y1<=EPS THEN
      PSET(I,J),2
    ELSEIF X2+Y2<=EPS THEN
      PSET(I,J),1
```

```
    ELSE
        PSET(I,J),0
    END IF
    NEXT J
    NEXT I
    END
```

19. The binomial theorem states that

$$(1 + x)^n = 1 + nx + \frac{n(n-1)}{2}x^2 + \frac{n(n-1)(n-2)}{3 \cdot 2}x^3$$
$$+ \frac{n(n-1)(n-2)(n-3)}{4 \cdot 3 \cdot 2} + \cdots .$$

When $n$ is a positive integer, this formula stops after $n+1$ terms. In this case, it was stated as our Theorem 3.4.2. Newton showed that it also works when $n$ is not an integer, provided $|x| < 1$. The infinite series obtained can be used to approximate the answer as closely as desired. Use the first five terms of this series to calculate an approximate value of $\sqrt{1.2} = (1 + .2)^{1/2}$. Compare your answer with that given by a calculator.

# Epilogue

We have now completed our study of three topics illustrating the power of mathematics. We have seen how mathematicians' dissatisfaction with Euclid's fifth postulate led to the discovery of alternative forms of geometry, which changed scientists' conception of the universe. We have learned how some ideas in elementary number theory, developed out of abstract curiosity, formed the basis for a method of cryptography that is now used by banks, governments, and the military. And we saw that the study of the behavior of simple functions under iteration, begun by mathematicians before 1920, leads to beautiful computer graphics as well as the idea of chaos (sensitive dependence on initial conditions).

In addition to these three principal examples, we have seen other manifestations of our theme of the unexpected applications of abstract mathematics. Thus we saw how the seventeenth-century astronomer Johannes Kepler was influenced by two ideas of Greek mathematics which had been developed two thousand years earlier; one of these enabled him to discover the laws of planetary motion. Finally, we learned how Alan Turing's study of abstract computable functions led to the development of the most revolutionary product of the second half of the twentieth century—the computer.

Along the way, we have learned a lot about mathematical proof and the nature of mathematics. The axiomatic method, most effectively presented by Euclid, was used also in Einstein's Special Theory of Relativity. But even Euclid's system was not rigorous enough for ninteenth- and twentieth-century mathematicians. Mathematics today is generally viewed as logical deductions about undefined terms. We saw how number theorists struggle to prove statements that can be understood by a schoolchild and have been verified in billions of cases. Yet the mathematician is not satisfied until the statement has been rigorously proved. We do not expect that many of our readers will go on to a career in mathematics, but the method of logical thought is important in everyone's daily life.

We have not taught calculus. We have dealt with topics, geometry, number theory, and computers, which are for the most part divorced from calculus. Yet we saw that ideas from calculus kept popping up. This included limits, slopes, derivatives, areas, and infinite series. We hope that seeing these ideas in these different contexts might inspire some of our readers to want to study calculus, which is a prerequisite for most of modern mathematics.

The reader should now appreciate that mathematics is more than the development of dry techniques for use in engineering or business. Of course, such applications provide a major reason for the importance of mathematics, but now you have seen the prettier side of the subject. You have learned about some of the great thinkers who have devoted their lives to the study of abstract concepts, because they were inexorably drawn by the ideas. These people are rewarded by the intellectual challenge involved in the study, and sometimes, many years later, the rest of the world is rewarded by practical applications of this work.

# Bibliography

Abbott, E. A. 1884. *Flatland: A Romance of Many Dimensions.* Reprint. New York: Barnes & Noble, 1963.

Adleman, L. M., R. L. Rivest, and A. Shamir. 1978. "A method for obtaining digital signatures and public-key cryptography." *Communications of the ACM* 21: 120–26.

Barnsley, M. 1988. *Fractals Everywhere.* San Diego: Academic Press.

Beck, A., M. Bleicher, and D. Crowe. 1969. *Excursions into Mathematics.* New York: Worth.

Becker, K.-H., and M. Dörfler, 1989. *Dynamical Systems and Fractals.* Cambridge, U.K.: Cambridge University Press.

Bell, E. T. 1937. *Men of Mathematics.* New York: Simon and Schuster.

Bonola, R. 1912. *Non-Euclidean Geometry.* Reprint. New York: Dover, 1955.

Bressoud, D. 1989. *Factorization and Primality Testing.* New York: Springer-Verlag.

Bronowski, J. 1973. *The Ascent of Man.* Boston: Little, Brown, and Co.

Burger, D. 1965. *Sphereland: A Fantasy about Curved Spaces and an Expanding Universe.* New York: Barnes & Noble.

Carslaw, H. S. 1916. *The Elements of Non-Euclidean Plane Geometry and Trigonometry.* London: Longmans, Green & Co.

Caspar, M. 1959. *Kepler.* London: Abelard-Schuman.

Cook, S. A. 1971. "The complexity of theorem-proving procedures." *Proceedings of the Third ACM Symposium on the Theory of Computing,* 151–58.

Coughlin, R., and D. E. Zitarelli. 1984. *The Ascent of Mathematics.* New York: McGraw-Hill.

Crutchfield, J., J. Farmer, J. Packard, and R. Shaw. 1986. "Chaos." *Scientific American* 255: 46–57.

Devaney, R. L. 1990. *Chaos, Fractals, and Dynamics.* Menlo Park, Calif.: Addison-Wesley.

Devaney, R. L., and L. Keen, eds. 1989. *Chaos and Fractals.* Providence: American Mathematical Society.

Devlin, K. 1988. *Mathematics: The New Golden Age.* London: Penguin.

Dewdney, A. K. 1987. "The Mandelbrot set and its cousins named Julia." *Scientific American* 257: 140–46.

———. 1989. *The Turing Omnibus: 61 Excursions in Computer Science.* Rockville, Md.: Computer Science Press.

Edwards, H. M. 1977. *Fermat's Last Theorem: A Genetic Introduction to Algebraic Number Theory*. New York: Springer-Verlag.

Ekeland, I. 1988. *Mathematics and the Unexpected*. Chicago: University of Chicago Press.

Eves, H. 1969. *In Mathematical Circles*. Boston: Prindle, Weber, and Schmidt.

———. 1972. *A Survey of Geometry*. Boston: Allyn & Bacon.

———. 1990. *Introduction to the History of Mathematics*. 6th ed. Philadelphia: Saunders.

Faber, R. L. 1983a. *Foundations of Euclidean and Non-Euclidean Geometry*. New York: Marcel Dekker.

———. 1983b. *Differential Geometry and Relativity Theory*. New York: Marcel Dekker.

Falconer, K. 1990. *Fractal Geometry: Mathematical Foundations and Applications*. Chichester, U.K.: John Wiley.

Feder, J. 1988. *Fractals*. New York: Plenum.

Fischer, P., and W. R. Smith. 1985. *Chaos, Fractals, and Dynamics*. New York: Marcel Dekker.

French, A. P., ed. 1979. *Einstein: A Centenary Volume*. Cambridge, Mass.: Harvard University Press.

Gaines, H. F. 1956. *Cryptanalysis: A Study of Ciphers and Their Solutions*. New York: Dover.

Gardner, M. 1962. *Relativity for the Million*. New York: Macmillan.

———. 1972. *Codes, Ciphers, and Secret Writing*. New York: Dover.

Gleick, J. 1987. *Chaos*. London: Penguin.

Gray, J. 1979. *Ideas of Space*. Oxford: Clarendon Press.

Greenberg, M. J. 1980. *Euclidean and Non-Euclidean Geometries*. New York: W. H. Freeman.

Hall, T. 1970. *Carl Friedrich Gauss: A Biography*. Cambridge, Mass.: MIT Press.

Handler, I., L. H. Kauffman, and D. Sandin. 1987. "On crossing the boundary of the Mandelbrot set." In *Computers in Geometry and Topology*, ed. M. C. Tangora, 151-77. New York: Marcel Dekker.

Hardy, G. H. 1967. *A Mathematician's Apology*. Cambridge, U.K.: Cambridge University Press.

Hawking, S. W. 1988. *A Brief History of Time*. New York: Bantam.

Heath, T. L. 1921. *A History of Greek Mathematics*. Oxford: Oxford University Press.

———. 1956. *Euclid's Elements*. New York: Dover.

Hellman, M. E. 1979. "The mathematics of public-key cryptography." *Scientific American* 241: 146-57.

Hilbert, D. 1956. *Grundlagen der Geometrie*. Stuttgart: Teubner.

Hilton, P. J. 1988. "Reminiscences of Bletchley Park, 1942-1945." In

*A Century of Mathematics in America*, Part I, ed. R. Askey, P.L. Duren, and U. Merzbach, 291–301. Providence: American Mathematical Society.

Hinton, C.H. 1980. *Speculations on the Fourth Dimension: Selected Writings of Charles H. Hinton.* Ed. R.B. Rucker. New York: Dover.

Hodges, A. 1983. *Alan Turing: The Enigma.* New York: Simon and Schuster.

Hofstadter, D. 1981. "Strange attractors." *Scientific American* 245: 16–29.

Hoyle, F. 1962. *Astronomy.* Garden City, N.Y.: Doubleday.

Jürgens, H., H.-O. Peitgen, and D. Saupe. 1990. "The language of fractals." *Scientific American* 263: 60–67.

Kahn, D. 1967. *The Codebreakers: The Story of Secret Writing.* New York: Macmillan.

———. 1991. *Seizing the Enigma.* Boston: Houghton-Mifflin.

Kline, M. 1972. *Mathematical Thought from Ancient to Modern Times.* Oxford: Oxford University Press.

Koblitz, N. 1987. *A Course in Number Theory and Cryptography.* New York: Springer-Verlag.

Koestler, A. 1959. *The Sleepwalkers.* London: Penguin.

Konheim, A.G. 1982. *Cryptography: A Primer.* New York: John Wiley.

Koyré, A. 1973. *The Astronomical Revolution.* Paris: Hermann.

Kranakis, E. 1986. *Primality and Cryptography.* Stuttgart: Wiley Teubner.

Lauwerier, H. 1991. *Fractals: Endlessly Repeated Geometrical Figures.* Princeton: Princeton University Press.

Lei, T. 1990. "Similarity between the Mandelbrot set and Julia sets." *Commun. Math. Physics* 134: 587–617.

Lewis, H.R., and C.H. Papadimitriou. 1978. "The efficiency of algorithms." *Scientific American* 238: 96–109.

Loweke, G.P. 1982. *The Lore of Prime Numbers.* New York: Vantage.

Mandelbrot, B. 1983. *The Fractal Geometry of Nature.* New York: W.H. Freeman.

Metropolis, N., J. Howlett, and G.-C. Rota. 1980. *A History of Computing in the Twentieth Century.* New York: Academic Press.

Milnor, J. 1987. "Self-similarity and hairiness in the Mandelbrot set." In *Computers in Geometry and Topology*, ed. M.C. Tangora, 151–77. New York: Marcel Dekker.

Neyman, J. 1974. *The Heritage of Copernicus.* Cambridge, Mass.: MIT Press.

Peitgen, H.-O. 1989. *Newton's Method and Dynamical Systems.* Dordrecht: Kluwer Academic Publishers.

Peitgen, H.-O., and P. H. Richter. 1986. *The Beauty of Fractals*. Berlin: Springer-Verlag.

Peitgen, H.-O., and D. Saupe. 1988. *The Science of Fractal Images*. Berlin: Springer-Verlag.

Peitgen, H.-O., H. Jürgens, and D. Saupe. 1992. *Fractals for the Classroom. I*. New York: Springer-Verlag.

Peterson, I. 1988. *The Mathematical Tourist*. New York: W. H. Freeman.

————. 1990. *Islands of Truth*. New York: W. H. Freeman.

Pickover, C. A. 1990. *Computers, Pattern, Chaos, and Beauty*. New York: St. Martin's Press.

Prenowitz, W., and M. Jordan. 1965. *Basic Concepts of Geometry*. Lexington, Mass.: Xerox.

Preston, R. 1992. "Mountains of pi." *The New Yorker* 68 (March 2): 37–67.

Rademacher, H. 1964. *Lectures on Elementary Number Theory*. Huntington, N.Y.: Krieger.

Ribenboim, P. 1988. *The Book of Prime Number Records*. New York: Springer-Verlag.

Rosen, K. H. 1988. *Elementary Number Theory and Its Applications*. Reading, Mass.: Addison-Wesley.

Rosenfeld, B. A. 1988. *A History of Non-Euclidean Geometry*. New York: Springer-Verlag.

Rucker, R. B. 1977. *Geometry, Relativity, and the Fourth Dimension*. New York: Dover.

Savage, J. H. 1976. *The Complexity of Computing*. New York: John Wiley.

Schwinger, J. 1986. *Einstein's Legacy: The Unity of Space and Time*. New York: Scientific American Books.

Stark, H. M. 1970. *An Introduction to Number Theory*. Cambridge, Mass.: MIT Press.

Stewart, I. 1989. *Does God Play Dice?* London: Penguin.

Trudeau, R. J. 1987. *The Non-Euclidean Revolution*. Boston: Birkhauser.

Weeks, J. R. 1985. *The Shape of Space*. New York: Marcel Dekker.

Westfall, R. S. 1980. *Never at Rest: A Biography of Isaac Newton*. Cambridge, U.K.: Cambridge University Press.

Winterbotham, F. W. 1974. *The Ultra Secret*. New York: Harper & Row.

Yaglom, I. M. 1988. *Felix Klein and Sophus Lie*. Boston: Birkhauser.

Zollner, J. C. F. 1888. *Transcendental Physics*, Boston: Colby & Rich.

# Credits

Page 8: Associated Press.

Page 16: The Bettmann Archive.

Page 20: The Bettmann Archive.

Page 39, left: *Science* ©AAAS, Gina Kolata, "Math Proof Refuted During Berkeley Scrutiny," 234 (December 19, 1986): 1498.

Page 39, right: Copyright ©1986/90 by the New York Times Company. Reprinted by permission.

Page 60: Johannes Kepler, *Prodomus dissertationum*, Tübingen, 1596. Rare Books and Manuscripts Division, The New York Public Library; Astor, Lenox and Tilden Foundations.

Page 63: Johannes Kepler, *Astronomia Nova*, Heidelberg, 1609. Rare Books and Manuscripts Division, The New York Public Library; Astor, Lenox and Tilden Foundations.

Page 64: The Bettmann Archive.

Page 65: Reprinted from *Excursions into Mathematics*, by A. Beck, M. Bleicher, and D. Crowe (1969) with permission of Worth Publishers, New York.

Page 95: The Bettmann Archive.

Page 104: The Bettmann Archive.

Page 125: Reprinted from *Flatland*, by E. A. Abbott, with permission of Princeton University Press.

Page 130: The Bettman Archive.

Page 152: B-26983 Circle Limit I, 1958, M. C. Escher, 1898-1972, National Gallery of Art, Washington, D.C., Cornelius Van S. Roosevelt Collection. ©1958 M. C. Escher/Cordon Art–Baarn–Holland.

Page 160: Detail of B-26983 Circle Limit I, 1958, M. C. Escher, 1898-1972, National Gallery of Art, Washington, D.C., Cornelius Van S. Roosevelt Collection. ©1958 M. C. Escher/Cordon Art–Baarn–Holland.

Page 178: The Bettmann Archive.

Page 200: The Bettmann Archive.

Page 251: Reprinted from *The Codebreakers: The Story of Secret Writing* by David Kahn, with permission of the author.

Page 260: From the collection of Donald Michie, Glasgow.

Page 276: Copyright ©1986/90 by the New York Times Company. Reprinted by permission.

Page 296: Benoit B. Mandelbrot.

Page 297: From *The Encyclopedia Americana*, 1989 edition. Copyright 1989 by Grolier Incorporated. Reprinted by permission.
Following p. 354 and cover illustrations: Color plates produced by Kenneth G. Monks.
Page 367: The Bettman Archive.
Page 368: Cambridge University Library.

# Index